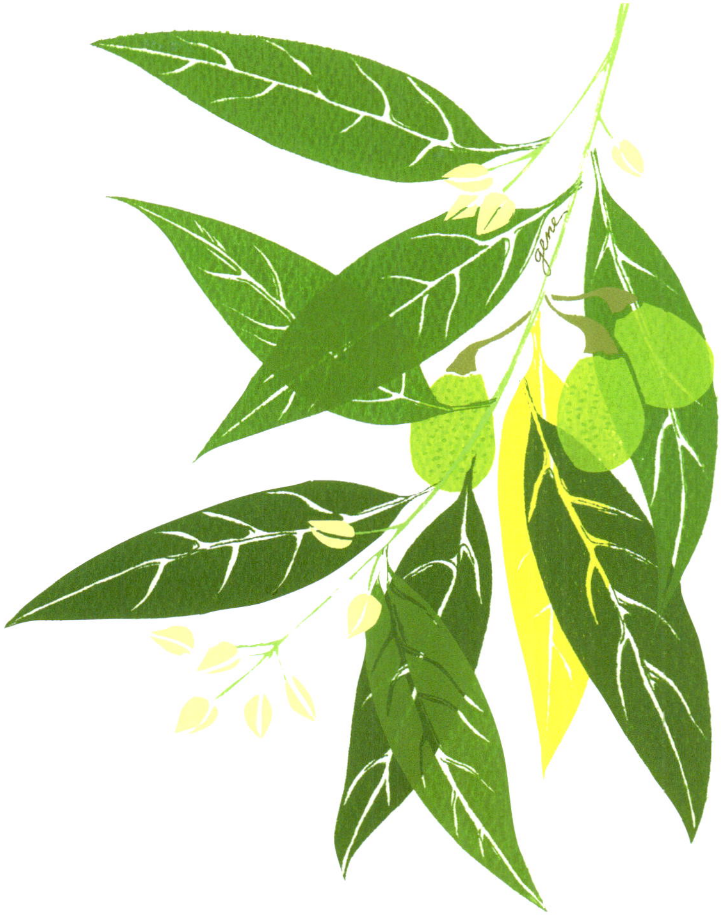

Botanical Serigraphs

THE GENE BAUER COLLECTION

ESRI Press
REDLANDS, CALIFORNIA

ESRI Press, 380 New York Street, Redlands, California 92373-8100

14 13 12 11 10   1 2 3 4 5 6 7 8 9 10
Printed in the United States of America

ESRI Press titles are distributed to the trade by the following:
In North America:
Ingram Publisher Services
Toll-free telephone: (800) 648-3104
Toll-free fax: (800) 838-1149
E-mail: customerservice@ingrampublisherservices.com

117035
DUAL3M11/09jh

Printed in the USA

Contents

Acknowledgments

IN 1997 MY HOME AND ORIGINAL ARTWORK WERE DESTROYED IN A WILDFIRE. I wish to thank these thoughtful people who returned their available copies to me:

Bill Miller *ESRI senior staff*
Phil Barker *Plant physiologist, United States Forest Service*
Max and Eleanor Dawes *Art collaborators*
Don and Nancy Albinson *Designers*
Corrine Bromberger *Horticulturist extraordinaire*
Joyce Dean *Archives curator, Redlands Horticultural and Improvement Society*

It has been a pleasure working with the staff at ESRI Press. I am most appreciative of the care with which my original artwork was handled. Everyone who worked on this project has been most gracious and creative. I especially want to thank the following ESRI staff members for their punctuality, thoroughness, and kindness afforded me during the preparation of this book:

Peter Adams *ESRI Press Manager*
Kathleen Morgan *Project Coordinator*
Candace Lyle Hogan *Editor*
Tiffany Wilkerson *Copy editor*
Jennifer Hasselbeck *Designer*
Kelley Heider *ESRI Press Projects Coordinator*

Introduction

THE SERIGRAPHS IN THIS COLLECTION represent my life's work in serigraphy, the art of original silk-screen printing. Each one originated from a pencil line drawing of the real plant (or animal) in front of me, sometimes in the wild, other times in one of California's many impressive arboretums or botanical gardens. My subjects include two birds and almost every type of plant: tree, shrub, perennial, cactus, saprophyte, vine, bulb, grass. A state larger than some countries, California offers many different environments for plants, so no type of region escaped me either. Thus the book covers high elevation, chaparral, coastal lands, bay lands, desert, and inland valley areas.

The serigraphs reproduced here were made from the last existing full collection in my possession. The other collections of originals I kept for myself were at home when the Mill Fire destroyed our house in 1997. I am grateful to the people who, hearing that all mine had been burned, sent some of their numbered prints back to me to make this book possible. The serigraphs reproduced in this book represent the covers and inside art of sixty-one individual booklets that I mailed to botanical garden enthusiasts all over California in the 1970s.

Originally I created these serigraphs in two periods of time, from July 1972 to May 1974 and from June 1976 to May 1978. I did so, respectively, for The Golden Native Collection while I was California native flora chairman of the California Garden Clubs, Inc., and for the Golden Botanical Gardens Collection while I was chairman of arboretums and botanical gardens for the same group. The serigraphs are the original illustrations for booklets I wrote—"a series of communications"—sent to the respective chairpersons of each of the twenty-six districts of the organization, urging them to use the information inside to educate their district members. I was under no illusion that I could add much knowledge to what had already been written about the vast array of California native flora, well known and loved worldwide (although I did favor those not-so-well-known in my choices). Rather, with each monthly message devoted exclusively to one plant, I hoped to better acquaint my fellow inhabitants of the state, as I wrote at the time, "with the marvelous treasures that exist here and to make them more aware of the necessity of preserving them for those who follow us."

I began with The Golden Native series (twenty-seven booklets, 1972–1974) followed by the Golden Botanical Gardens (twenty-nine booklets, 1976–1978). I created five additional booklets for special events. These are included in the last section of the book, "Living with Nature," along with the stories that inspired them, experiences I have had enjoying the wilderness around my home in the San Bernardino Mountains.

I chose different types of plants, both native and introduced to California, for the serigraphs on the booklet covers, considering several different things before the final choice of plant. I selected plants from six continents with strong emphasis on those from Asia and Australia. The choice for each cover was made the day I visited the garden, and it was always a plant that was particularly attractive to me on that day and not necessarily what was typical or characteristic of the garden. I intended to show that these gardens are interesting every day of the year.

I personally visited fifty-six public gardens in California in search of my subject matter, then created a booklet in which to record what I had found, and sent it by mail to a selected number of garden club officers. I printed by hand about eighty numbered copies each, and assembled the pages of each booklet at the fold with knotted yarn. I sent every monthly issue through the U.S. Postal Service, which of course gave me the opportunity to include one more serigraph on the envelope and to ponder meaningfully over my selection of the other pictorial accompanying it, the postage stamp.

I gave a lot of thought to selecting the postage stamps, considering color as well as subject matter and message. For example, when I used the state flower *(Eschscholzia californica)* for a cover, I selected the stamp of the butterfly chosen as the state insect of California. For the

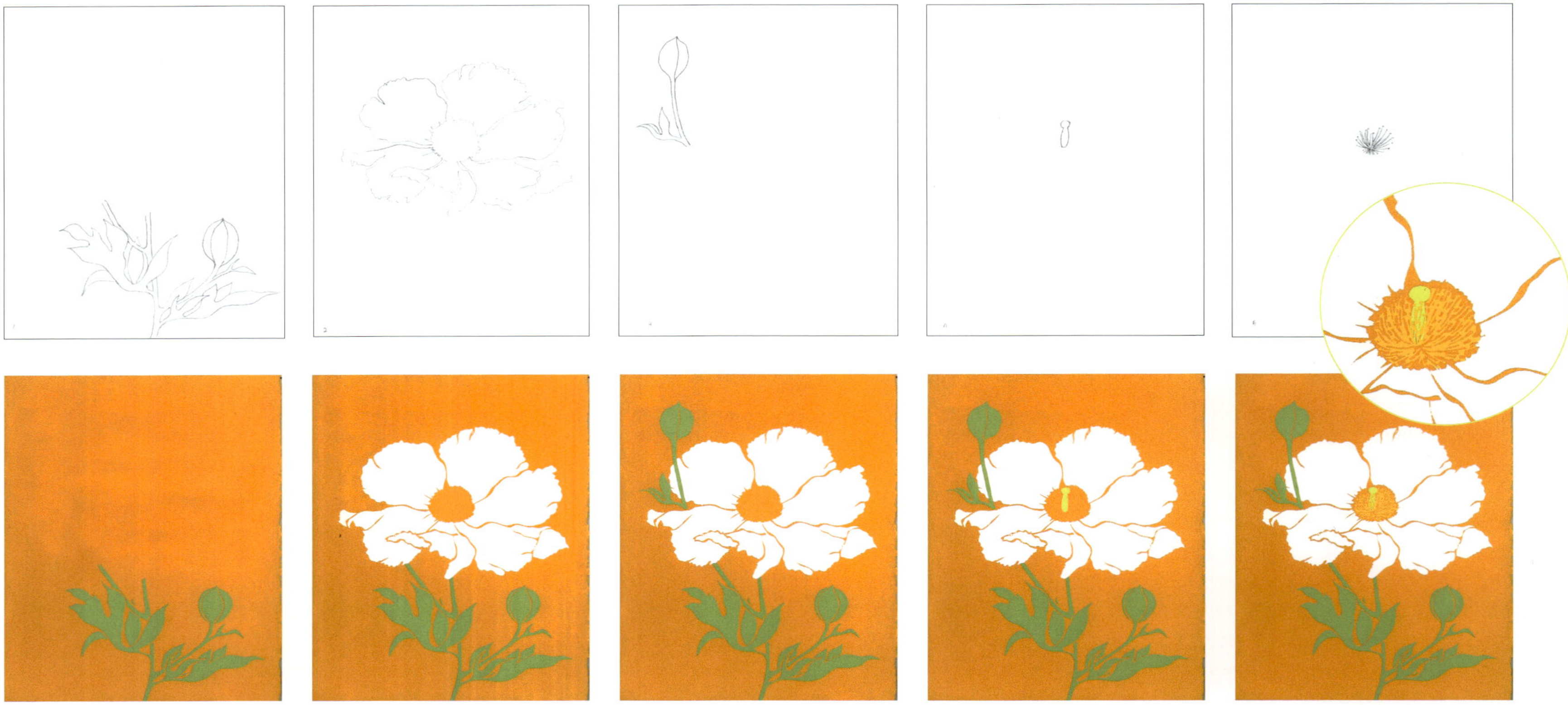

booklet exploring the Santa Clara Valley District, I chose the stamp honoring the first civil settlement in Alta California, the town of San Jose that was prominently featured in the issue. The U.S. Postal Service was very considerate in issuing stamps that I particularly desired, even one commemorating the dance just in time for me to use for the dancing daffodils.

The silk-screen printing process is a varied one, having been used to make large commercial signs, commercial labels, textiles, stationery, as well as the more personal hand-printed work suitable for framing. There is a fine distinction between prints called silk-screen prints and those called serigraphs. Those considered as fine art—and usually hand-printed—are serigraphs. I created the serigraphs reproduced here by a rather complex process described simply as the hand-cut lacquer film technique. In recent years, this procedure has been replaced by a photographic method. But in my process, I used no mechanical device of any kind in creating the originals.

This is how I create a serigraph:

I start from a line drawing made with a soft pencil. For each stage of printing the serigraph, I need to cut a stencil out of lacquer film. Lacquer film is solidified lacquer on one side and waxed paper on the other. I cut only the lacquer, not the waxed paper, using a craft knife with a 1/16-inch blade and three small line cutter tools. Using a solvent, I adhere the lacquer film to silk fabric that has been stretched over a wooden frame. Now the waxed paper backing can be peeled off. I choose the type and color of paper with care and mix the oil-based inks. Then I use a squeegee to press the ink through the open mesh of the stencil onto the paper. A completed serigraph may require five to eight, or more, color runs. On average it took five to seven days—from pencil line drawing to completion of printing the cover—to finish one complete run.

I had many interesting experiences visiting California's public gardens. I never took with me an elaborate easel or even a drawing board, but I did carry a small ten- by twelve-inch piece of heavy illustration board attached to a piece of drawing paper the size of a booklet cover. I also carried a small pencil box with about five different pencils and an eraser. This seemed a minimal and inconspicuous amount of gear to handle, but when I was drawing it apparently gave me an air of attachment to the garden. Sometimes I could hardly complete the drawing because of so many interruptions. People approached me as if I was a garden employee, and suddenly becoming an information bureau, I was expected to know everything about the garden. Some of the questions I was asked were, "Where are the dawn redwoods?" "How do you pronounce that name?" "How old is this tree?" "Where is this tree from?" People really want to know about plants, but they don't want to spend time sitting in a library reading. They would rather be told about plants. I tried to be helpful, but escorting people to various areas of the garden and explaining botanical nomenclature sometimes came dangerously close to taxing my patience, especially when my purpose there was to complete a drawing. Still, their curiosity did fortify my intent at the time.

After all, my purpose in exploring the arboretums and botanical gardens of California was to better acquaint people with them and what very special features may be found in each one. In the booklets I included maps, the correct address, the visiting hours, the facilities to be found within, and the general type of plant material growing in each establishment—all important information now available online in a most up-to-date fashion, so you will not find that here. In deciding what to reproduce in this book, I did not update the content and I deleted most of my extensive descriptions of the establishments themselves. While I may have done them justice then, these gardens have had thirty years more to grow, and even though there are fewer flora in the wild to see, their garden-bound abundance has increased, so better to leave you to have your own botanical garden adventures.

I do encourage you to visit the arboretums and botanical gardens wherever you live. The ones in California are spectacular. Many of them are well known—even world famous—while others are little known. Some were new at the time I wrote about them, while others were planned and developed many years before. Some are very small, while others encompass large acreage. Some are privately endowed, and others are supported by public funds. Many are closely allied to, if not actually part of, educational institutions. Regardless, each is unique, projecting its own distinct character.

We now know botanic gardens began in Europe in the form of monastic gardens, where healing plants and herbs (called "simples") were cultivated. Eventually they came out into the open as city gardens, whose purpose was to beautify and offer congested urban populations the enjoyment lost when the large areas left to nature became smaller and less available to everyone. Almost every European city already had a garden by the time the science and academic establishments decided on a further purpose for them: development as botanical gardens for research and education.

In a way, when visiting a botanical garden, you are seeing nature's dance choreographed, pathways set for you to walk about on, placements and settings planned so as to give you the best view and good light. And so too a book, organized in such a way as to lead you to the best vantage point, should then leave it to you to experience what you will on your own. Years ago, I made the booklet form a device by which to take my fellow garden club enthusiasts with me on the adventures of discovering each plant. This book in your hands is probably the closest you can come to having the whole stack of booklets on your lap. You will find the serigraph of the plant material (the cover of the booklet), described at the time as best I could without being a botanist, then get a look at the subject from a different angle (the booklet's tissue-paper serigraph), now that you are really looking. As I intended with the booklets, I'm hoping this book will give what a botanical garden—rooted both in simples and science—traditionally offers: inspiration to invoke in ourselves both the potential for self-healing and our powers of observation.

Along with every booklet, each serigraph was created with the same amount of care and affection, which I believe comes across in the book you are holding now. I will end its introduction with the last thing I wrote in the last booklet of the final series. I called that last issue a summary of the two years, 1976 to 1978, at the end of which I had visited fifty-six public gardens in the state of California and produced twenty-nine booklets and their incumbent serigraphs for the Golden Botanical Gardens series:

"Four parts of my body—brain, eyes, heart, and hands—have been utilized almost to their limit these past two years. Without any one of these it would have been impossible to complete this series. At times I have resorted to that standard phrase of mine, usually associated with outdoor physical work, 'I hurt *all* over.' After standing all day, mixing inks and printing, I *do* hurt *all* over. Another phrase that I have overworked is, 'Oh, I can't wait. I simply can't wait.' This is usually uttered late in the evening when all is in readiness to print but must be delayed because I do not like to work under artificial light. After a night's rest and the breaking of a new day, I can at last see how it will look.

"The first color is nice but rather uninteresting; the second color is nicer and adds interest; the third color usually begins to build up the form and from then on, although it is time-consuming, it is most enjoyable work. About the time the fifth color is printed I usually am overcome with affection and exclaim, 'Oh, I love you.'"

Gene Bauer

TO

Laura and Jack Dangermond
for your vision in collecting my botanical artwork
for a book under the imprint of ESRI Press.

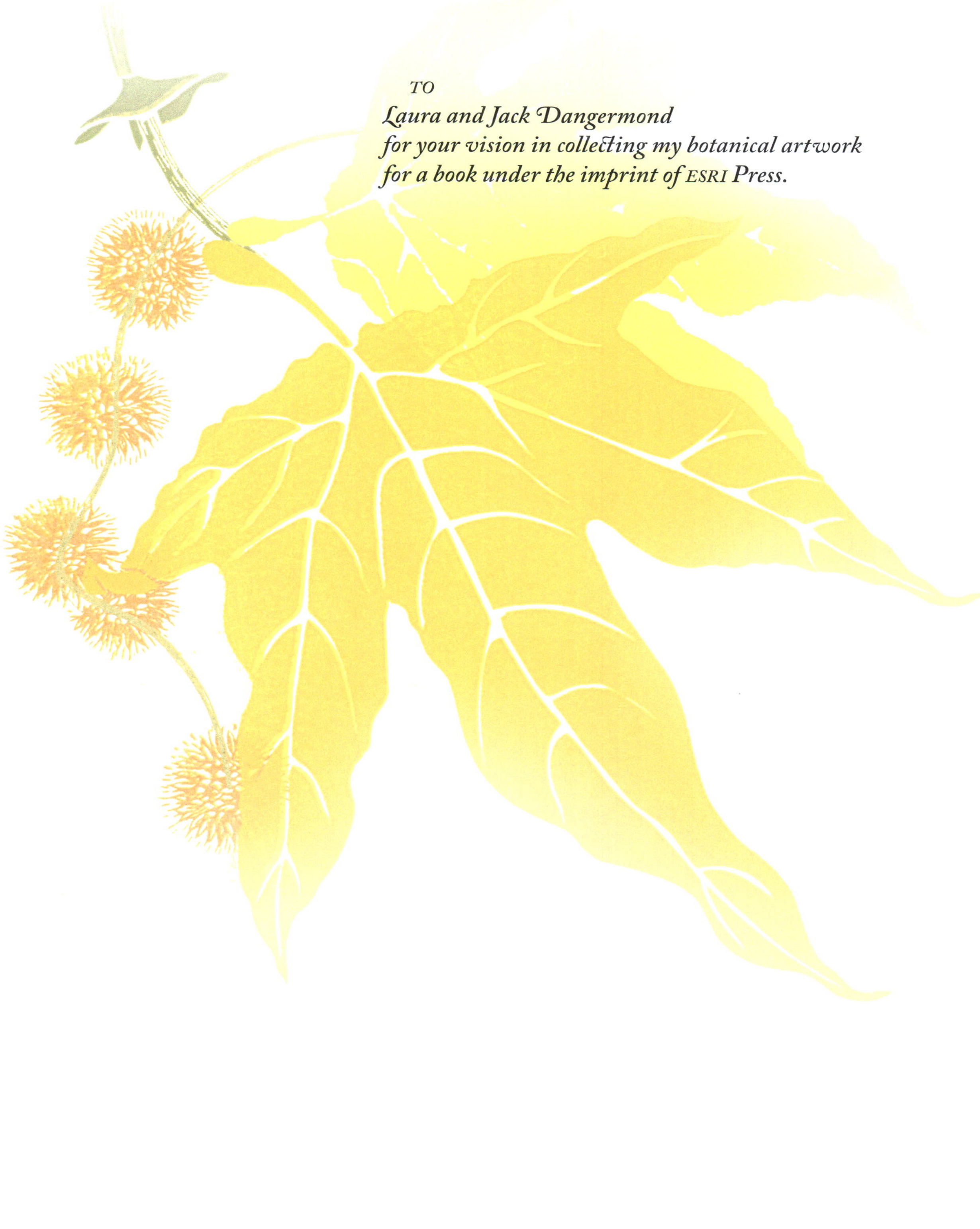

Collection 1

the Golden native 1972–1974

Lily *(Lilium humboldtii* var. *bloomerianum)*

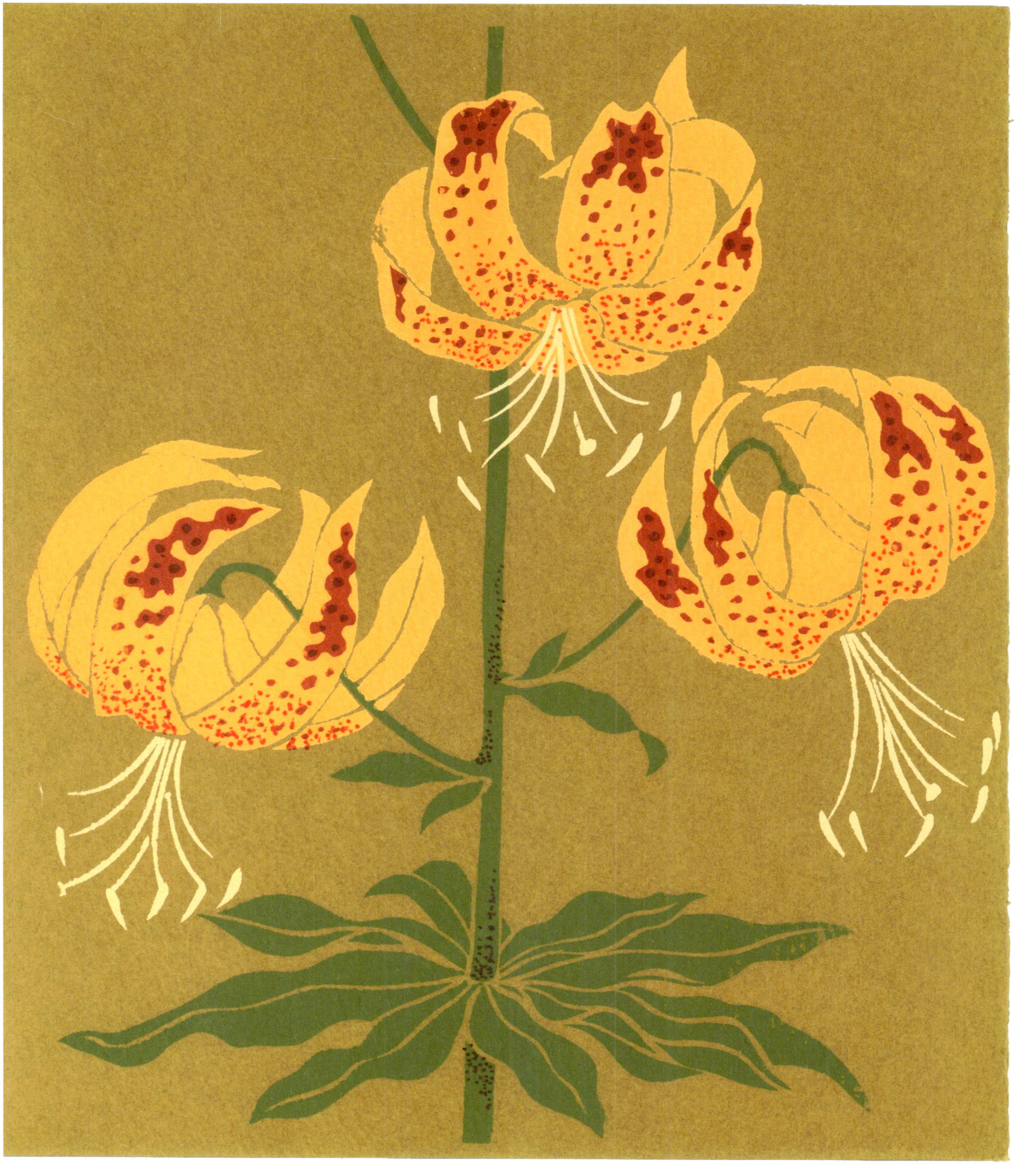

July 1972

The plant group that is possibly both the least known within the state of California and the most cherished outside the state is the group of bulbous plants known as Pacific Coast lilies. Lilies have always fascinated me, and I vividly recall seeing my first wild lily. Actually, it was the first lily I had ever seen, wild or tame. I was a child and on a family outing when we accidentally discovered several magnificent ones growing in a very isolated, remote area of Silverado Canyon in Orange County. In all my life—all ten years of it—I had never seen anything like those lilies. Of course I was told they were "tiger lilies," a popular name then and now. They were exquisite, and now I know that they were *Lilium humboldtii* var. *ocellatum.* I've often wondered if they are still there. This variety is much more spectacular than either the typical form, *Lilium humboldtii,* or variety, *bloomerianum,* that I drew. *Ocellatum* grows taller, and the flowers are larger and more richly colored. The name means "with little eyes," which describes the rings around the spots on the petals. *Ocellatum* grows from San Luis Obispo County to Orange County, thus having a much more southern distribution than the type that has been described as the "glory of the Sierra Nevada range of Central and Northern California (Tehama County to Fresno County), to which it is limited ...."

Lilies are one of the many genera of plants confined to the northern hemisphere. Among the eighty-seven known species of lilies in the world, forty-nine are native to Asia, twelve are native to Europe, and twenty-six are native to North America. Of the latter, fifteen are native to California. I have seen nearly all fifteen growing in the wild, thanks to Mr. Frank Ford, who guided my husband and me to all their secret hiding places in the month in which they bloom (July). I still cannot decide which species is my favorite—each in its own environment is enchanting—but I can speak about the one I drew and the one I would have drawn had it been in bloom, a species that is distinctively and almost solely native to the district in which I live.

Lilium humboldtii variety *bloomerianum*

Reproduced here is my version of the alpine form *Lilium humboldtii* variety *bloomerianum,* native to my home area at 5,500-foot elevation in the San Bernardino Mountains. I find it especially in the remote canyons where human beings have not dug the bulbs. I have grown it from seed quite easily, but mine resides within exactly the same conditions it enjoys in the wild—the wild area where it grows, after all, is less than a half mile from my home, so it is no wonder I am enchanted by it.

Lilium humboldtii variety *bloomerianum* grows at high elevations in Southern California and is smaller in all its parts than the type and the variety *ocellatum.* The plant grows four to five feet tall and has between ten to twenty blooms. It is a true Turks cap flower, a medium orange with crimson blotches on the tips of the petals and spots of the same color. The foliage is very lovely with wavy edges and arranged in whorls.

I am always a bit distressed when someone says to me, "Oh, you have some tiger lilies." This common name is reserved (or should be) for that lily known botanically as *Lilium tigrinum,* a native of China, Korea, and Japan. I am so proud that mine are California natives and natives of my own surroundings that I don't like them referred to by any other name.

Lilium parryi

I can't resist telling you about this lily. It is both elegant and rare, growing at high elevations (from 5,000 to 9,000 feet) in Riverside and San Bernardino Counties, and first discovered in the San Bernardino Mountains in 1876. One of the few remaining areas where it may be found in the wild is a very high, moist mountain meadow, surely one of the most beautiful areas in all Southern California and only about fifteen miles from where I live. The flowers of *L. parryi* are composed of a funnel-shaped bloom of an exquisite lemon yellow and lightly spotted brown. It can reach up to six feet tall with as many as fifteen blooms. Commonly known as lemon lily, it is delightfully fragrant.

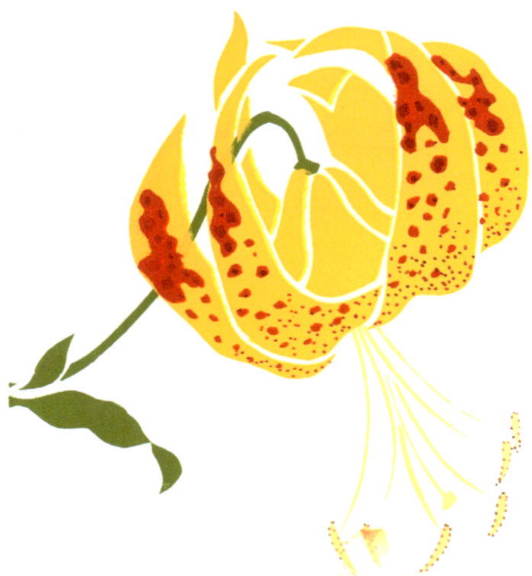

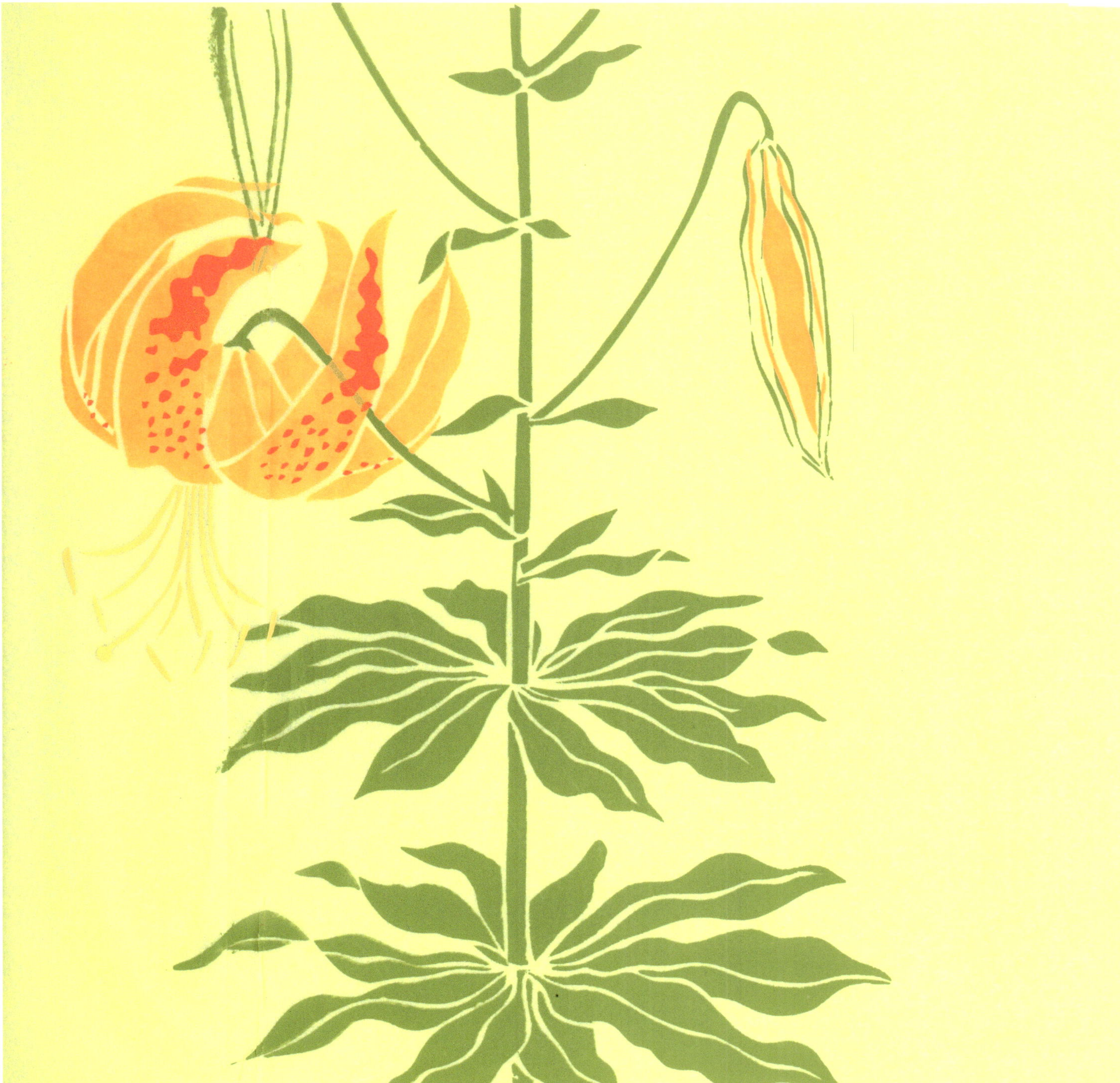

Wild buckwheat *(Eriogonum fasciculatum)*

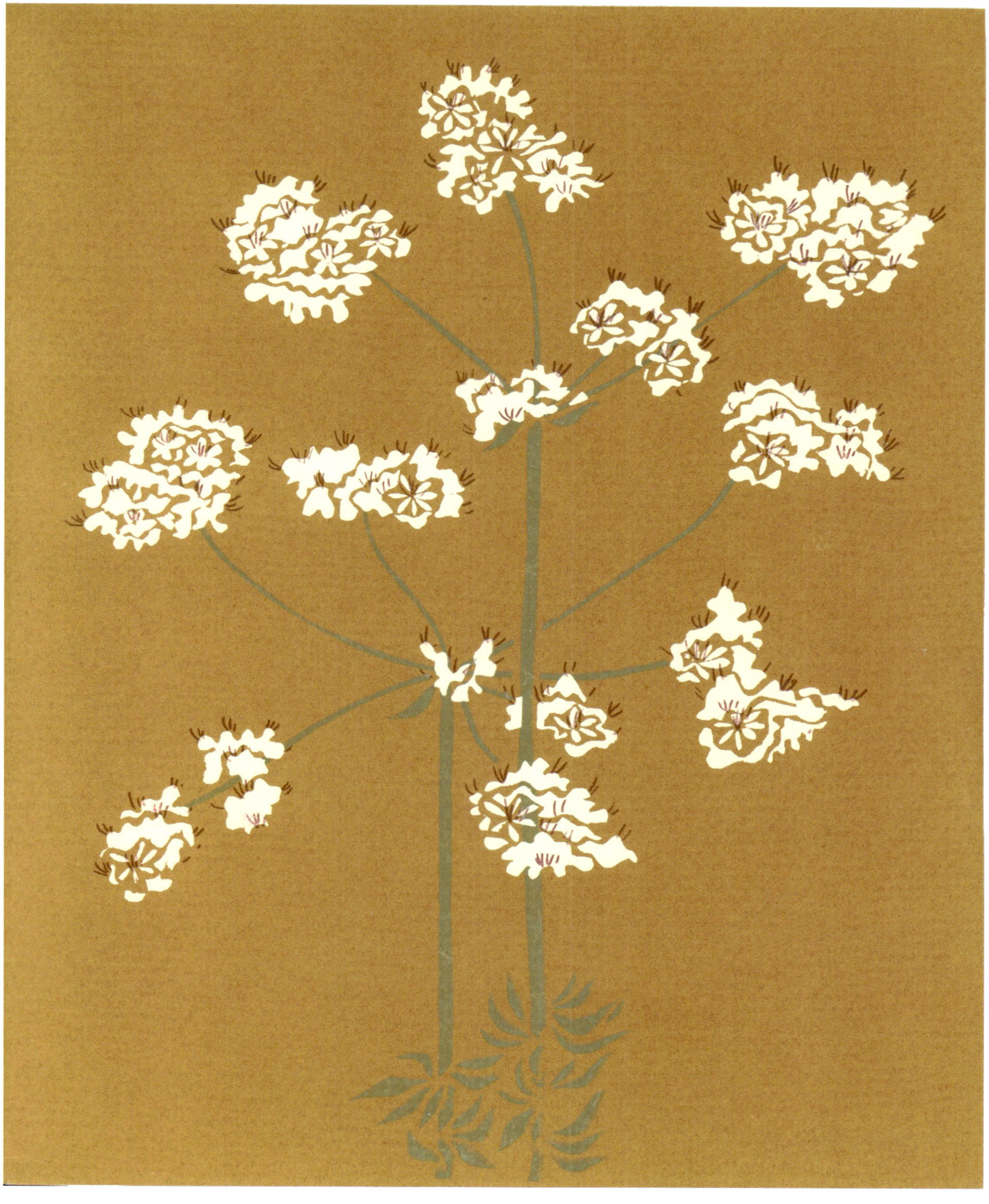

August 1972

Among the most abundant and least appreciated group of plants in California are those of the genus *Eriogonum*. More than one hundred of the total one hundred sixty species in the genus are native to the state, so perhaps we take them for granted. There is an *Eriogonum* species native to almost everywhere in California, including coastal, desert, valley, foothill, and high mountain areas. In fact, while one particular species of these natives has never been found growing in a native state on the mainland of California, it does grow wild offshore: *E. giganteum* is native only to the Channel Islands, particularly enjoying the windy exposure it likes best on Santa Catalina and San Clemente islands where it reaches a height of seven feet on dry slopes that are steep and rocky.

The name *Eriogonum* is derived from two Greek words, *erion* (wool) and *gonu* (knee or joint); both refer to the hairy joints and stems of some species. The group is commonly called wild buckwheat, a name that always evokes in me visions of creamy whites and rich, earth colors—russet, the siennas, the umbers. (They are in no way connected with the herb cultivated as a food plant whose seed is ground to make buckwheat flour, however; that is the Eurasian native, *Fagopyrum sagittatum*.) On most of the *Eriogonums* the flowers are in globular clusters and are usually white, yellow, or bronze, though some are pink. Many bloom in the heat of summer, and especially the white flowers that blanket large areas offer a refreshing illusion—a vista covered with snow under the bright August sun. Then the summer months fade and the tiny white blossoms take on the mellow russet tones that are so typical of a California autumn.

Ranging as they do from one-inch dwarfs to shrubs six feet tall or more, plants in this genus are an interesting group, attractive both in their native habitat and in cultivation. What other California native plant as large as *E. giganteum* (known as St. Catherine's lace) could possibly produce such an airy, lacy effect during its blooming season? Leathery, oblong leaves, soft, gray green above and white beneath, grow to the top of its branches, above which the naked stem rises high to proudly hold its flat, wide, and intricate bloom. *E. inflatum* or desert trumpet, with its oddly shaped, inflated stems topped by a lacy and delicate branching pattern, provides quite a contrast to its harsh desert environment. Imagine General John C. Fremont's relief, while crossing the Mojave Desert in 1844, when he discovered it: with oddly bulging stems that inflate below the flower and are hollow and moist inside, this desert inhabitant could become something of a thirsty man's canteen. Then there's *E. parishii,* called mountain mist because of the ethereal effect of its diffuse branching—a dense, rounded mass of very slender branchlets and miniscule flowers. As the plant ages it turns a gorgeous claret red, making one glad that wherever the seeds of this species fall, little plants pop up.

Eriogonum fasciculatum

Called wild buckwheat, California buckwheat, or flat-top, this low spreading perennial grows on dry hillsides and canyons near the coast. Two subspecies grow in the foothills, mesas, and at high elevations, which is why I happened upon my occasion to draw it. Native to the mountain areas of Southern California, variety *foliolosum* grows in abundance not far from my home. While I was prowling around investigating this area I suddenly came upon a vast array of beehives, and though quite startled, once I regained my wits I thought, how smart of whomever it was to facilitate the collecting of buckwheat honey this way. *Eriogonum fasciculatum* is very attractive to bees and is the third most-valued native bee-plant, after white sage and black sage.

Another important attribute of *E. fasciculatum* is that it grows in hot, dry soils and is thereby extremely valuable in controlling erosion. Its flowers are usually white and very small, but they form large, feathery, flat-topped clusters that sit atop long, bare stems standing stiffly upright all over the bush. These flowers turn a rusty brown in late summer. The leaves are narrow, lanceolate, and quite green.

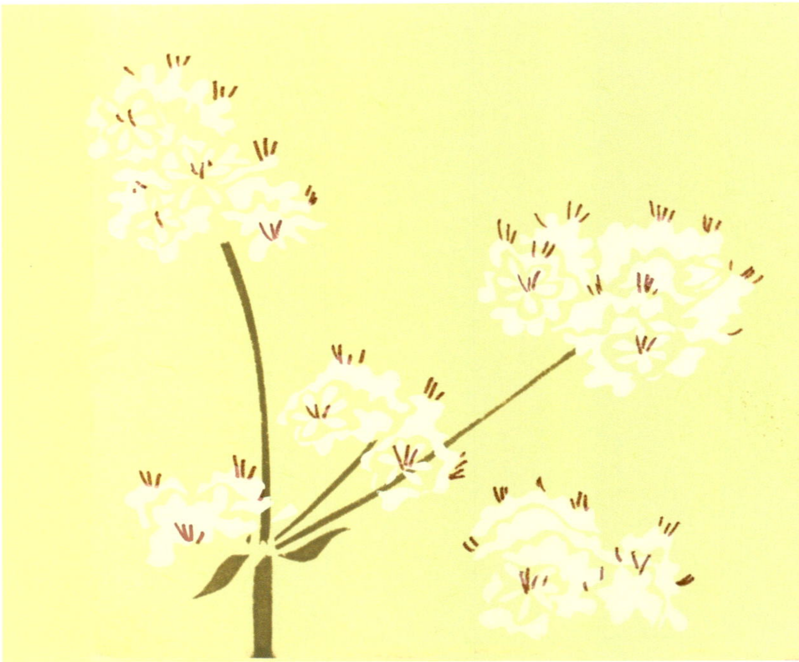

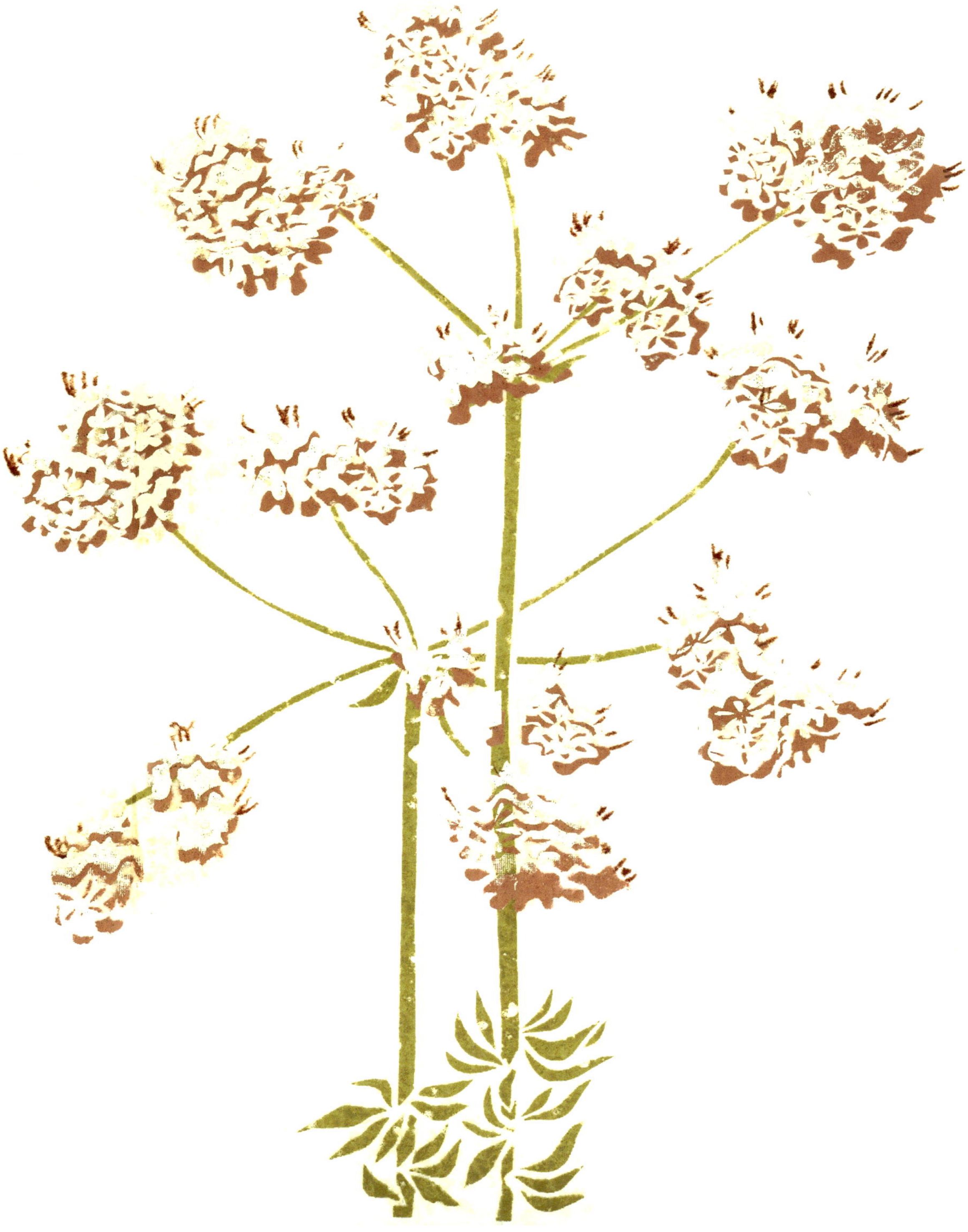

Bush chinquapin *(Castanopsis sempervirens)*

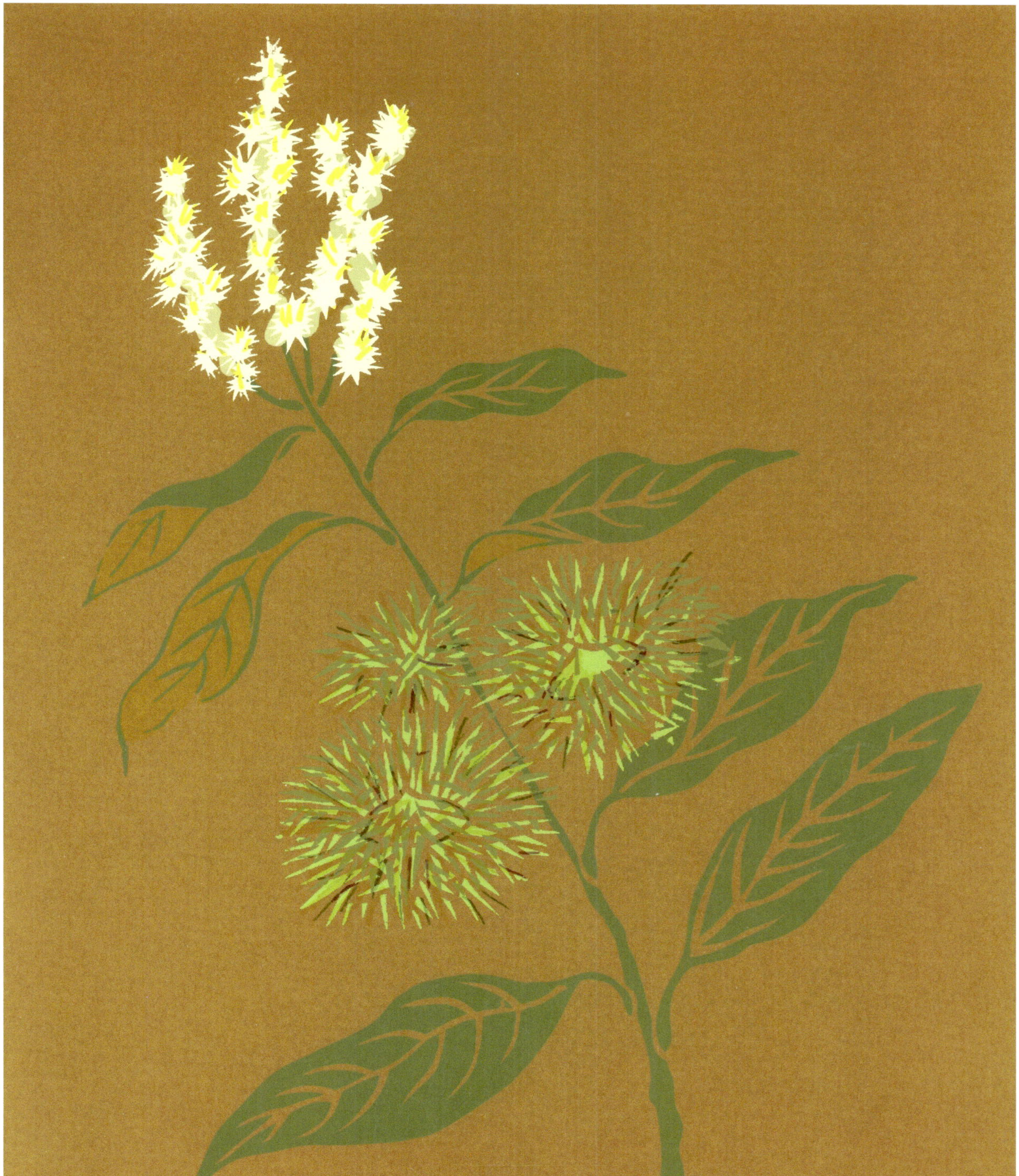

September 1972

The genus *Castanopsis*—from the two Greek words *kastanea,* meaning "chestnut," and *opsis,* meaning "resemblance"— includes about thirty species, all but two of which are widely distributed throughout Asia. The two that are not are found instead in the United States, both native only to the Pacific Coast: *C. chrysophylla* is the tree form and *C. sempervirens* is the alpine shrub form you see here. The tree is known by the common name of chinquapin (also golden chinquapin, western chinquapin, and golden-leaved chestnut). The flower of the chinquapin is very much like that of the chestnut in appearance and odor.

C. chrysophylla is rare among trees of the world in that it is found only in the very limited area of the Northwest to which it is indigenous. The trees that exist today range in height from fifty to seventy-five feet. Occasionally one is found that exceeds one hundred feet. This species reaches its greatest size in the coastal valleys of northwestern California and southwestern Oregon.

The glory of this tree is its dark, lustrous leaves—evergreen like the native shrub—and three to six inches long with golden yellow, fuzzy undersides. This gives the tree a green and golden look about it, an especially lovely sight in summer when it is in full bloom. Small creamy white flowers borne in clusters of fluffy spikes several inches long compose the blooms, particularly pleasing against a background of dense, glossy foliage. When in full bloom, the only tree in its range that resembles it is the tan oak, whose flowers are very similar. The chinquapin, however, can always be identified by its distinctive leaf with its golden undersurface.

Another distinctive feature of the tree is its fruit—a nut enclosed in a prickly burr much like a small chestnut burr. They are about 1 to 1 ½ inch in diameter, the burr often a little shorter than the nut so that the nut tip protrudes a bit. The nut is sweet and matures at the end of the second season. But it fruits sparingly, and in fact this weak seed production has been thought to be the reason why the tree is relatively rare compared to its closest forest companions, the redwood, tan oak, madrona, and Douglas fir.

Castanopsis sempervirens

The tree *C. chrysophylla* stands like a giant sentinel reaching for the sky, yet its fellow California native species makes an equally majestic display by hugging the ground. The alpine shrub form of the genus *Castanopsis, C. sempervirens,* makes a big impression on me even from a distance. Its leaves, flowers, and fruit are very similar to its tall cousin. Known as bush chinquapin or Sierra chinquapin, it even shares in part of its common name. But there the similarities end. *Sempervirens* is the alpine species that grows more to the east and to the south than *C. chrysophylla* and generally at higher elevations. It forms thickets on dry mountain slopes or rocky ridges and composes an important part of the very high elevation chaparral in the San Bernardino and San Jacinto mountains. In these areas where I live it grows at elevations of 8,000 to 10,800 feet (where I gaze long and admiringly upon it); in the Sierra Nevada it grows at elevations between 6,000 to 9,500 feet.

At such heights the mountains are clothed in a mantle of tight, dense, rounded mats about three feet tall, each one spreading out six or eight feet. *C. sempervirens* forms these mats with *Ceanothus* and manzanita with which it usually grows. Together they make the mountain look like a marvelous patchwork quilt of irregular shapes in rich shades of blue, green, and yellow green. The green is a species of manzanita, the blue is a species of *Ceanothus,* and the yellow-green of course is the shrub chinquapin. I can always detect from a distance where the chinquapin predominates because of that golden quality emanating from the undersides of its leaves.

I could gaze upon this covering all day and never tire of it. I always marvel at the exquisite placement of each plant and its handsome coloring in relation to the other plants and their coloring. It constantly amazes me that man had nothing to do with the arrangement of these particular colors and textures. Such infinite variety already exists in nature! Of course then my thoughts go to how unfortunate it is that so many plantings made by man are in straight lines and the space between is evenly measured. We could benefit so much from nature's better eye if only we would be a bit more observant and willing to learn.

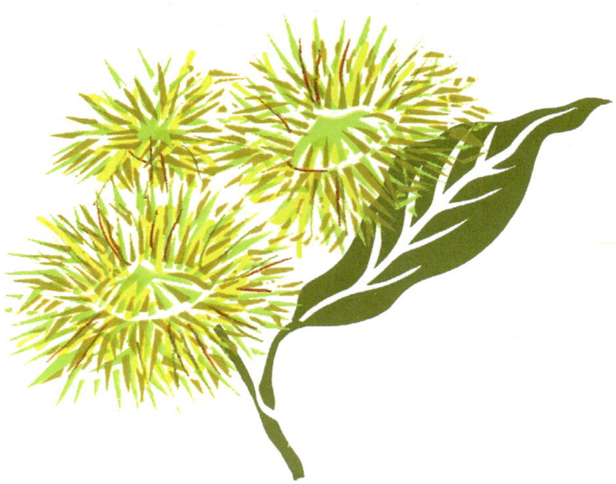

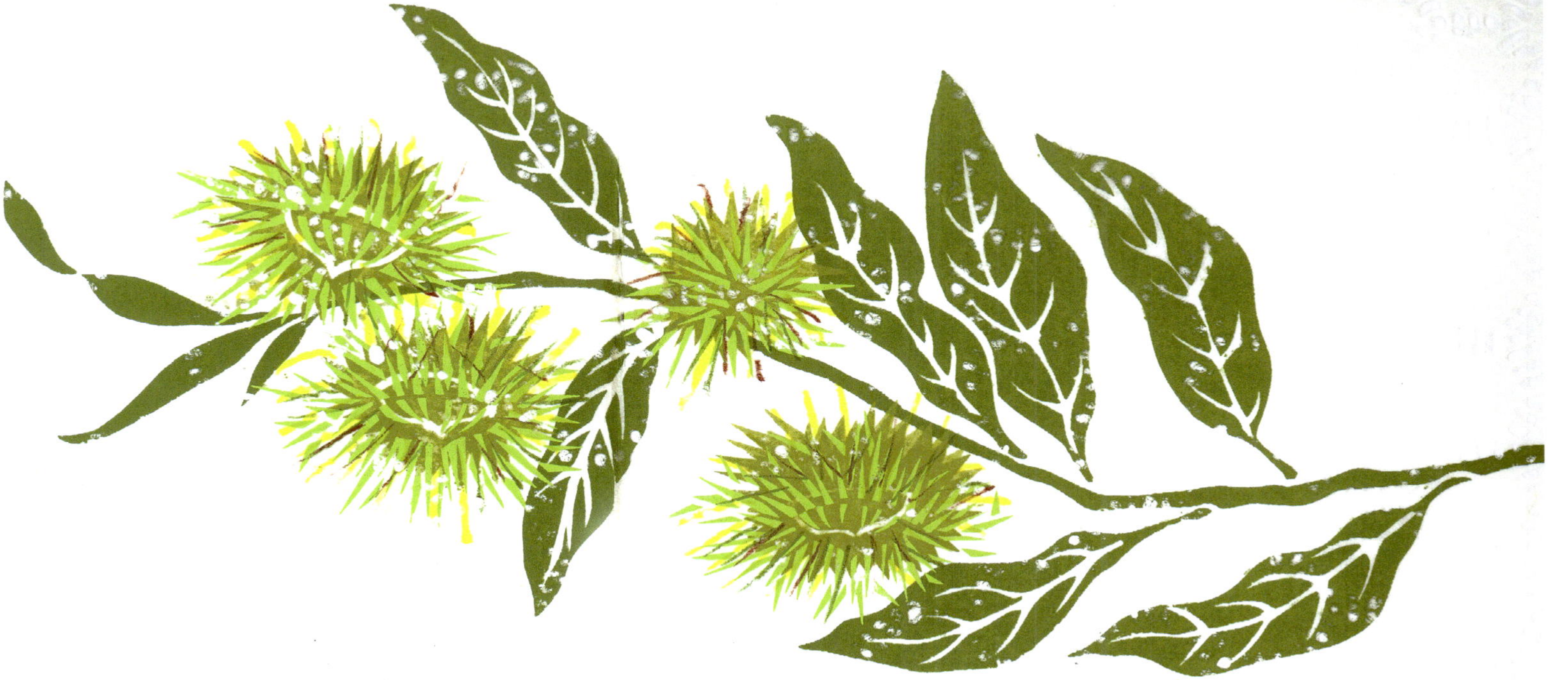

Vine maple *(Acer circinatum)*

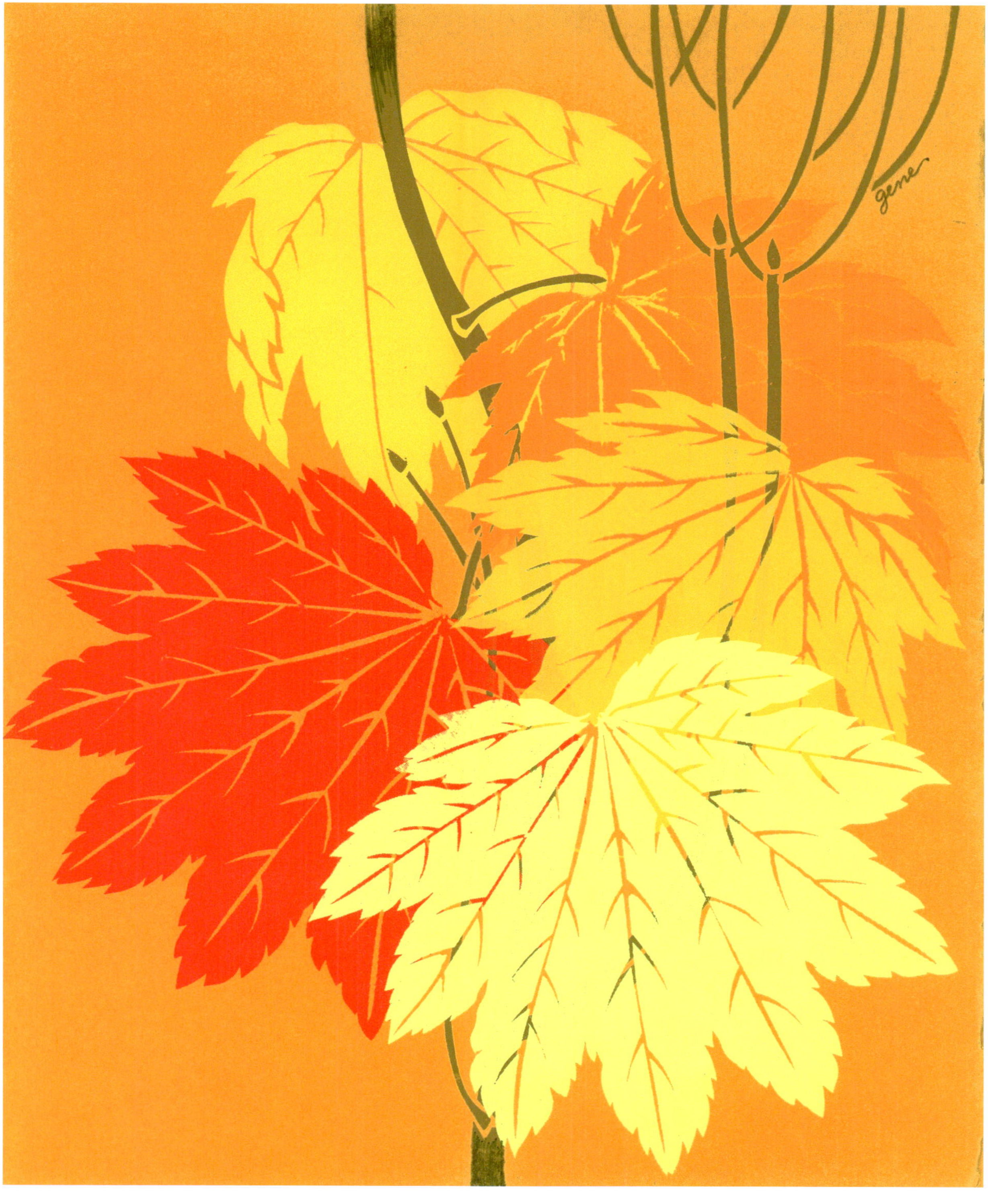
gene

October 1972

To me, each day, week, month, and season in the year has its own particular splendor. But the month of October and with it the autumnal season has always held a special enchantment for me. It is at this time of year that all of nature's warmth, mellowness, and goodness is felt in the abundant bounty that this earth produces. It is a time of maturity and a time of harvest.

October is the month of shorter days and cooler nights. It is the time when deciduous trees, like the maple, display their most lavish leaf color before shedding them for winter dormancy. California forests may never be as flamboyant as eastern forests. But that truth can't diminish the soft golden glow that permeates our forests at this time of year.

Acer is the classical Latin name for the maple tree. Most people associate the maple with the hardwood forests of the eastern United States and Canada (the maple is as dearly loved by Canadians as the oak is by the English), yet the genus *Acer* includes approximately one hundred twenty species widely scattered over the northern hemisphere. Twenty-eight species are native to North America; four of these are native to California.

Even though western maples may be overshadowed by their eastern cousins in general, they do tend to stand out where they are: the forests of the eastern United States are basically hardwoods, while those of California are coniferous or evergreen. So wherever a maple is found in California, it is usually framed by an exquisite green background and does not have to compete with ten other species of hardwoods for recognition.

The largest and most commercially important hardwood tree of the Pacific Coast—in fact its *only* large California native maple—is *Acer macrophyllum,* commonly known as big-leaf maple, broad leaf maple, and Oregon maple, and found scattered or in small groves from Alaska to Southern California. At a height of one hundred feet, with a trunk diameter of twelve to twenty inches, such a tree, its interlacing limbs forming a forest ceiling of moss-covered arches, makes what John Muir called "lofty sky gardens." Its leaves (six to fourteen inches wide), the largest of any maple, turn a clear, golden yellow in the fall.

Still, of all the trees native to California, only two could compete with the exquisite autumnal tints of eastern trees. One is *Cornus nuttallii* (see page 51), considered among the most handsome of all North American trees, and the other is the one I have chosen here, *Acer circinatum.*

Acer circinatum

From the Latin word *circinatus,* meaning "ring-shaped" or "circular," the *circinatum* earned its name by the circular shape of its leaves. This is a maple that is often vine-like or prostrate, thus its common name, vine maple. Half tree, half shrub, it grows from five to thirty-five feet tall with a trunk diameter of only three or four inches. Often the slender trunk will curve out and take root forming a vine maple thicket or clump. It is usually found as an understory tree in very dense shade, but it can grow in the open, as is evidenced by its presence in places where the cover trees have been destroyed. The tree's natural habitat is from British Columbia south into California.

The leaf of the vine maple, from four to six inches long, has from seven to nine sharply pointed, cut-toothed lobes, which form something of a semicircle in outline. The leaf is thin in texture and particularly delicate in appearance, attractive from its first unfolding in spring into a soft shade of green. Often the petioles or stems of the leaves are scarlet, and the vigorous young shoots usually carry a tinge of scarlet, too. Its true glory is revealed in October when it provides the only real scarlet in our forests. It has some of the same luminosity of the exquisite *Acer saccharum* (sugar maple) of the East. While other native trees do provide some autumn color, none are as beautiful as this tree.

Acer circinatum has been widely used in landscaping, especially in Northern California, Oregon, and Washington. It grows slowly, but always looking neat and refined, it is often used quite appropriately as a substitute for Japanese species. It deserves to be even more widely used. It likes rich humus, summer watering, and shady situations. If you can provide the conditions it requires, do plant this exceedingly attractive California native. It will give you much pleasure.

Excellent specimens of *Acer circinatum* can be seen in the California Native Plant section of Strybing Arboretum and Botanical Garden in Golden Gate Park, San Francisco.

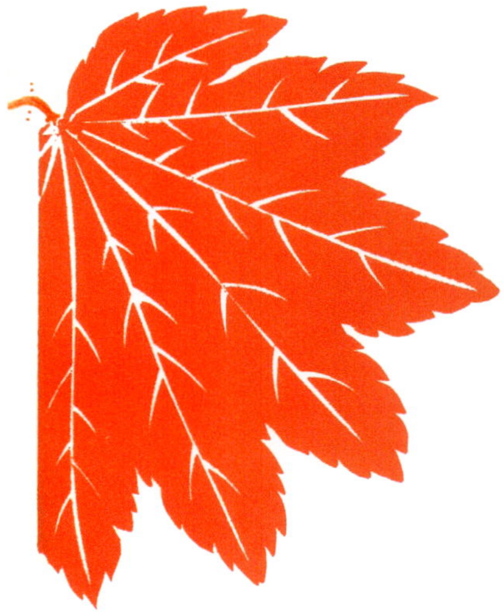

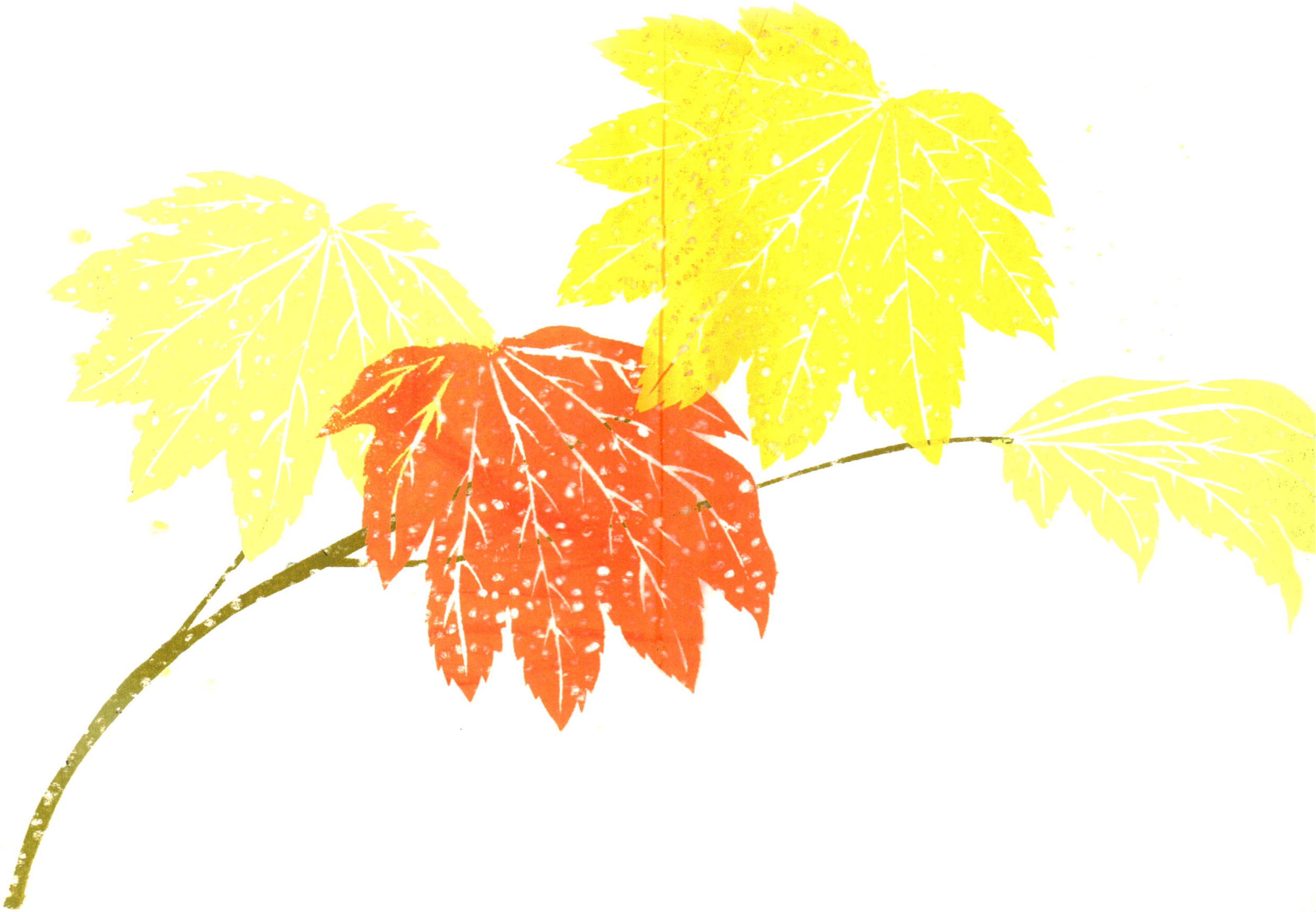

California fan palm *(Washingtonia filifera)*

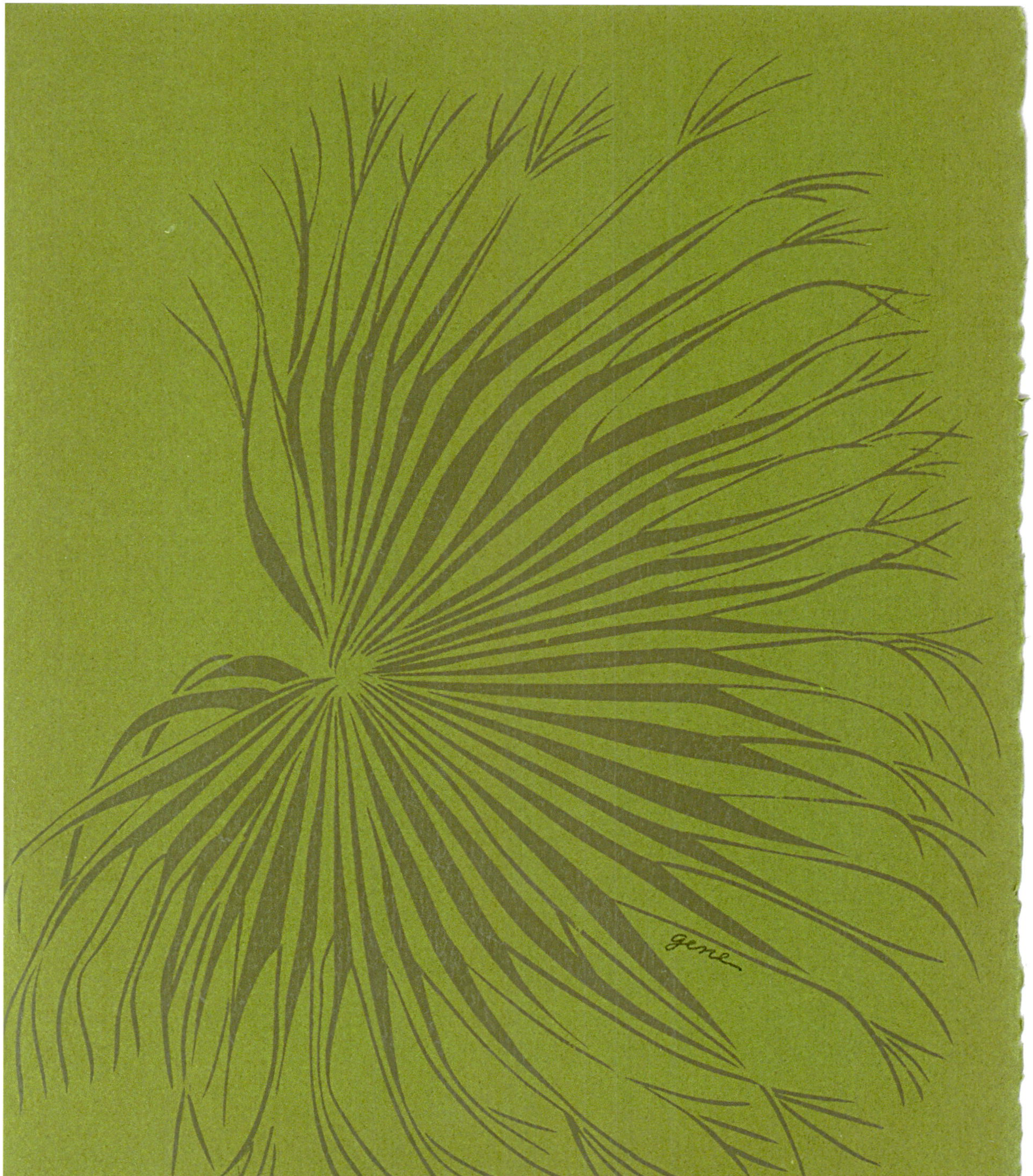
gene

November 1972

California is known as the land of palm trees, yet the only palm native to it, *Washingtonia filifera,* grows in the wild in a very limited area in the state's southern portion. So rare is any tree at all in the desert that a dozen or more towns or scenic regions in this small area have celebrated the romance of the palm in their name (Twentynine Palms, Thousand Palms, Hidden Palm Canyon, and Palm Desert). Commonly known as the fan palm, desert palm, or Washington palm, it grows wild mainly in Palm Canyon near Palm Springs. Yet of all North American palms, it has traveled the farthest under cultivation, and as an ornamental, is an integral part of urban landscaping in cities along the Gulf Coast, Florida, Hawaii, and the Mediterranean.

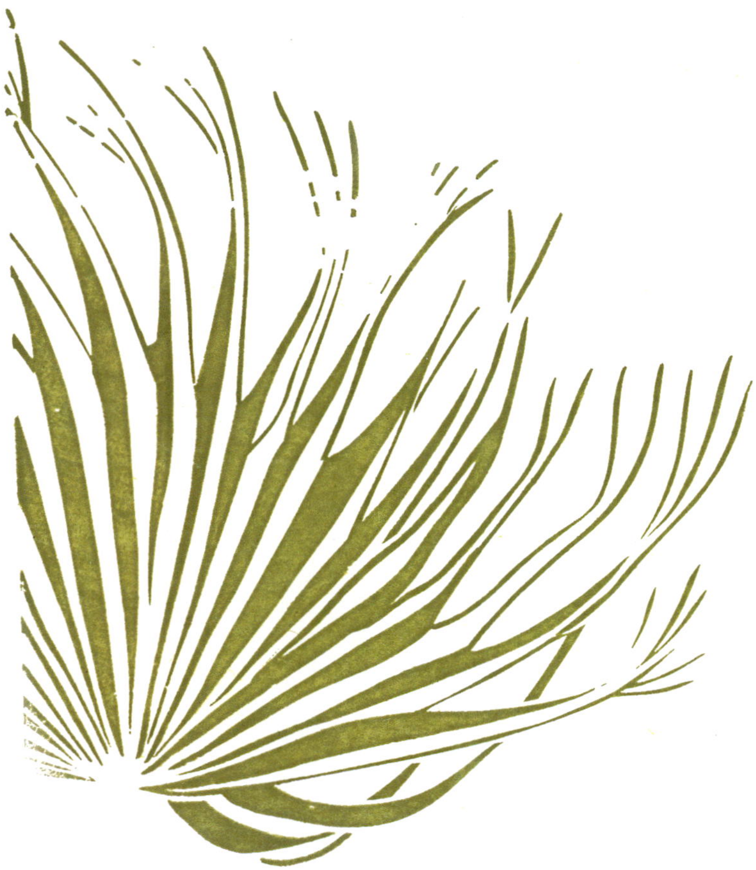

Washingtonia filifera

Filifera is one of two species in the genus *Washingtonia* and the only one found within the United States. (The other, *robusta,* is native to Mexico.) Being a monocotyledon, washingtonians are unique among the Golden State's native trees insofar as their woody tissue is not arranged in annual concentric rings, but rather scattered irregularly throughout the trunk. While other trees have two or more seed leaves, monocotyledons bear but a single seed leaf when the seed is newly germinated. The veins of the leaves are usually parallel as in leaves of grasses rather than the feather and net-like veinings of most tree leaves.

The fronds (leaves) of all palm trees are shaped either like a long feather or a round fan. The species indigenous to California, the *filifera* is a fan-type palm. In fact, it has the largest and strongest leaves of all fan palms—some are six feet in length and as wide as they are long. The petioles (stems or stalks) are as long as the leaf or longer, both edges armed with irregular, sharp, and often hooked spines or teeth. In Latin, *filifera* means "bearing filaments or threads," and refers to the way the leaves are deeply slashed into stiff, ribbon-like divisions with the edges and ends frayed into many long, threadlike fibers. Usually forty or more of these fan-shaped fronds form the tree's crown.

It is interesting to watch new palm leaves open. They start closed, thin and upright like green swords, then gradually open until they are completely pleated. They are solid on the lower half with the upper half slit into long fibers. The newest leaves appear at the top of the tree. Starting upright, slowly they begin to lean toward the other fronds, eventually spreading out at various angles. Some of the older leaves bend backward, and as they continue to age, becoming yellow and dry. Then turning brown, they hang down and hug the trunk.

For this reason, trees growing in their natural habitat of desert canyons and ravines appear somewhat different than those in cultivated areas. Untrimmed, all the dead fronds hang down ultimately and form a skirt around the trunk. This dense, thatched mass of dead leaves can become a column as thick as eight feet, reaching from the ground to the top of the trunk, and so solid that even the desert's strong winds can't tear the leaves away. It requires sawing to remove each frond, an important thing to do in cities where they are real fire hazards.

In May and June, small whitish flowers are borne in large clusters eight to ten feet long. Birds and other animals eat the small berry-like fruit that develops when it turns black and ripe in autumn. Birds use threads from the fronds for nest-building. Native Americans used the fibers for basket weaving, the leaves to thatch houses, and sometimes the trunks as timber for building.

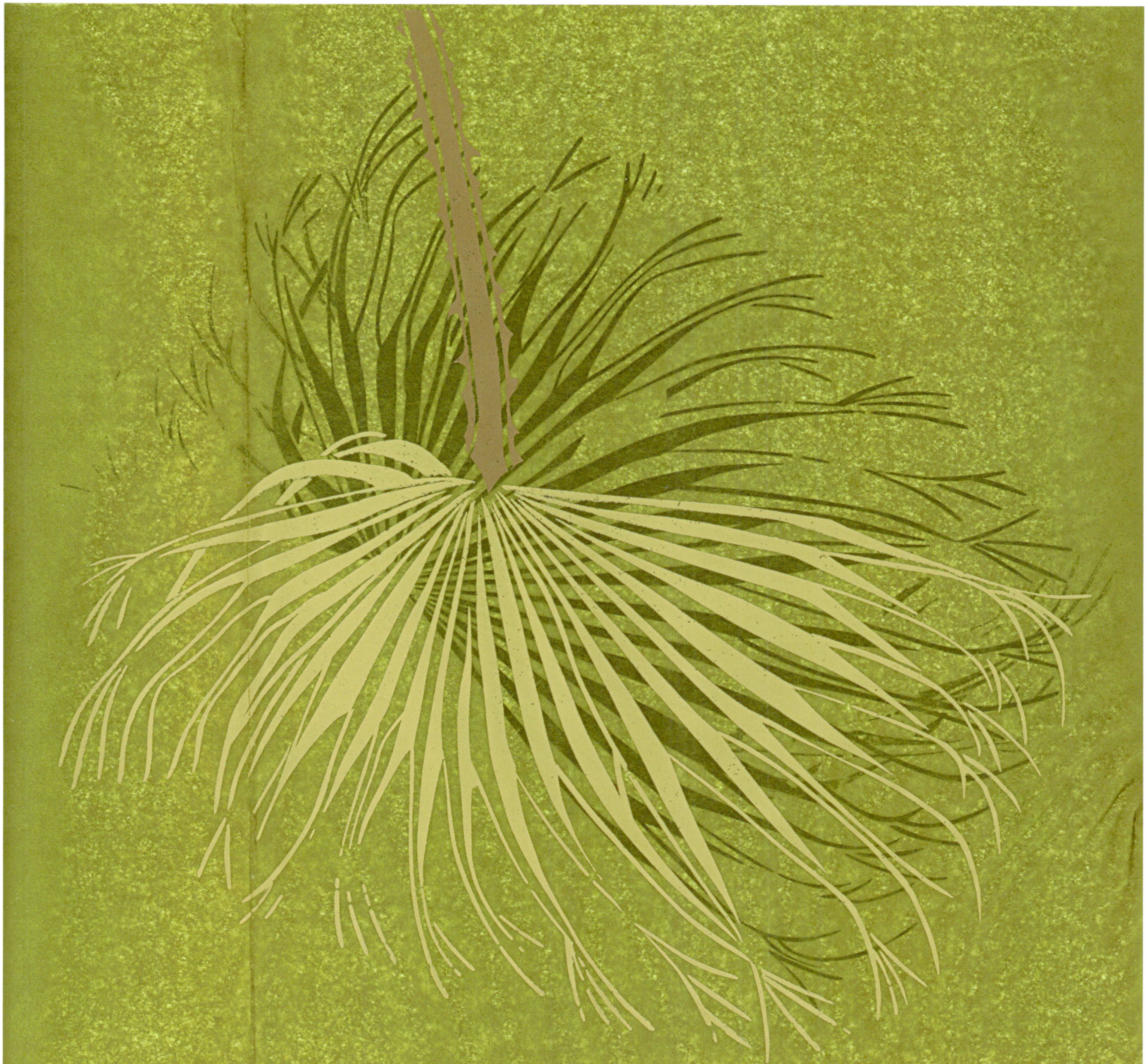

Big-cone Douglas fir *(Pseudotsuga macrocarpa)*

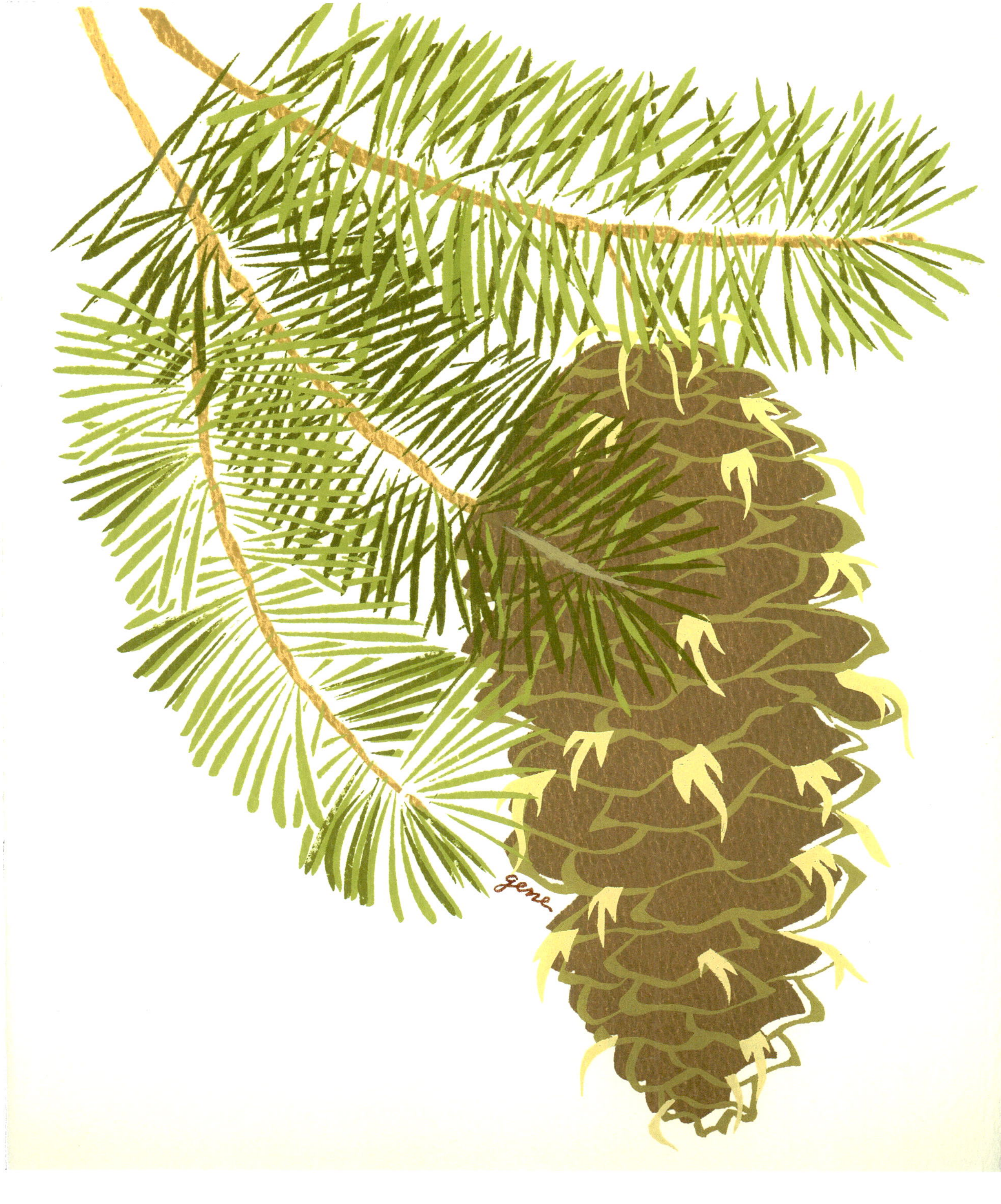
gene

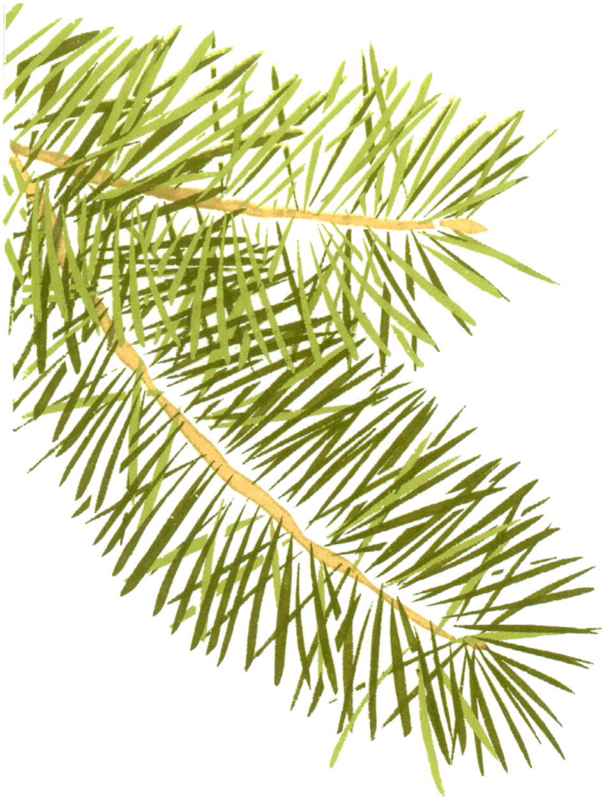

December 1972

The genus *Pseudotsuga* includes five species, two of which are native to western North America. Of those two, one is native to Southern California and one is native to Northern California. The northern species, Douglas fir, is famous of course. The most abundant tree of the Northwest and the state tree of our neighbor to the north, Oregon, Douglas fir is renowned for its extremely strong wood. The southern species shown here is little known and is found nowhere else in the world but here.

Pseudotsuga macrocarpa

This California native, indigenous to nearly all forested areas of Southern California, is particularly symbolic of the season for me, and not just because it brings to mind the proverbial Christmas tree. *P. macrocarpa* is usually the first coniferous tree one encounters when ascending the higher elevations in both the San Gabriel and San Bernardino mountains. Like the genus overall, these trees have broad pyramidal crowns and are particularly lovely and graceful in their youth.

This species grows on many of the near-vertical cuts that resulted when State Highway 30 was constructed as the road up to the San Bernardino Mountains. I've always had a very special fondness for this tree in how it manages to grow in so many rugged situations. On my way home one day I was delighted to see one about twenty feet above the highway surface, and admired how it was growing on a very steep cut in decomposed granite and sharp jagged rocks. Looking up at so many exquisite young trees that flourish on seemingly nothing, I often wonder how seeds arrive there at all, how they find enough mineral soil on which to germinate; how in the world do they survive?

Some years ago this particular tree, which stands not far from the official elevation sign at 5,000 feet, was about six feet tall. Nothing grew near it because the cut was so precipitous, and I couldn't imagine that any human being could get near it. However, four days before Christmas I was rather startled and amused to see that it was wearing holiday finery—someone had climbed up and hung half a dozen shiny baubles on it. The ornaments adorned the tree well into early spring, but gradually they disappeared one by one. Still, I always had a warm feeling when I passed by the tree, remembering how I came to realize that someone else admired it, too.

The Christmas after that, however, my heart nearly broke. Apparently someone had climbed up and cut the tree. I could not understand then, and still cannot, this kind of human behavior. Fortunately though, it was cut above a whorl of branches, so it formed a new leader and by now has almost grown back to its former lovely self.

Most of the time *P. macrocarpa* are nestled in ravines at lower elevations, their upper limit being about 6,000 feet. This is an extremely long-limbed tree, with its lower branches swooping out horizontally to great lengths. In this respect, it is often confused from a distance with *Pinus lambertiana* (sugar pine). From the Greek, the generic name means "false hemlock." Indeed, all *Pseudotsuga* trees combine some of the characteristics of hemlocks, spruces, and true firs and are often mistaken for them as well as for pine trees.

Adding to the confusion about the genus in general, *P. macrocarpa* is known as big-cone spruce, big-cone fir, and big-cone Douglas fir. The name of the species, *macrocarpa,* does mean "big cone," and at four to seven inches in length, its cone is considerably larger than that of the Northern California species. In fact, cones are the most distinctive feature of *Pseudotsuga* trees in general, and examining them is the way to distinguish these trees from the others for which they are often mistaken.

Though their cones are pendulous like hemlocks and spruces, the cones of *Pseudotsuga* trees differ from those of other coniferous trees in that they have three-pointed scale-like bracts protruding between the scales. With scales rather thin in texture and rounded, the cones seem loosely put together and exhibit a wavy, graceful quality. To me the cones of the southern species are among the most beautiful of the world's cone-bearing trees. Knowing the cones is the most accurate way to identify these trees, but the needles help as well. Flat and ½ to 1½ inches long, the needles grow singly around the branch, forming long feathery sprays that lend a drooping effect to the tree, further distinguishing them from the stiff appearance of the fir.

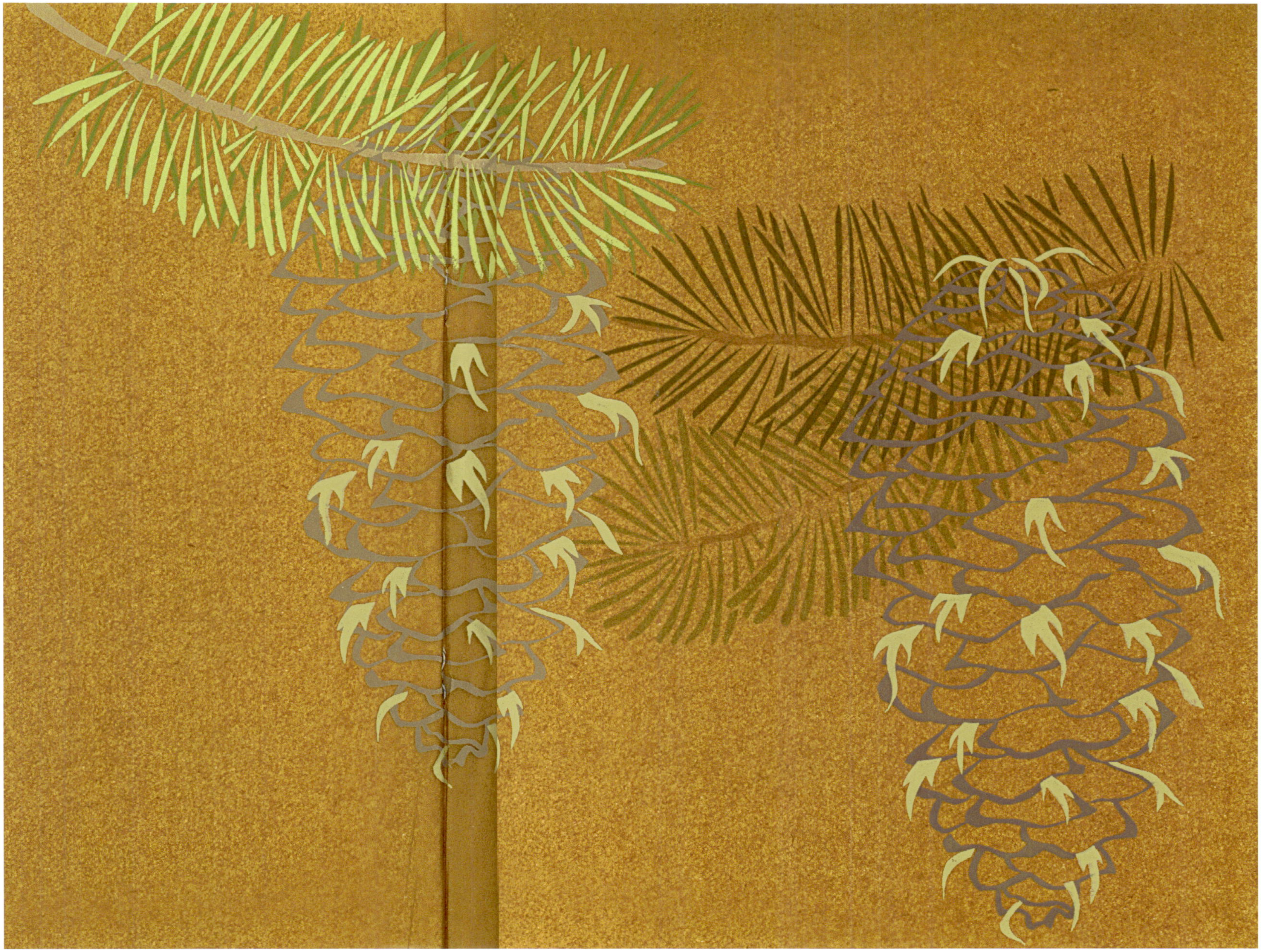

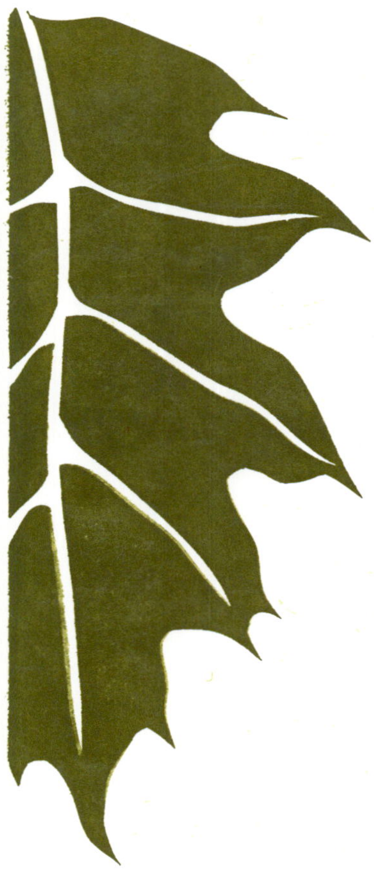

Oracle oak *(Quercus morehus)*

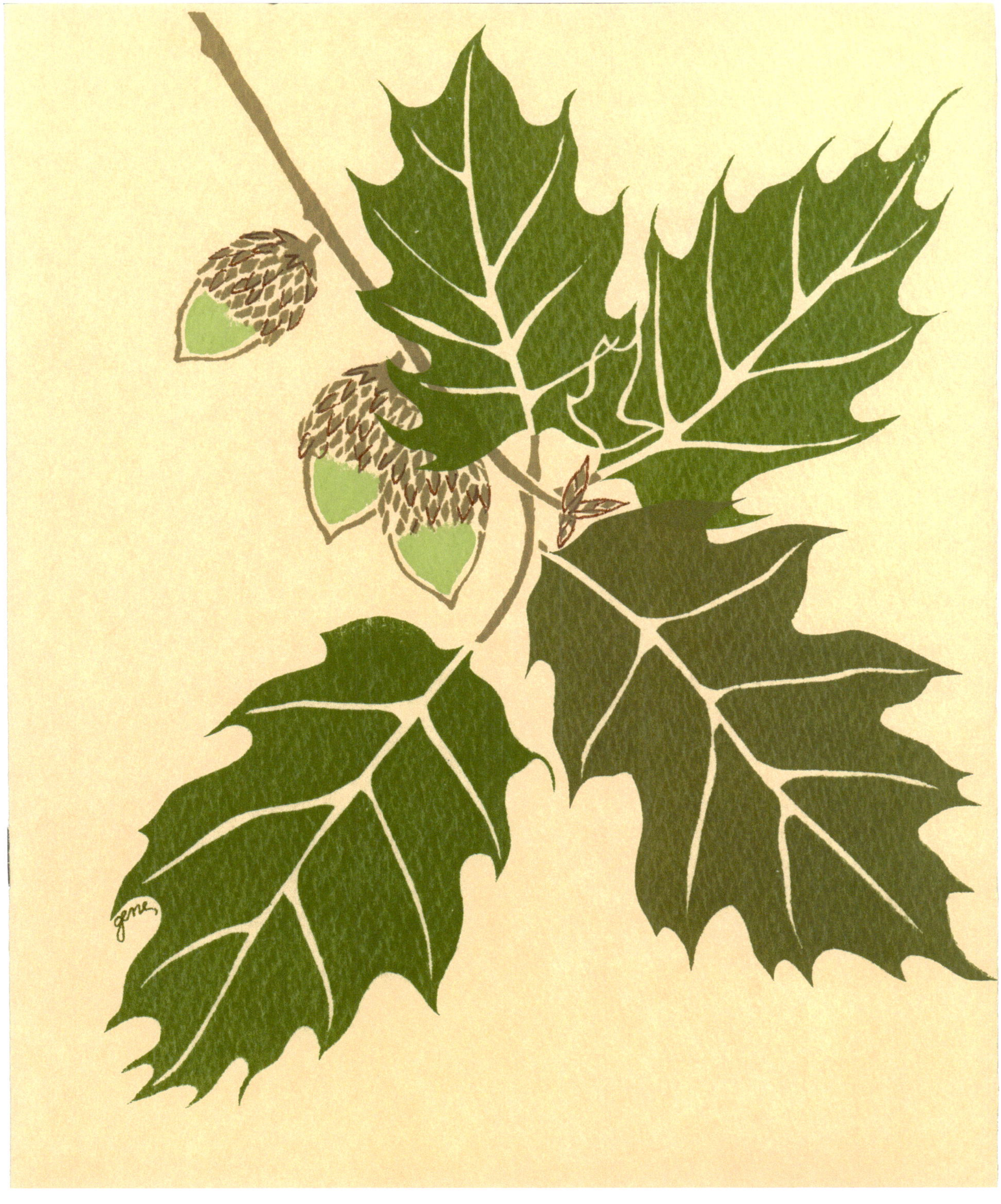
gene

January 1973

Since the beginning of time as we know it, the mighty oak has been symbolic of strength and durability, one of the most venerated of all sacred plants. Ancient mythologies tell us that many people believed the oak to be the first tree created and that humankind sprang from it. It was the celestial tree of the Celtic Druids, and druidic rites took place with the aid of the oak and its satellite, the mistletoe. *Quercus,* the classical Latin name of the oak, understates it well: a composite of the Celtic *quer,* meaning "fine," and *cuez,* meaning "tree," indeed, a fine tree it is. We have it in every part of California except for the desert areas in the southeast and the extremely high elevations. In fact, a superb assemblage of oak species from around the world is located in Shields Grove, part of the arboretum on the campus of the University of California at Davis. This is the largest oak collection in the western United States, possibly the world.

The genus *Quercus* is very large, numbering well over three hundred species widely distributed throughout the northern hemisphere and into the mountains of the tropics. More than fifty species are native to the United States, and fifteen of these (including one hybrid) are native to California. The genus includes both evergreen and deciduous species, and while many grow into enormous, rugged-looking trees, others remain dense shrubs. The most accurate way to distinguish the various species is to identify the leaf and the nut-like fruit of the tree, which is of course the acorn.

Many oak trees owe their existence to the bad memory of squirrels, who hoard the acorns, bury them for their winter food, then forget (often) where they have stored them. The earliest inhabitants of the state regarded this handsome, unique fruit of the oak as the staff of life. The favorites were those from *Q. lobata* and *Q. agrifolia.* After leaching them to remove the bitter taste, Native Americans pounded the acorns into flour, which was then cooked as a gruel or baked into loaves. Oak bark was used in California's early days for medicinal purposes and for curing hides and making dyes. Oak wood was highly prized as a fuel. In fact, on this very day when it is a blustery 18 degrees outside, this is the quiet, slow-burning wood keeping me warm.

The trees from which I did this drawing are located off the main highway in the general vicinity of Oak Glen, an area in which grow at least three different species of native oaks. The species I chose is a rare and unusual oak, and an especially interesting one at this time of year.

Quercus morehus

The leaves of *Q. morehus* are a rich, glossy green of medium texture, 2½ to 4 inches in length. Most oaks with leaves like these are deciduous; the leaves fall in October and November. Yet here I am in January, by which time most deciduous oaks are completely bare. Because it retains its green leaves well into winter, *Q. morehus* would seem like an evergreen oak, but it is not. It drops its leaves just before the buds burst open with new leaves in the spring. With this leaf dropping, the tree appears to be forecasting when spring is coming. So I think its common name is very appropriate: oracle oak.

In fact, after much investigation, I learned that both the specific Latin and common names of this species may be traced back to its 1863 discoverer, Dr. Albert Kellogg, an MD who liked sketching and writing about trees he thought might be new to science. Another nineteenth-century personage, Edward L. Greene, wrote about him, "A vein of religious feeling, which was deep in him, frequently obtained expression in his botanical writings." Greene described Kellogg's terminology as "somewhat original and his way of making Latin adjectives still more so, insomuch that grammarians have been obliged to correct the endings of many [species names] before giving them further currency." It wasn't easy, but further investigation has revealed that the biblical "Oak of Moreh" was connected with a sanctuary whose priests gave oracles on questions asked by worshippers.

Even today new specimens of *Q. morehus* are being found because it is a widely distributed species, although nowhere is it really abundant. This oak is most common in the Sierra Nevada foothills, and usually found growing in the company of *Q. wislizenii* and *Q. kelloggii,* of which it is a hybrid. The leaves of *Q. morehus* are a little smaller and less deeply lobed but very much like *Q. kelloggii,* a deciduous tree. Its acorns are very much akin to those of *Q. wislizenii,* an evergreen tree.

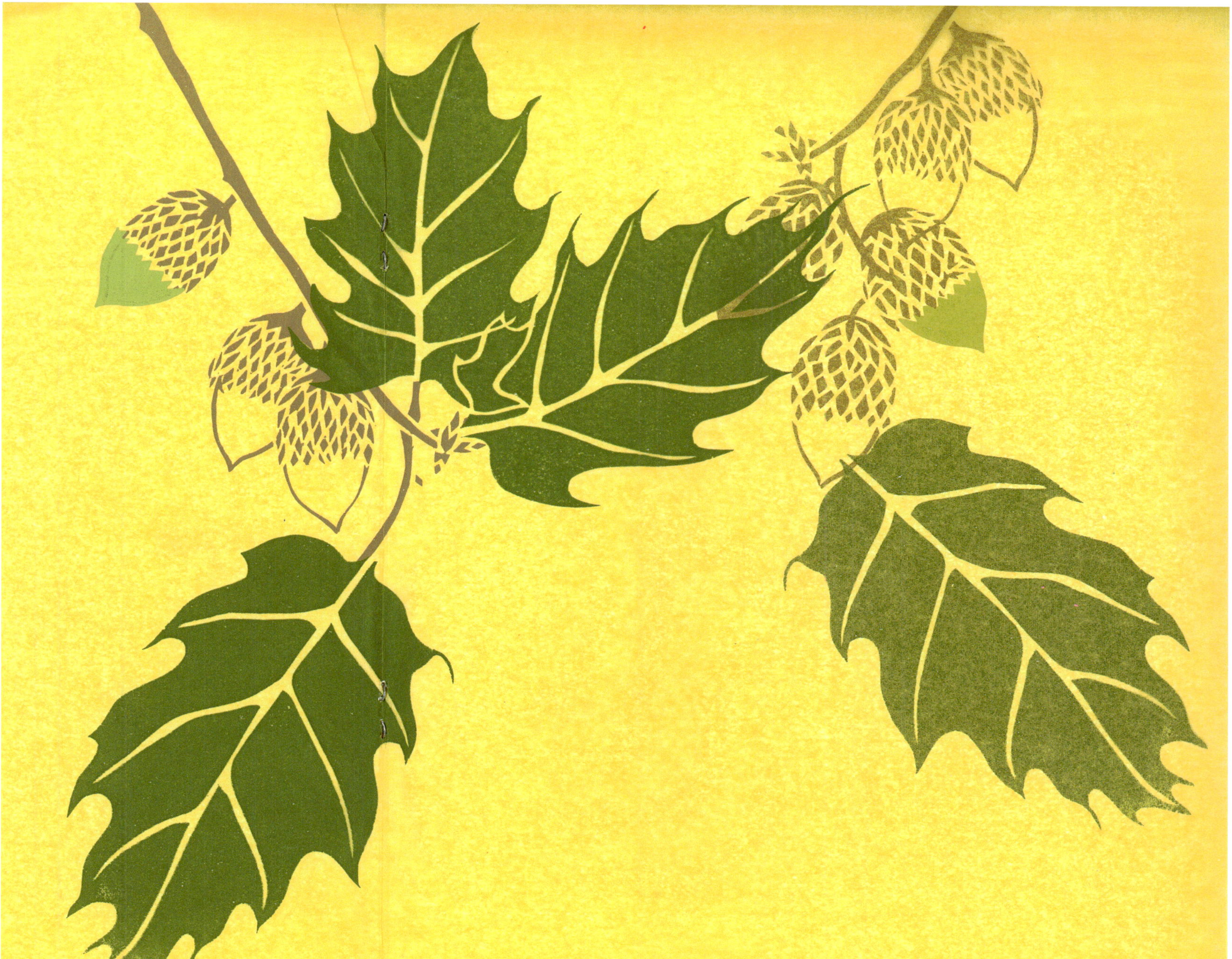

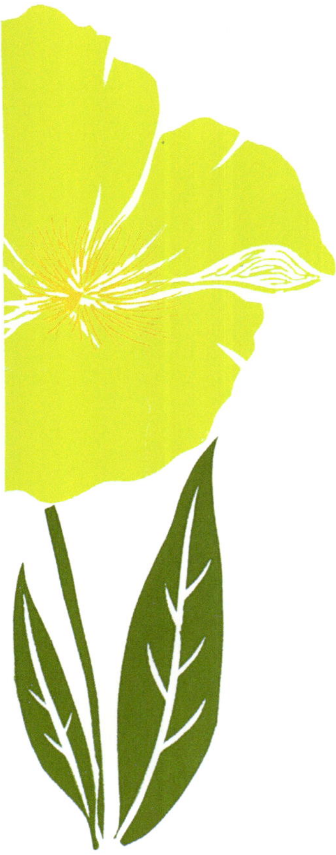

Island tree poppy *(Dendromecon harfordii)*

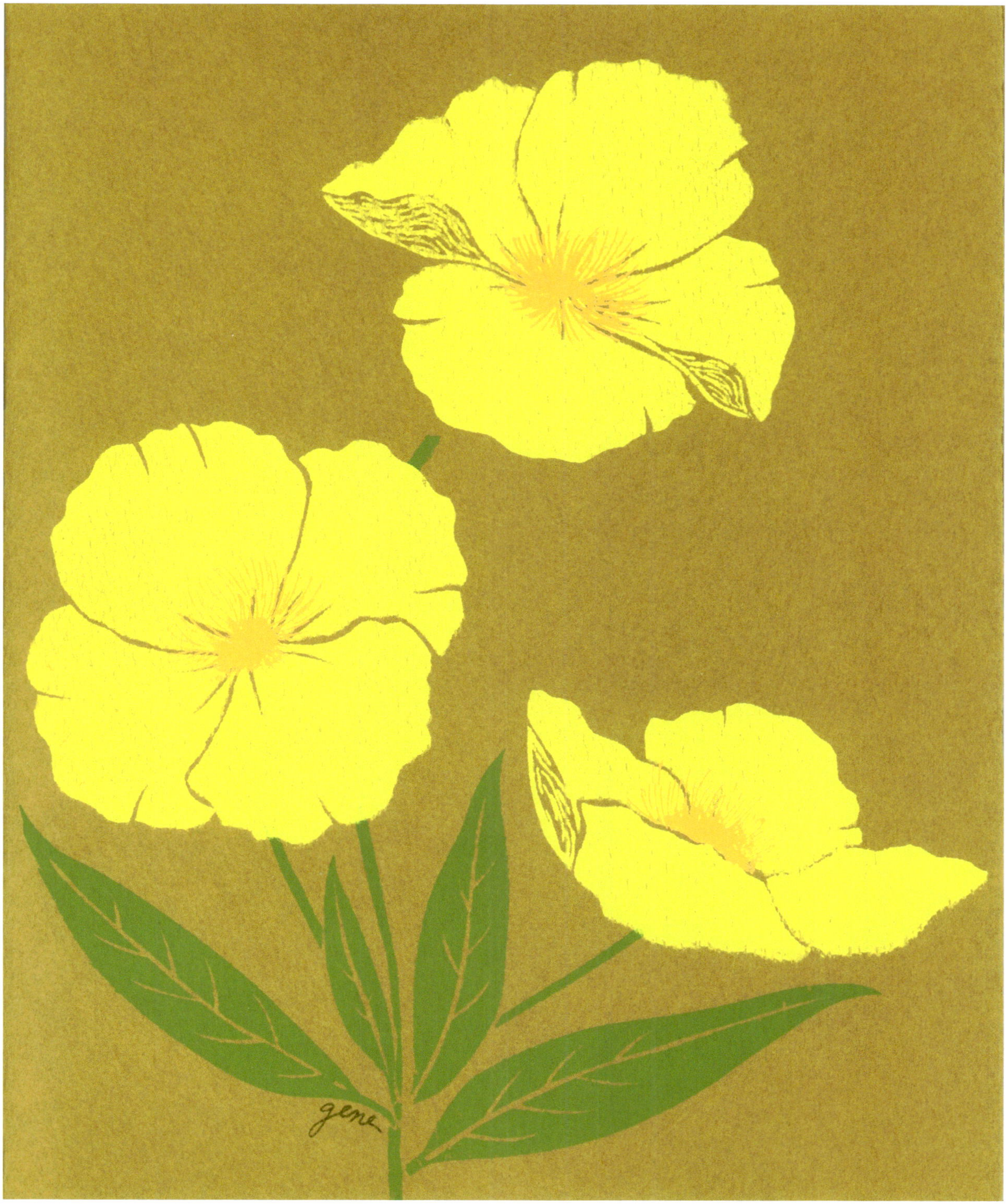
gene

February 1973

One of the most decorative and popular of our native shrubs is the tree poppy, the only native shrub member of the poppy family and, when in bloom, certainly a contributor to the scene John Muir described looking at the flowing foothills of the Santa Clara Valley when he first arrived in California around the mid-1800s: "They are covered with flowers growing close together in cloud-shaped companies, acres and hillsides in size, white, purple and yellow, separate, yet blending like the hills upon which they grow." From the Greek *dendron* (tree) and *mecon* (poppy), the genus *Dendromecon* contains two species, both native to California. One is found on the mainland of California and the other on the Channel Islands.

Dendromecon rigida

The flowers of the mainland's *Dendromecon rigida* are from one to three inches in diameter and composed of four petals of a lovely clear yellow. Basking in the sunshine, the flowers open wide to display a mass of deep golden stamens at their center. Most commonly in flower from April until July, the plant might be considered ever-blooming because it produces some flowers nearly all year long. This may explain why its popularity is so far-reaching. The *rigida* species name refers to the stiff quality of the plant's foliage structure. It is an evergreen shrub growing from two to ten feet high, with willow-like leaves from one to four inches long that are leathery and lanceolate shaped. *D. rigida* has light-colored, shreddy bark, and as with so many drought-resistant California plants, its foliage is gray green. This plant seems to prefer open, sunny hillsides of the chaparral belt and dry washes between 1,000 to 3,000 feet. Often found in areas of the wild with no rain from February to December, it makes a likely candidate for the drier garden, as long as it's grown on hot, stony, gravelly banks where its roots may seek out their own source of deep moisture. Meanwhile, the parts above ground seek plenty of sunshine.

Dendromecon harfordii

Dendromecon harfordii is the species found in a native state only on the Channel Islands, and so it's called island tree poppy. Very much like *D. rigida,* it is even more likely to produce flowers all year long and is a bit freer in flowering—the flowers it produces are even a little larger. *D. harfordii* grows a little taller, too, and is more treelike and less stiff in appearance than *D. rigida*. Overall, the dendromecons are extremely variable, especially in the size and shape of the leaves, and it was once thought that there were as many as twenty distinctive enough to be recognized as a species. But in fact there are but these two, and if you are considering cultivating them, growing the dendromecons from seed can be a very rewarding experience.

When I sent out a questionnaire to the twenty-six district representatives of the California Garden Clubs asking for their overall selections, I was struck by a comment from the California native flora chairman of the Santa Clara Valley District. Hers is a region in the central part of the state, a fertile valley bounded by some foothills of moderate elevation. First for comparison, here is how John Muir described the area when he first set foot on it, heading for the Sierra Nevada in 1868:

> "We crossed the bay by the Oakland Ferry and proceeded up the Santa Clara Valley to San Jose. This is one of the most fertile of the many small valleys of the coast .... It was now spring-time, and the weather was the best we have enjoyed. Larks and streams sang everywhere; the sky was cloudless, and the whole valley was a lake of light. The atmosphere was spicy and exhilarating...a sky not simply pure and bright, and mixed with plenty of well-tempered sunshine, but it possessed a positive flavor, a taste that thrilled throughout every tissue of the body. The foothills of the valley are in near view all the way to Gilroy, those of the Monte Diablo range on our left, those of Santa Cruz on our right; they are smooth and flowing, and come down to the bottom levels in curves of most surpassing beauty. They are covered with flowers...."

And this is what the native flora chairman had to say about the same area 104 years later: "Unfortunately Santa Clara Valley has been practically taken over by subdivisions, freeways, and shopping centers so that our native flora is almost gone."

Tree poppy seeds are a bit difficult to germinate but surely worth the extra effort. Knowing where and how they grow in the wild and that new plants are abundant after a fire could provide some helpful clues. Rancho Santa Ana Botanic Garden conducted a study wherein good germination was achieved by planting the seed in a light soil mixture in a flat. A layer of straw or pine needles was placed over the soil and burned. After this, the flat was thoroughly watered and about a month later the seedlings appeared. Remember that both species require perfect drainage and sunshine—and very little water during the summer months.

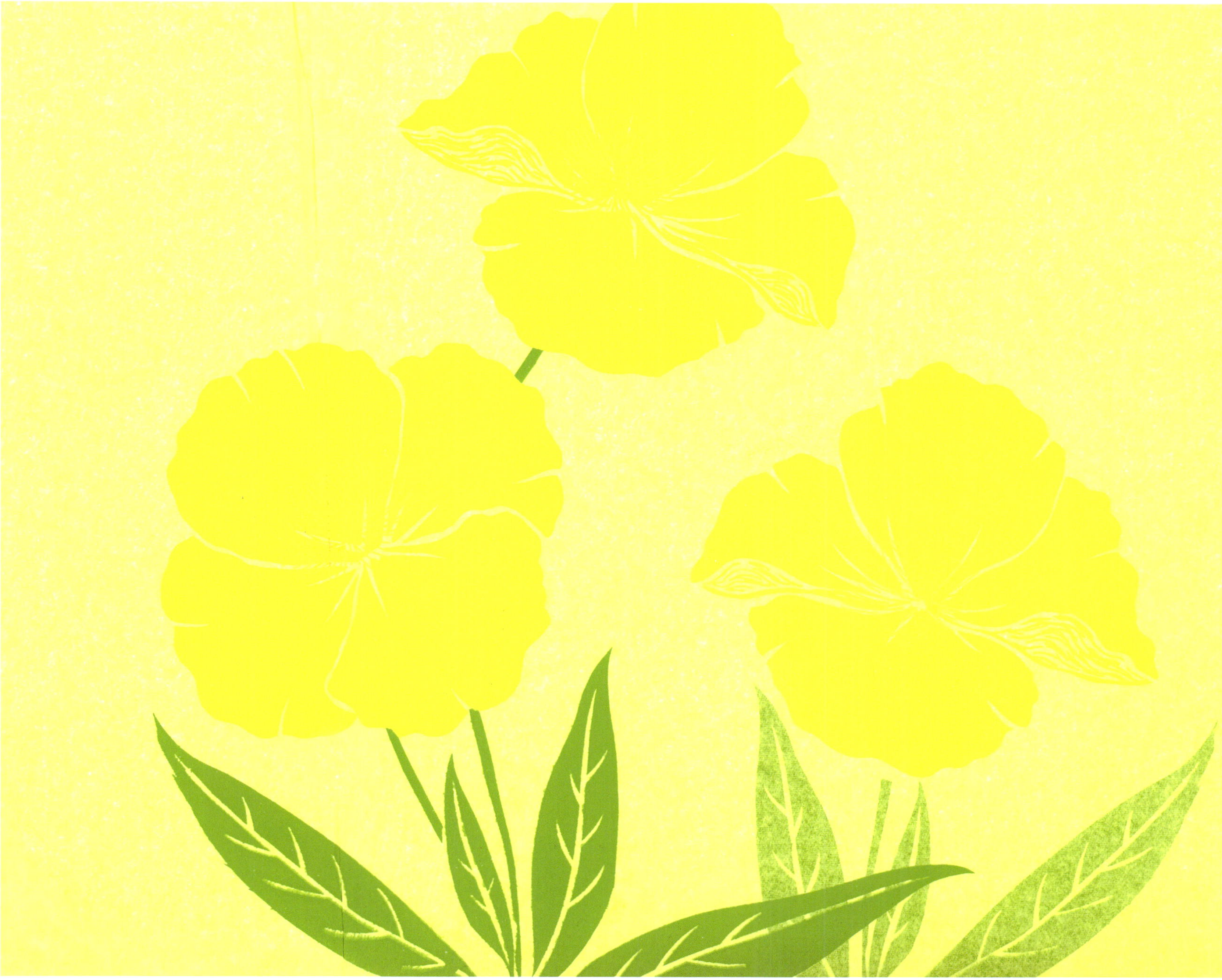

Manzanita (*Arctostaphylos uva ursi* 'Point Reyes')

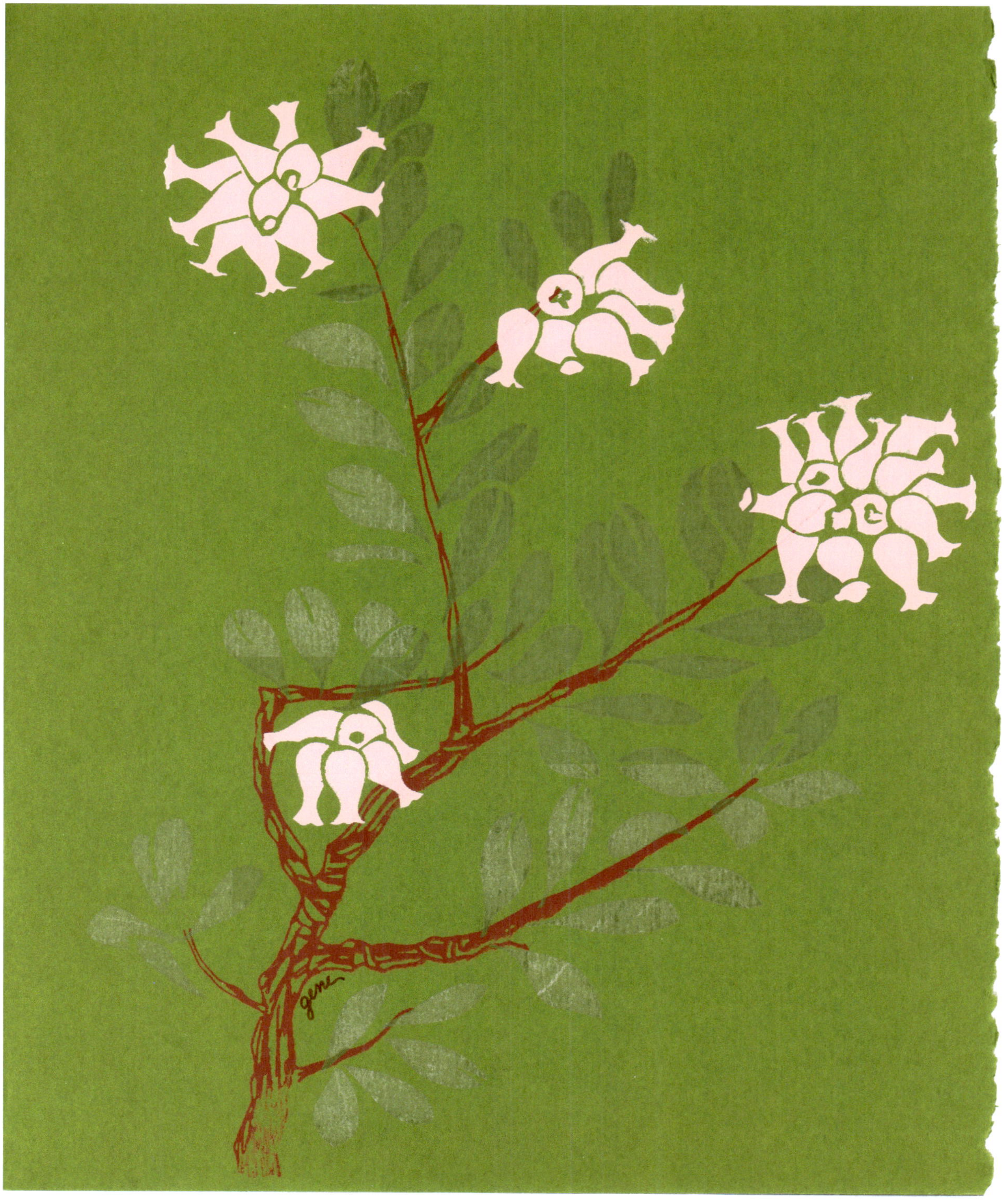
gene

March 1973

As a child on our annual trek to the mountains I was always intrigued with the crooked shrub that had smooth, shiny red bark. Though I was not aware of its name or that it bloomed and set berries, the plant was always familiar to me because I recognized that smooth, red bark. Now I know it is the manzanita from the genus *Arctostaphylos,* which contains about fifty species distributed mainly in Central and North America but reaches its highest development in the West and especially in California. Widespread throughout the state, thirty-eight species and many varieties and forms of it are native to California.

The manzanitas are extremely variable in form ranging from low, dense creepers to medium-size shrubs, and some are small trees that grow to between twenty and thirty feet tall. One of the easiest ways to recognize a manzanita is by its bark. Variously described as reddish brown, purplish red, mahogany, chocolate- or cinnamon-colored, the bark is smooth and polished looking. It is also very thin, and as the plant grows, the bark cracks and curls back. This shredding and flaking bark eventually falls to the ground exposing a fresh new covering that is soft and green but soon takes on the rich, handsome tans and reds that are so characteristic of this plant.

Manzanitas are evergreen with leaves that are thick and leathery. Often the leaves are standing up edgewise to minimize their exposure to the sun and thus conserve their moisture. The wood is very hard and always crooked, and early Californians found many uses for it, including using manzanita pegs to substitute for nails in building. The plant's fruit, a red-brown berry, was also utilized for food and medicinal purposes, much like the acorn.

The manzanitas are among the earliest of the native shrubs to bloom, with some species blooming as early as November and others during the winter months. The flowers—white, white tinged with pink, or all pink—must be seen up close to be fully appreciated. Usually close-set in clusters, the individual flowers are urn-shaped, little bells that have a waxy purity about them. The most perfect description I've read is from *Yosemite Trails* by J. S. Chase when he says they look "like little classic vases cut in alabaster." The flowers later develop into the red-brown berries, or "bear berries" as they are called in some regions. The genus name is from the Greek words *arktos,* meaning "bear," and *staphule,* meaning "grape," and derives from the observation that bears are very fond of this fruit. Throughout California, however, it is better known by the lovely name Spanish Californians gave it, manzanita, the Spanish word for "little apple."

Species identification is complicated by the fact that it is common for several species in an area to overlap and freely hybridize, creating many intermediate and perplexing forms. In identifying a species, several things must be determined: presence or absence of a burl at ground level; height and general habit (creeping, procumbent, or erect); stickiness of the leaves, twigs, and fruit; position of flowers (erect or drooping). Because many species of the *Arctostaphylos* genus are members of the chaparral community they are often subjected to fires. The species that have a burl or rounded crown at ground level sprout after a fire; others don't after a fire or having been cut.

Arctostaphylos pechoensis

I made the drawing from a plant of *Arctostaphylos uva ursi* 'Point Reyes', but I want to be sure to mention *A. pechoensis,* both for its aesthetic value and because it is a superb plant for holding soil in place. In a dense, unbroken mass, it is like a leafy carpet forming mounds over dry slopes and ridges. Occurring in the low mountains from 500- to 2,600-foot elevation in the west central portion of the county, the species is also known as Pecho Mountain manzanita, named after the medium-high mountains in San Luis Obispo County. A low-growing, dense shrub at one to five feet high, it has a wide and spreading root crown with many stems. When situated close among other shrubs or otherwise sheltered, the stems stand upright; but when exposed on windy mesas, they spread out and are more mat-like. They can extend horizontally six to ten feet or more.

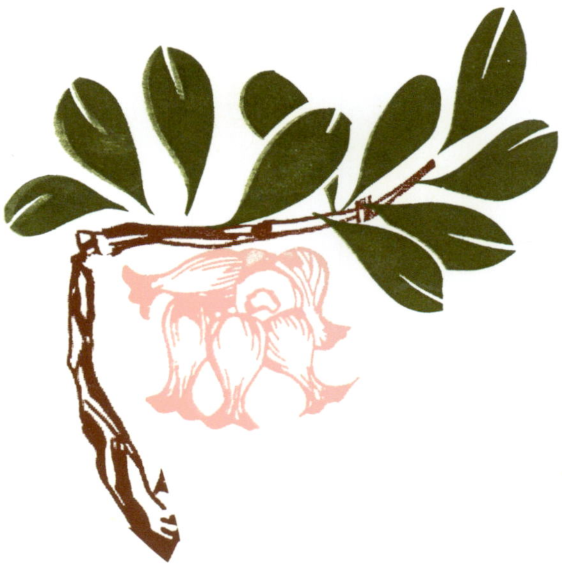

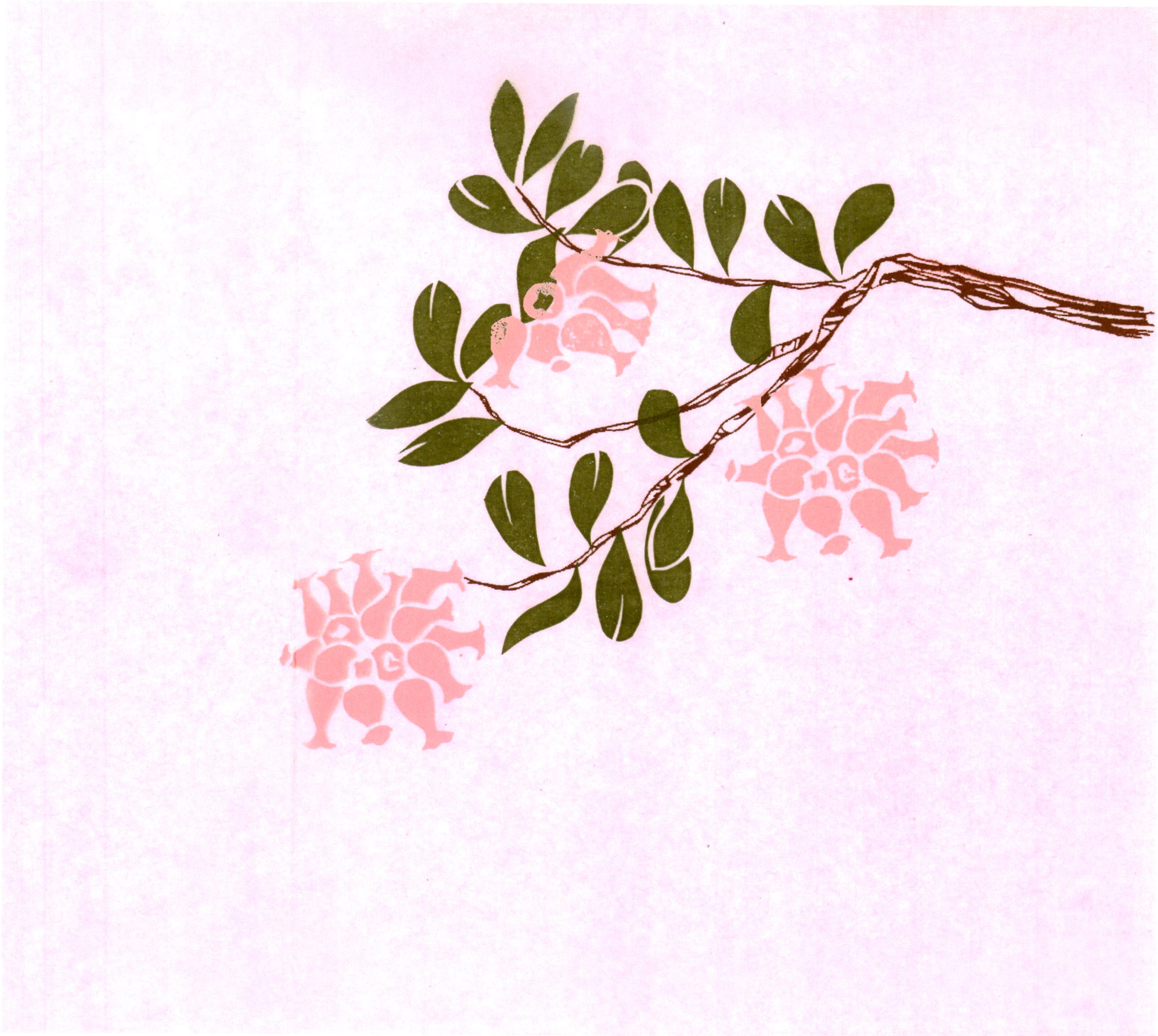

Western redbud *(Cercis occidentalis)*

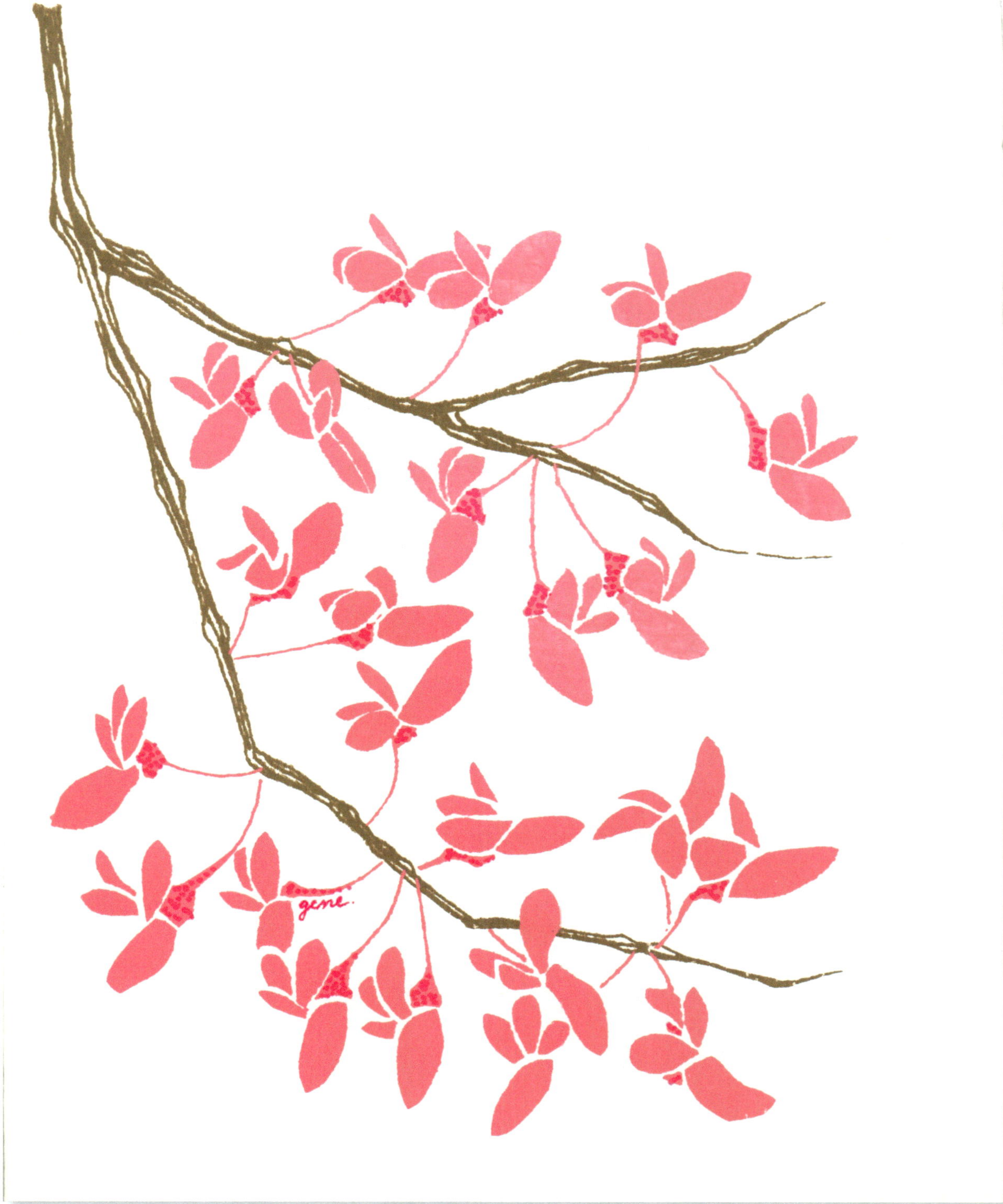
gene

April 1973

A winter that is a consistent series of cool, overcast, rainy days, like ours has been, leads to an outstanding spring for wildflowers in California. Here is an early species that blooms in March and April, its flowers bursting open on bare branches before the leaves open. The genus *Cercis* is from the Greek *kerkis,* the name of a weaver's tool and referring to the appearance of the 2½- to 3½-inch seedpods of the shrubs and small trees in this group, commonly known as redbud. The six species of *Cercis* are distributed in Asia, Europe, and North America.

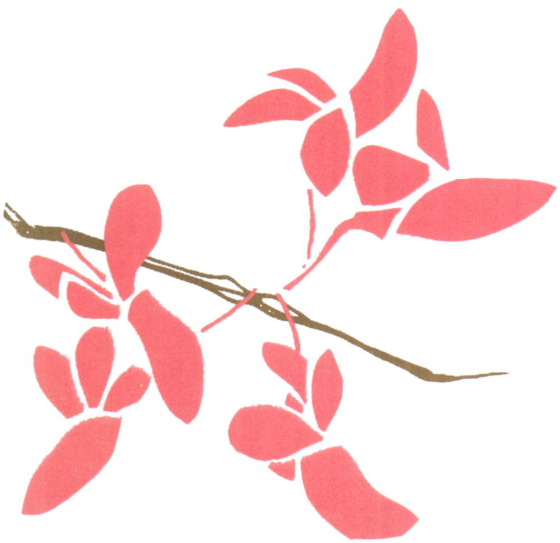

Cercis occidentalis

The one species native to California is *occidentalis,* meaning "western." This deciduous shrub or small tree (eight to twenty feet high) prefers sunny hillsides and requires little moisture in the summer. In its native habitat, western redbud grows with California buckeye, *Ceanothus,* and manzanita on dry, gravelly banks and slopes. It is quite prevalent on dry slopes and canyons in the foothills of the Sierra Nevada and the inner Coast Ranges north of San Francisco. Western redbud is also found on the desert side of the Laguna and Cuyamaca mountains in San Diego County. Anyone traveling in these areas in early spring cannot miss this tree looking like a fragile, rosy cloud.

The outstanding feature of *C. occidentalis* is the unusual color of its blossoms, a rosy magenta that is unique among our native flowering trees and shrubs. Its many stems growing from the base tend to create a round, spreading shape to the small shrub or tree. In early spring, the flowers burst open before the leaves, so that the otherwise bare branches rise up to create an illusion of billowy, rosy clouds. The flowers are numerous but they are small and would be completely hidden by the leaves if they opened together—it is fortunate that they don't defeat their own beauty by waiting for the leaves. The leaves themselves are a most attractive bronze color upon first opening and soon develop into their permanent blue green. Somewhat heart-shaped at the base, the leaves are smooth and roundish, about 2 to 3½ inches wide, with a tendency to appear folded down the center.

The genus *Cercis* belongs to the second largest family of true flowering plants in the world—Leguminosae or the pea family—containing about 550 genera and many thousands of species. The flowers of *C. occidentalis* as well as of many other members of the pea family are classed as papilionaceous, in reference to the butterfly-like quality of the flower. Each flower is composed of five petals, distinguished as banner, wings, and keel. The banner is the single upright petal with two larger lateral petals (the wing petals) on either side. The remaining two petals are larger and slightly joined, forming a keel. By July the flowers of the western redbud have developed into long, reddish-brown seedpods. They hang down conspicuously, lending color to the tree until the following spring.

Even during the winter, the western redbud is attractive, its bare branches possessing a refined quality, dainty and fragile. In its native habitat it is a slow grower, and as such oftentimes has a finesse and refinement about it that fast growers rarely attain.

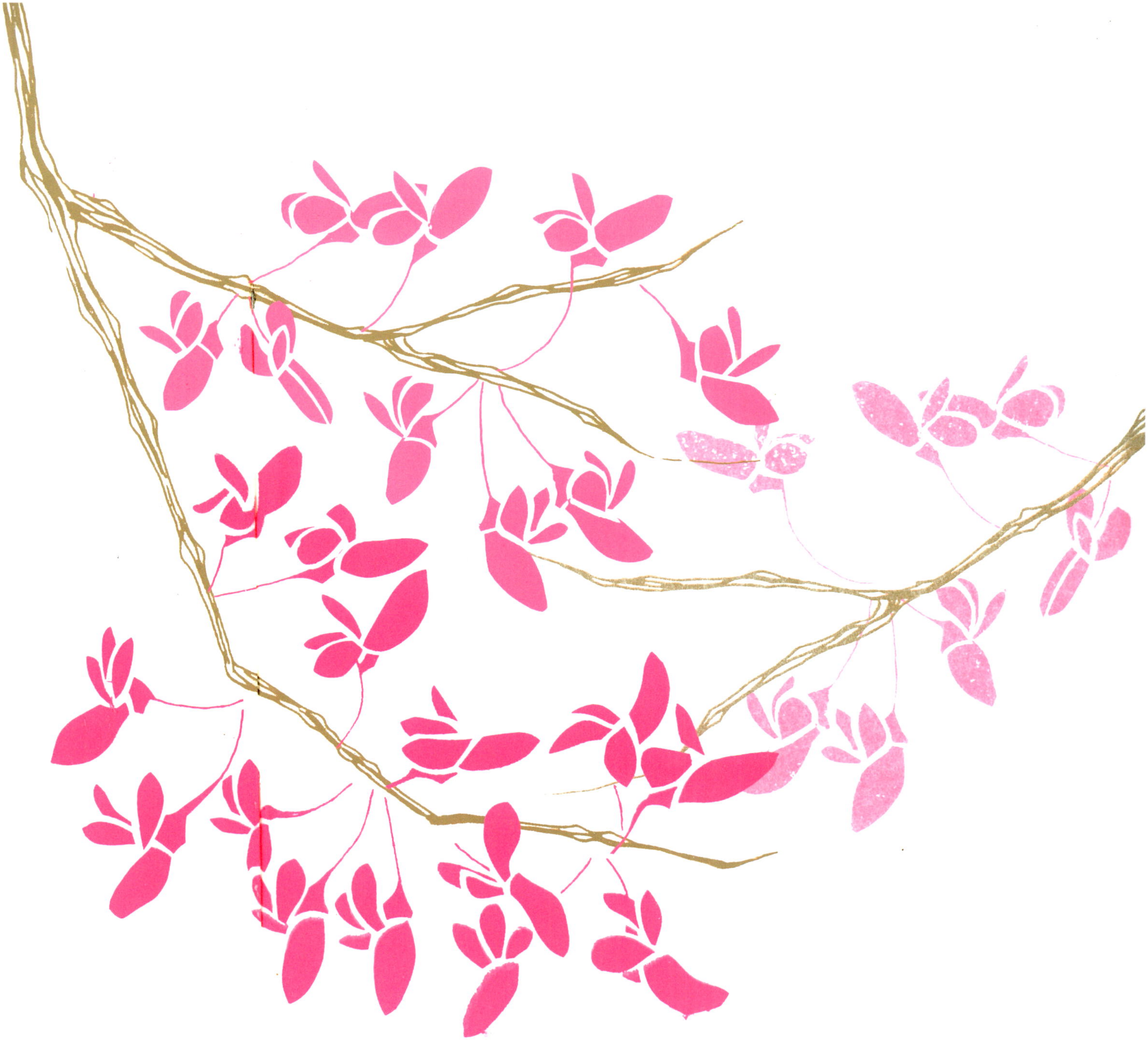

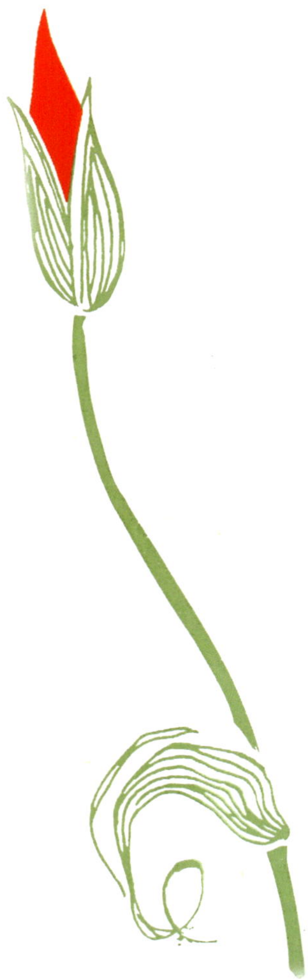

Desert mariposa *(Calochortus kennedyi)*

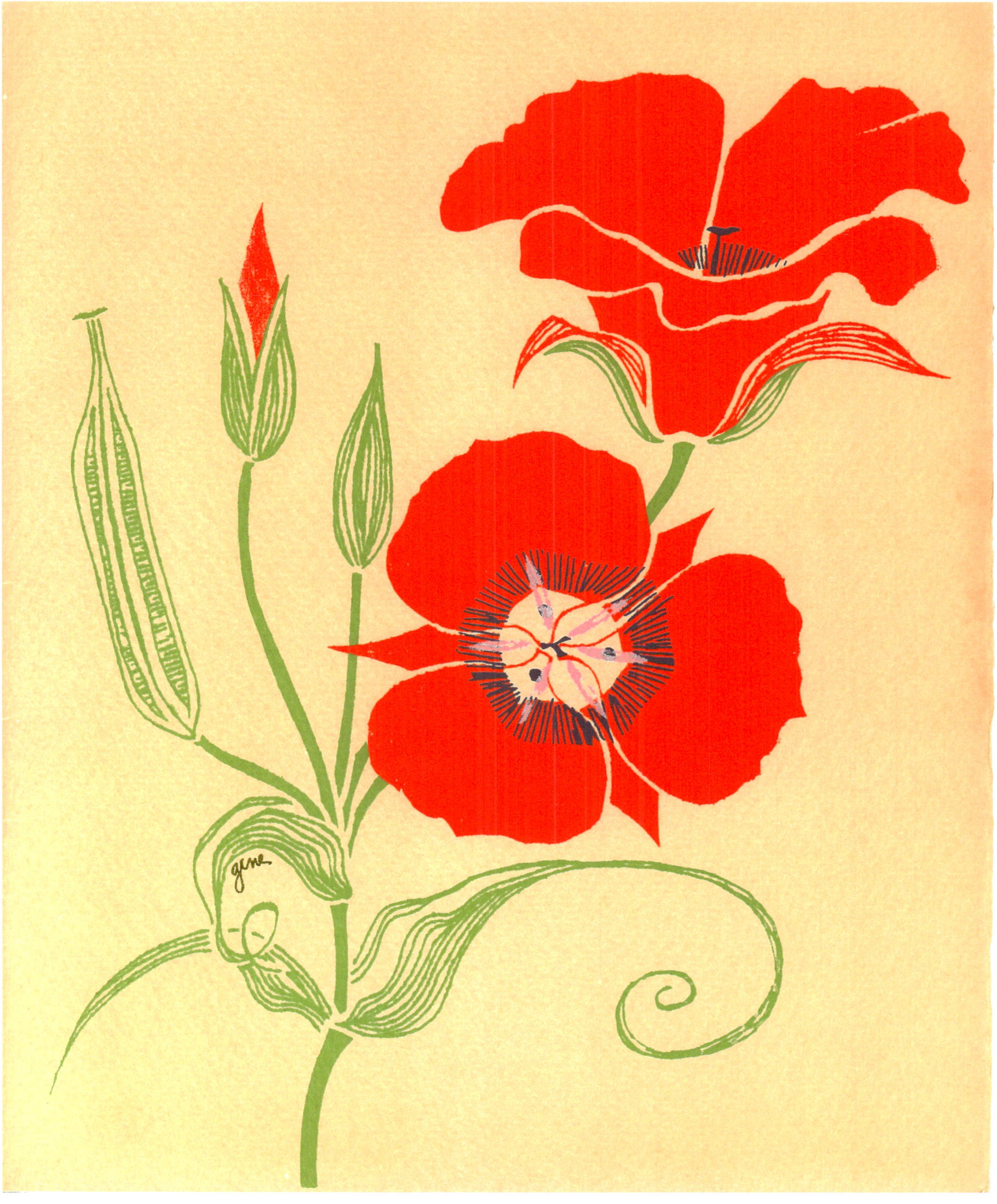

Early May 1973

When the California poppy was selected as the official state flower in 1890, the only other flower receiving any votes at all was the *Calochortus*. The genus includes about sixty species growing only in temperate North America (mostly in the West), of which thirty-seven are native to California. *Calochortus* plants are found in nearly every part of the state of California, with each area cherishing its own particular species. When I was planning this series on California natives, I wanted to include the genus *Calochortus,* and because the desert species is so utterly magnificent, I decided to use *C. kennedyi* to represent the inland desert area.

The two Greek words that compose the genus name (*kalos,* meaning "beautiful" and *chortos,* meaning "grass") describe the flower and the grass-like foliage of *Calochortus*. From a small bulb with a fibrous tunic come the stems, thin, extremely strong and rigid, possessing a somewhat wiry quality. The length of the stem can be anywhere from two inches to two feet tall, depending on the species and where it is found. The foliage is usually solitary, linear, and basal, and oftentimes it has dried up by the time the flower opens. The three different shapes of flowers of the genus *Calochortus* include hanging or nodding, globular flowers with nodding seed capsules; erect, star-like flowers with nodding seed capsules; large, erect, cup-shaped flowers with erect seed capsules. Members of this genus do go by the shapes they take; however, these common names, like mariposa lilies, mariposa tulips, globe tulips, star tulips, and sego lilies create much misunderstanding. *Calochortus* are neither tulips nor lilies, though they have often been referred to as the tulips of the New World. Upon very close examination they are, indeed, very much like the tulips of the Middle East. The flower consists of three petals and three sepals, just like a tulip. The sepals of the *Calochortus,* however, do not develop to as full color and maturity as those of the tulip.

Calochortus kennedyi

Desert mariposa, the common name of *Calochortus kennedyi,* alludes to the exquisite butterfly markings on the petals of this high-desert native, the most colorful of all the desert wildflowers. Characteristic of an area that includes the eastern portion of Kern County and the northern portion of San Bernardino County, this species, unlike the genus itself, is restricted to the desert.

I am very familiar with and dearly love the two species of *Calochortus* that are native to the mountains where I live, and I had seen many photographs of *C. kennedyi* but never the real flower until April 30, 1973. I prefer making my drawings from living specimens, so I was especially eager to see the real thing. So on that Monday morning when I got a call from Mrs. Berniece Meade, a past director of the Desert Empire District of the California Garden Clubs, offering to show me where she had seen a few *C. kennedyi* blooming, I was more than happy to go with her. (I admit I had extra motivation: it was snowing where I live, and clear, mild weather awaited me at our destination.) Leaving the snow at home, I braved the fog-shrouded Cajon Pass to meet Berniece in Ridgecrest, 150 miles away in the Mojave Desert. She packed me and a lunch into her vehicle and off we went, and after a somewhat bumpy ride, we came to the spot where she had seen some blooms just a few days earlier. But this day, there were hundreds—very possibly thousands—of *C. kennedyi* in full bloom. I found it nearly impossible to walk without stepping on them. On the distant hills we saw great flame-colored vistas of thousands more blooms. I was almost speechless. My only and very repetitive words were, "I can't believe it! I simply can't believe it!"

Words are inadequate to describe the beauty of this flower. The petals are from 1½ to 2 inches long and a brilliant vermilion. They do not have a "bloom" on them, but their color is so intense and so rich that these petals appear to be made of velvet. Deep inside the flower there is a charming design or pattern. Each petal, with a purplish-brown spot at its base, has a hairy gland, and even the sepals are complete on the inside with the vermilion color and gland. The two-inch seed capsule is exceedingly handsome as well, all decked out with its longitudinal stripes.

Under difficult conditions, the desert mariposa may bloom only a few inches above the ground, but usually it is from eight to fifteen inches tall. Sometimes it will stretch to twenty inches, growing right up through a small desert shrub in order to bloom over it. This plant may be found growing singly or in colonies; in heavy soil of open and brushy places at an altitude of between 2,000 and 6,500 feet; from Inyo County to San Bernardino County. It blooms at different times depending upon the elevation: at lower elevations in April; higher up, it blooms in May and June.

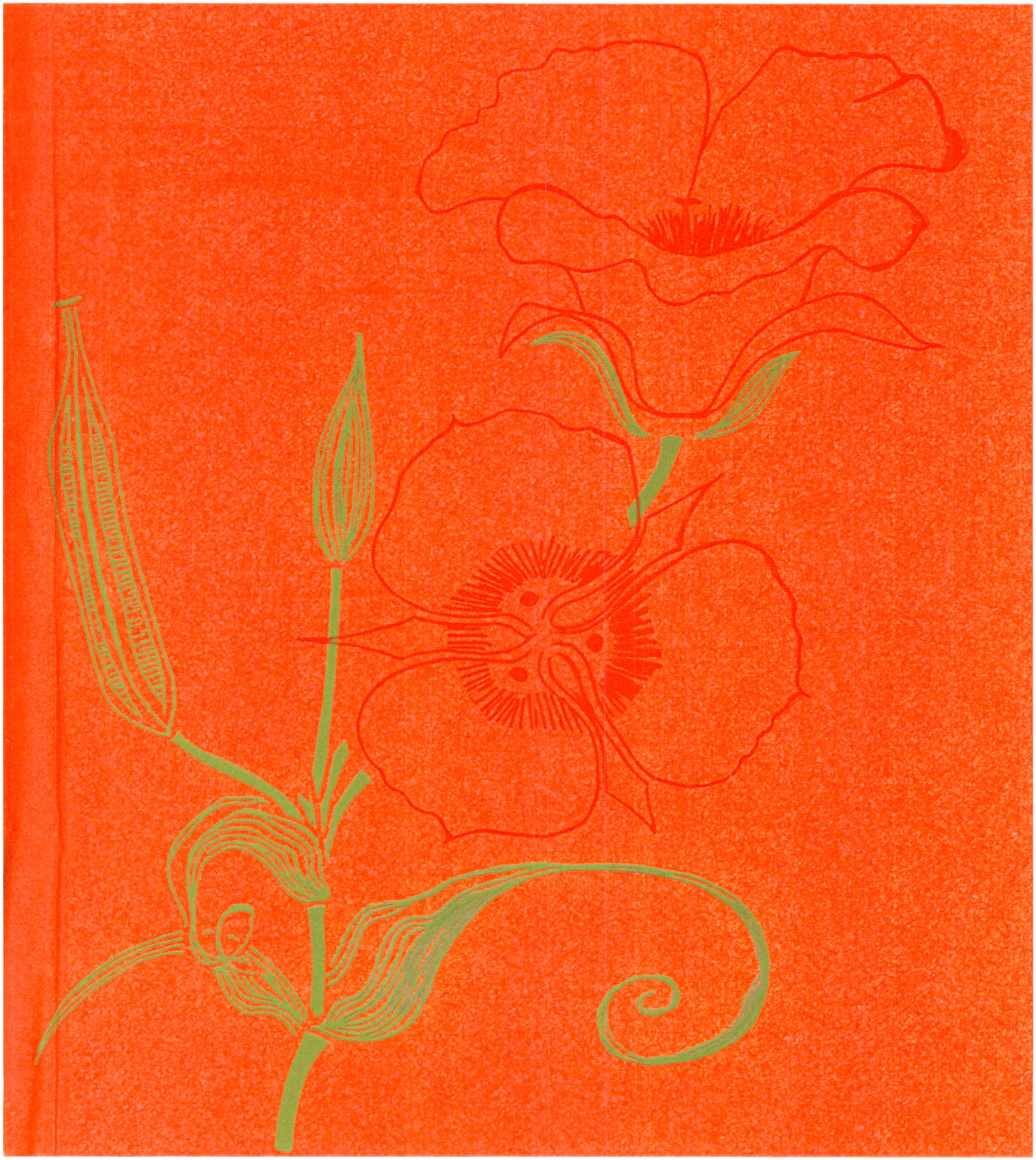

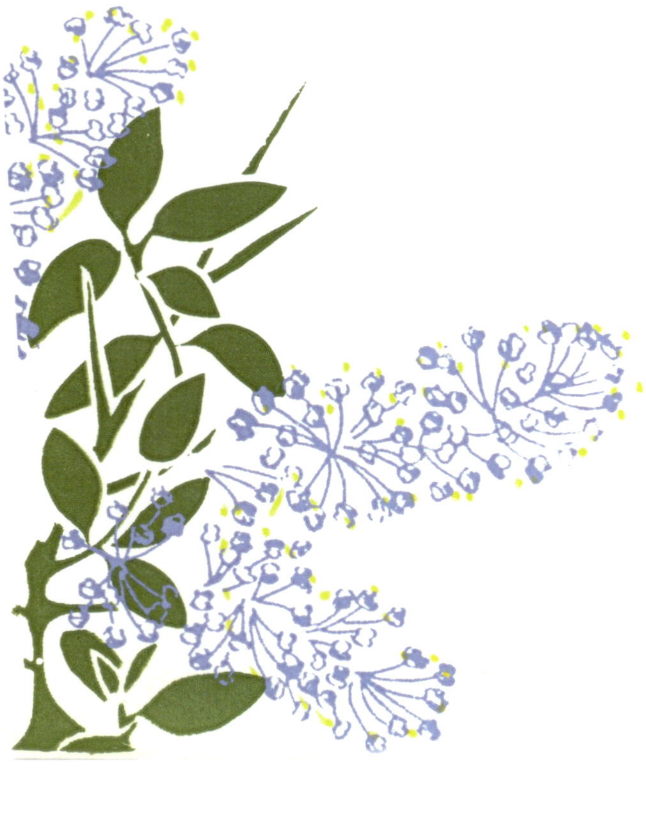

Buckthorn *(Ceanothus leucodermis)*

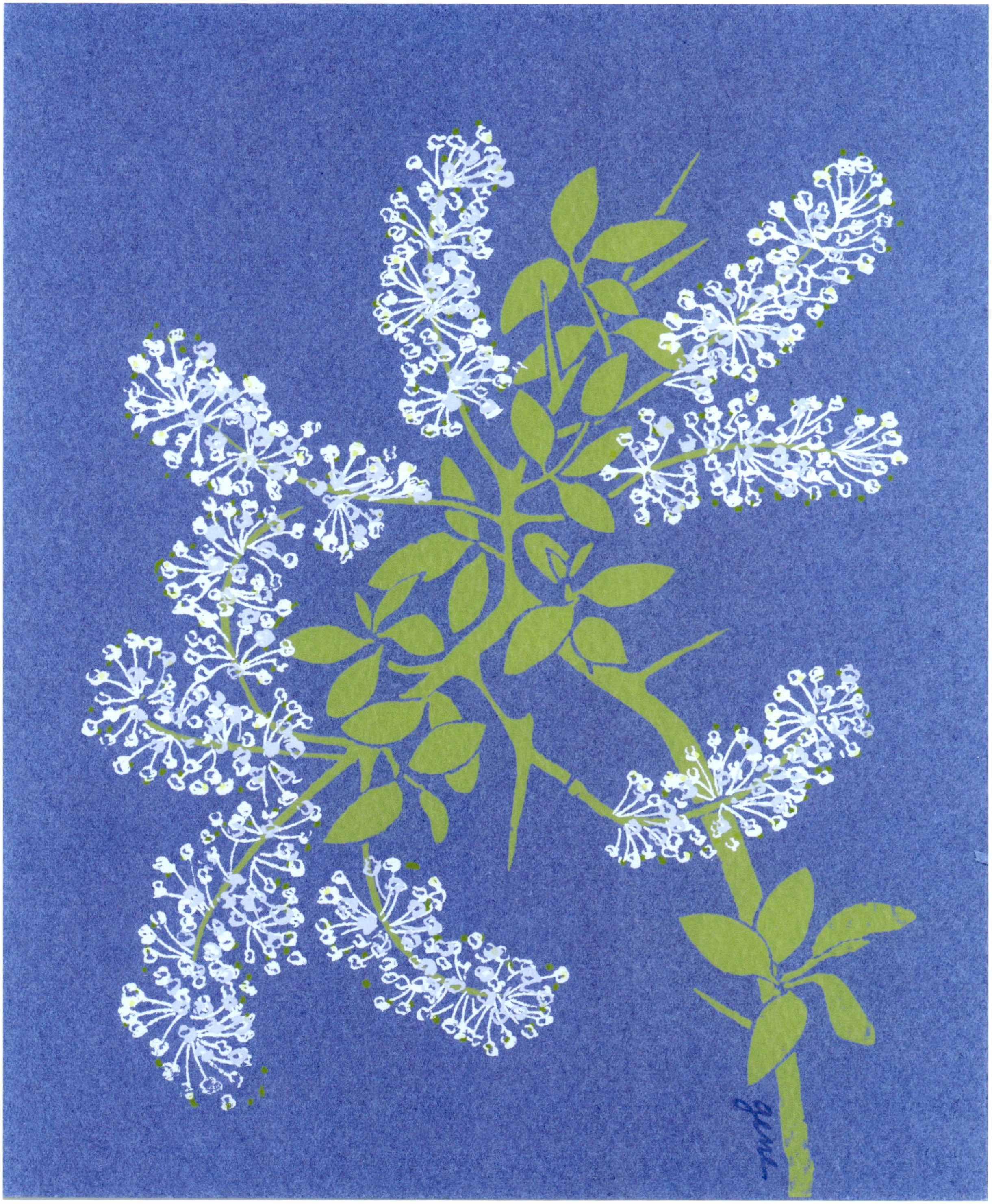

Late May 1973

Not just off its western border but also throughout the state of California from midwinter until midsummer, somewhere there is always a sea of blue. The genus *Ceanothus* (buckthorn) is to thank for that; soft, billowy clouds of every shade of blue from the palest to the deepest, even tinged with lavender and pink. Stand above and look down on the land and as far as the eye can see this ocean of blue goes on and on. Is it any wonder that this genus is among the best known and most popular in California? (On the questionnaire I sent to each district's garden club representative, the genus mentioned more than any other was *Ceanothus*.) Also mixed in with this exquisite blue—and sometimes alone because they usually appear earliest—are the whites: white-white, off-white, creamy white.

For the other half of the year, these areas where *Ceanothus* grows are little noticed and practically impenetrable, a nondescript, solid mass of many different plants growing together in harmony under like conditions. In varying textures and not very exciting shades of green (blue green, gray green, yellow green), these plants form a dense covering that provides protection and food for animals and birds, and they also serve as erosion control. With the first hint of warm weather, the blossoms begin to open and one of the glories of a California spring is well under way. The transformation happens quickly.

No shrubs are more typically Californian than those of this genus. Found in nearly every part of the state, from the coast through the chaparral belt to high mountains, top to bottom wherever each *Ceanothus* species grows, it is called by a name borne of local terminology: deer brush, buckthorn, buck brush, tobacco brush, mountain balm, lady-bloom, blue myrtle, musk-bush, white thorn, Jim brush, blue-blossom, downy lilac, squaw mat, mahala mat, sweet birch. The most prevalent of the common names are wild lilac and California lilac. The name lilac is really quite misleading because these shrubs have nothing to do with the lilac *(Syringa vulgaris)* of the eastern United States.

The majority of the sixty or so species of *Ceanothus* are native to California, and there are many varieties. Identical species may be separated by many miles, so they tend to be divided into types as well: coastal, chaparral, foothill, and high mountain. Most of the *Ceanothus* species are evergreen, but a few are deciduous. They range in size from low, mat-like creepers to medium-size shrubs, and a few attain tree height.

The individual flowers are very small and in clusters. In some species the clusters clothe the entire stem hiding the leaves and looking like a solid rounded mass of color. Many of the white flowering species present this appearance. In others, the little clusters appear at the top of the branches in panicles, a likeness with the eastern lilac that may have prompted the common name of wild lilac. The flowers of all *Ceanothus* species are fragrant. Some are extremely sweet smelling—a rather earthily warm, musty sweetness difficult to describe but entirely different from the fragrance of *Syringa vulgaris*. When the sun-kissed flowers are crushed and rubbed in water they produce a cleansing lather, a quality that has earned it another name—soap bush.

Leaf size varies from quite small, one half inch on some species, to three inches long on others. Some have soft, thin leaves, while others bear stiff, prickly leaves. Many species are armed with thorns as long as one to three inches, and these can be extremely vicious. More than once, returning from forays into areas where it grows with our arms covered in scratches and our clothing torn, have we muttered, "Down with *Ceanothus!*"

Ceanothus leucodermis

Pictured here is the pale blue blossom of *Ceanothus leucodermis*. This species grows near my home, where I marvel at such a cool, celestial color emanating from an earthbound plant. It was impossible for me to find a bloom of *Ceanothus cyaneus,* commonly known as San Diego lilac, but this species is worth mentioning not only because it is widely grown and well known but also because of how it got that way. It is an excellent example of how to succeed in sharing a new discovery without causing damage to the native plant. First discovered around 1924 in hills of a 1,500-foot elevation above Lakeside in San Diego County, it was shown to Miss Kate Sessions and was named by Miss Alice Eastwood. Seeds were collected and were grown and propagated by Miss Sessions. It was readily made available, and now the beauty of this plant graces many gardens and areas far from its original home. The native stands were in no way damaged. The proper procedure of discovery, naming, propagation, and distribution was followed and all to the benefit of the plant.

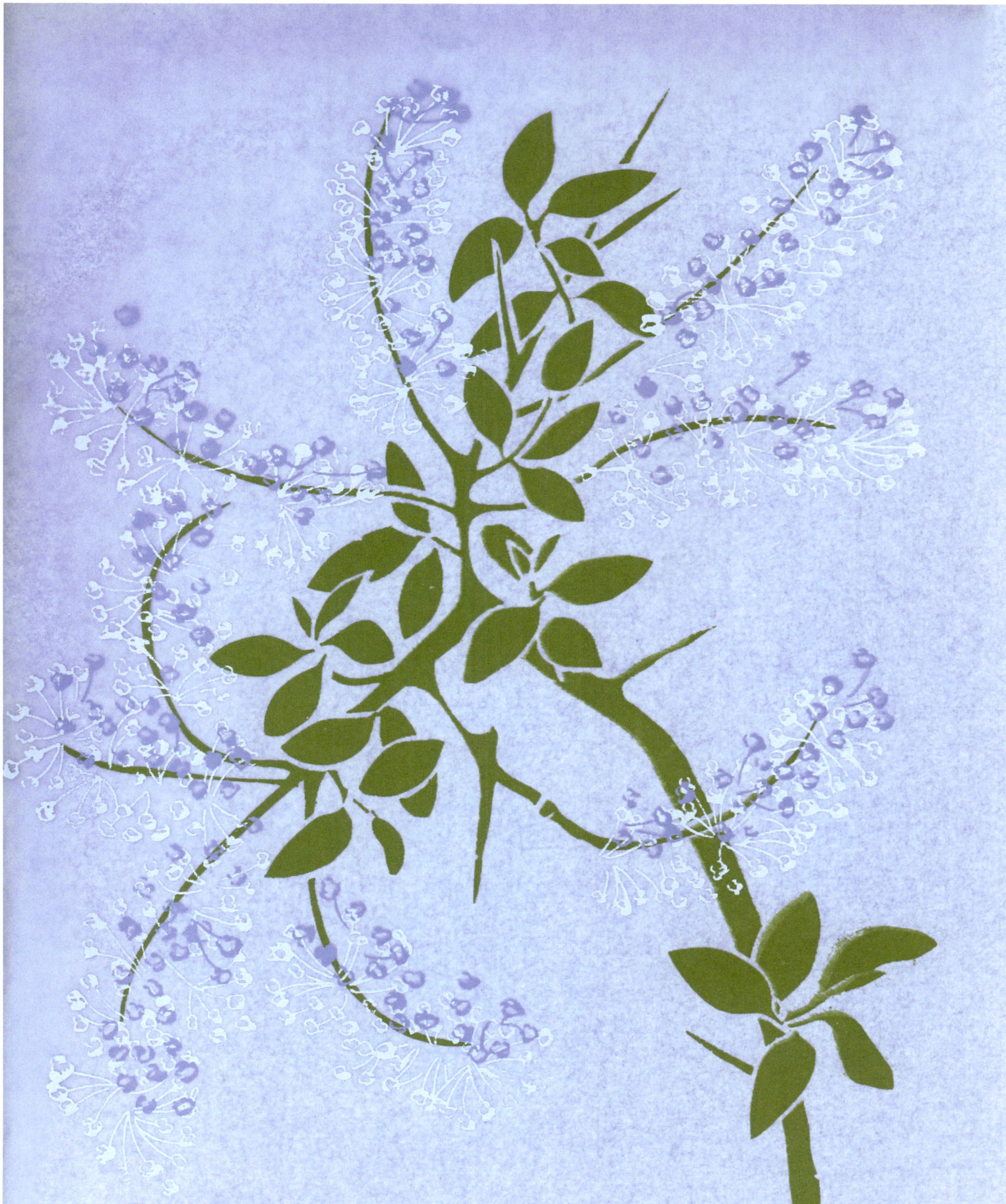

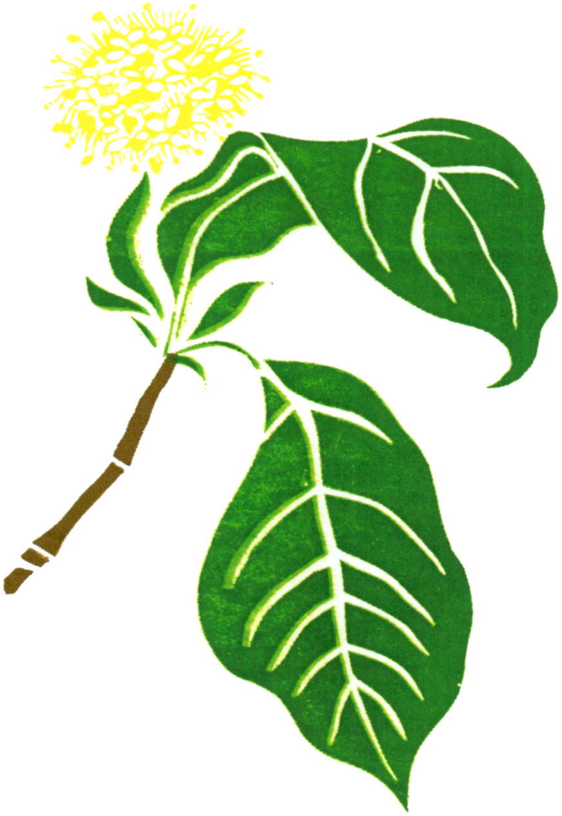

Pacific dogwood *(Cornus nuttallii)*

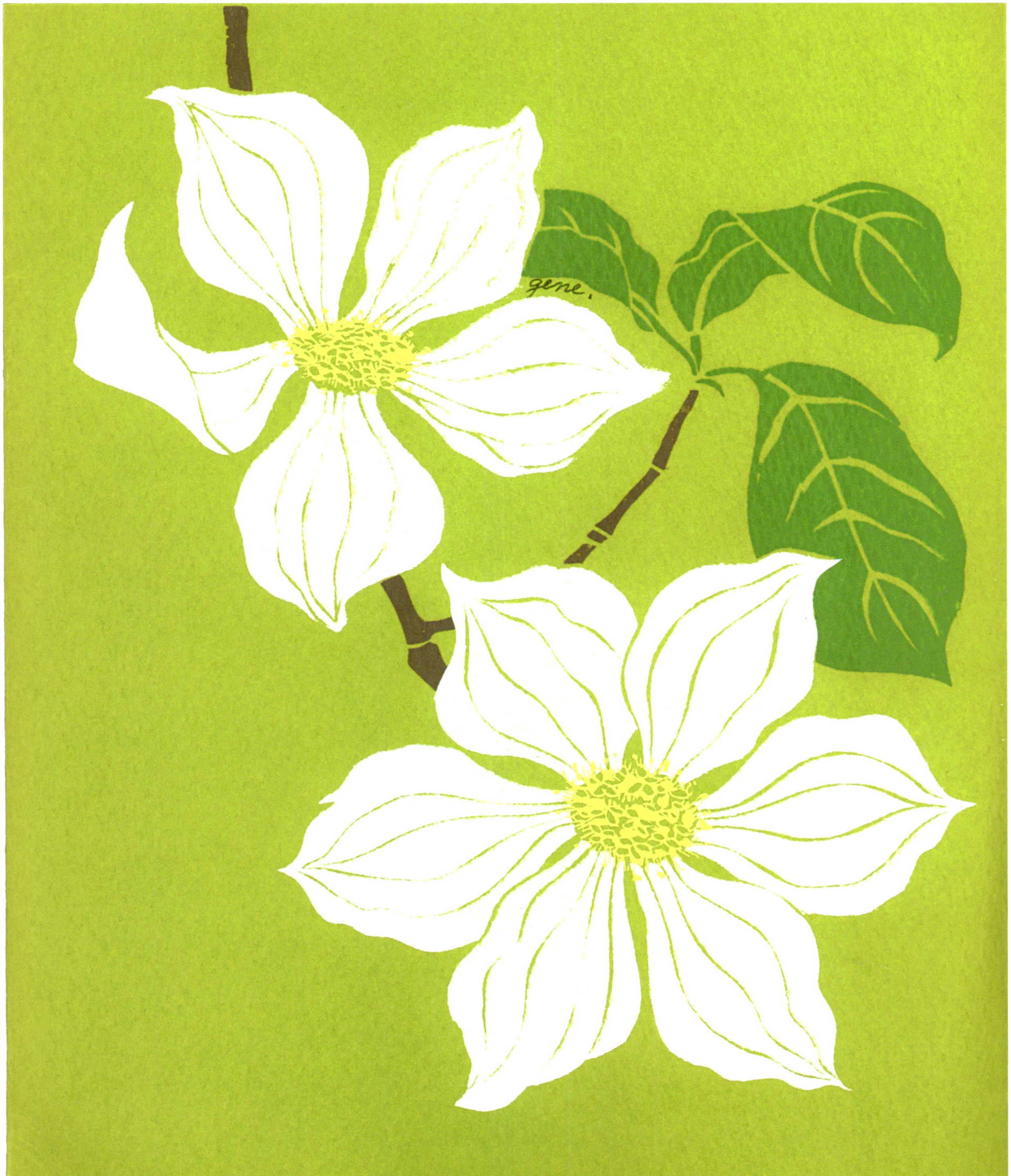
gene.

June 1973

Cornus is commonly known as dogwood. It is astonishing that a genus producing such lovely shrubs and trees could be saddled with such an unromantic name. Two theories have been advanced as to why: one, its astringent bark was used as a cure for mange in dogs; two, the hard wood was considered especially good for making skewers or "dags," and the word "dagwood" soon became "dogwood." I love it anyway. In fact, one of its six species native to California—and the only one that grows to tree size—*Cornus nuttallii* happens to be a personal favorite.

As for the genus itself, *Cornus* (meaning "horn" in Latin and referring to the hardness of the wood) includes fifty species. All but one (growing in Peru) are native to the northern hemisphere, mainly found in Europe, Asia, and North America. Of the six Golden State natives, one is a diminutive ground cover, *Cornus canadensis,* affectionately known as puppy-wood or bunch-berry. Usually only three to nine inches high, it grows from Mendocino County north across our border with Oregon and on through Washington and British Columbia. Four of the six California natives are shrubs found in moist areas and along stream banks. Though they have attractive foliage and interesting plant structure, their spring blooms are tiny and white, not at all showy. The shrubs remain inconspicuous, seemingly content to serve their purpose of controlling stream erosion. Quite conspicuous indeed is the one tree of the genus, *C. nuttallii,* a four-season beauty in a region that enjoys verdant meadows, valleys, and high mountains at the top of California.

Cornus nuttallii

California, mountain, Pacific, and western dogwood, by whatever name, *C. nuttallii* is recognized by many botanists as among the most handsome of all North American trees. Unfortunately, it is not at all happy growing outside its native habitat. Many botanic gardens throughout the world have tried to cultivate this tree, often unsuccessfully. A deciduous tree and usually small—from twenty to forty feet tall—it grows to magnificent proportions in the Puget Sound area, reaching heights of from fifty to sixty feet, some specimens with a diameter of more than two feet.

It is most conspicuous in the spring, usually in late April and May when the blooms open before the leaves appear, transforming shady green forests into lilting, ethereal fairylands. In summertime the tree is clothed in ovate leaves from three to five inches long. A particularly fresh, clean, clear green, these leaves are difficult to describe—could luminescent be the word? No other tree in our forests has leaves of this translucent quality.

Come fall, whole areas again are transformed into magnificent wonderlands, this time when the leaves turn color—every one imaginable from yellow orange, orange, pink, pinky rose, then rose to clear reds. At times, when all these colors appear on one tree and that tree happens to be adjacent to a cousin still in its stage of fresh summer green, you get to see the best of autumn colors against the backdrop of your best summer memory—as if the rainbow were backlit by luminous green. Look fast when the clusters of shiny red berries appear—their brilliant coloring in contrast against the autumn leaves—robins and band-tailed pigeons adore this fruit. I have seen as many as fifty robins feasting on these berries, literally stripping the trees. Each berry, or to be correct, each individual drupe is about a quarter of an inch long, fleshy, and contains a stone with a seed. The seeds are not harmed when devoured by the birds; in fact, this is a magnificent way to disperse them. With something for every season, these dogwood trees round out the winter by dressing their graceful, naked branches in the many half-inch, button-size flower buds with which to open the show again next spring.

C. nuttallii also grows in Oregon, Washington, and British Columbia. In 1956, the latter chose it to be the floral emblem of the province. It is abundant on Vancouver Island and used as a street tree in the town of Nanaimo, where it sometimes blooms again in the fall, its white bracts framed by scarlet foliage.

In the mind of many easterners, the name dogwood conjures up a flower with four-notched petals, an image traditionally associated with the eastern dogwood *(C. florida)* and, for some, with the cross in Christianity. Used on china, draperies, towels, placemats, and so on, the image has nothing to do with California's *C. nuttallii.* One day, a favorite restaurant not far from my home—and situated on the crest of a canyon lined with *C. nuttallii*—disappointed me when I looked at their new menu. The restaurant was within five hundred feet of those magnificent trees and yet its menu cover featured the four-notched petals of the eastern dogwood! It was as if the owners had never even seen the glorious blossoms that were right in front of them.

In all fairness, the true flowers—what matures into the red drupes with the seed inside—of all the species of *Cornus* are in the center of what appears to be the blossom. The large, white petals that surround the flowers are really bracts and not part of the true flowers. Still, if they had looked, they would have seen that our state's species tends to have six bracts, each larger than the eastern dogwood's four. The bracts of the *C. florida* are notched, while *C. nuttallii* are pointed. Furthermore, the buds of *C. florida* are a quarter of an inch in size and look much like a caper. The immature bracts completely cover the true flowers, like this: 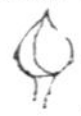. At half an inch, the *C. nuttallii* buds are larger and flatter in size, and the bracts do not cover the flowers: 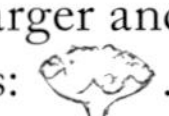.

So when you hear the word "dogwood," try to erase the four-notched-petals image from your mind and really look at this elegant western beauty. It's an individual worthy of your admiration.

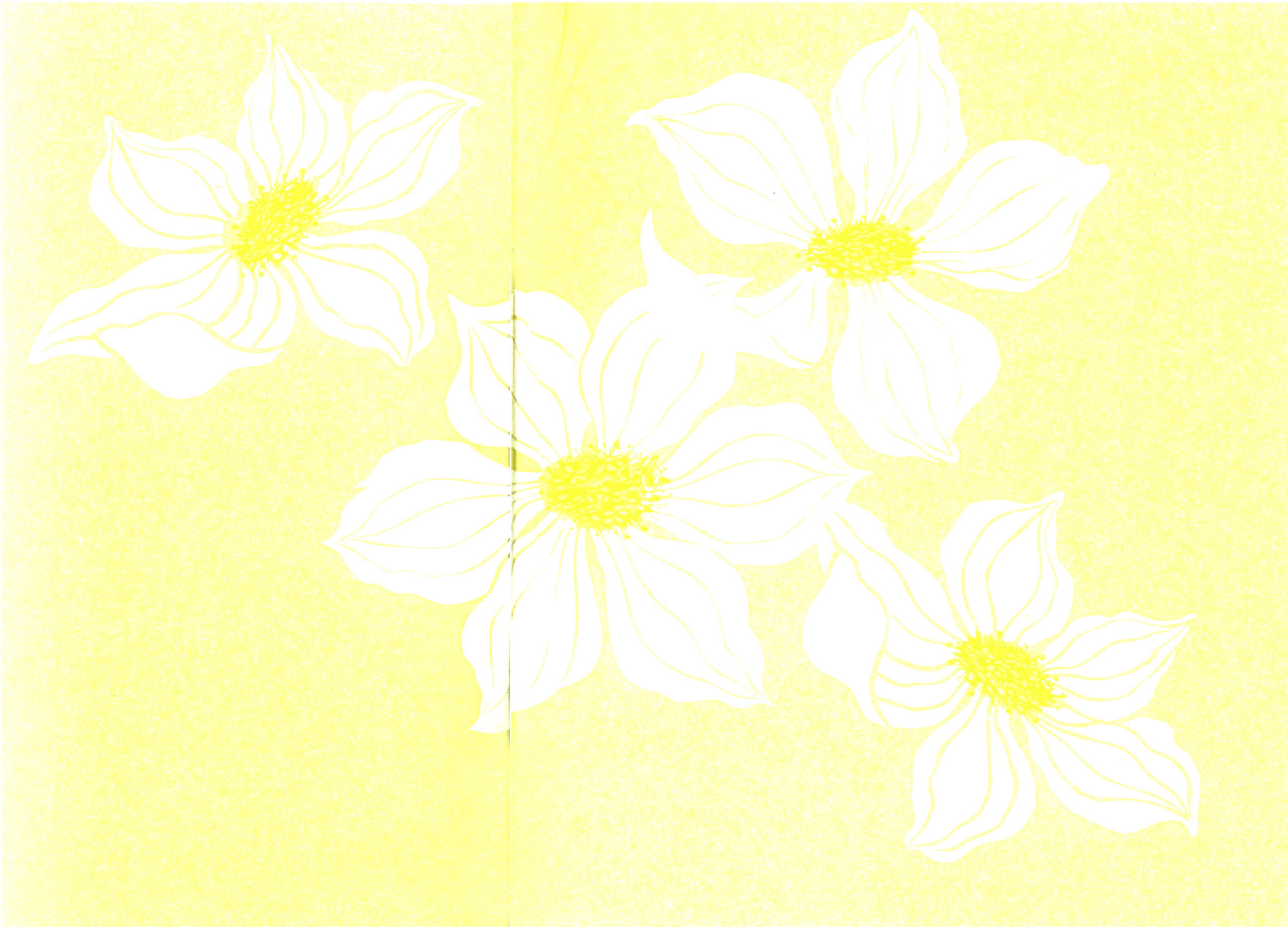

Snow plant *(Sarcodes sanguinea)*

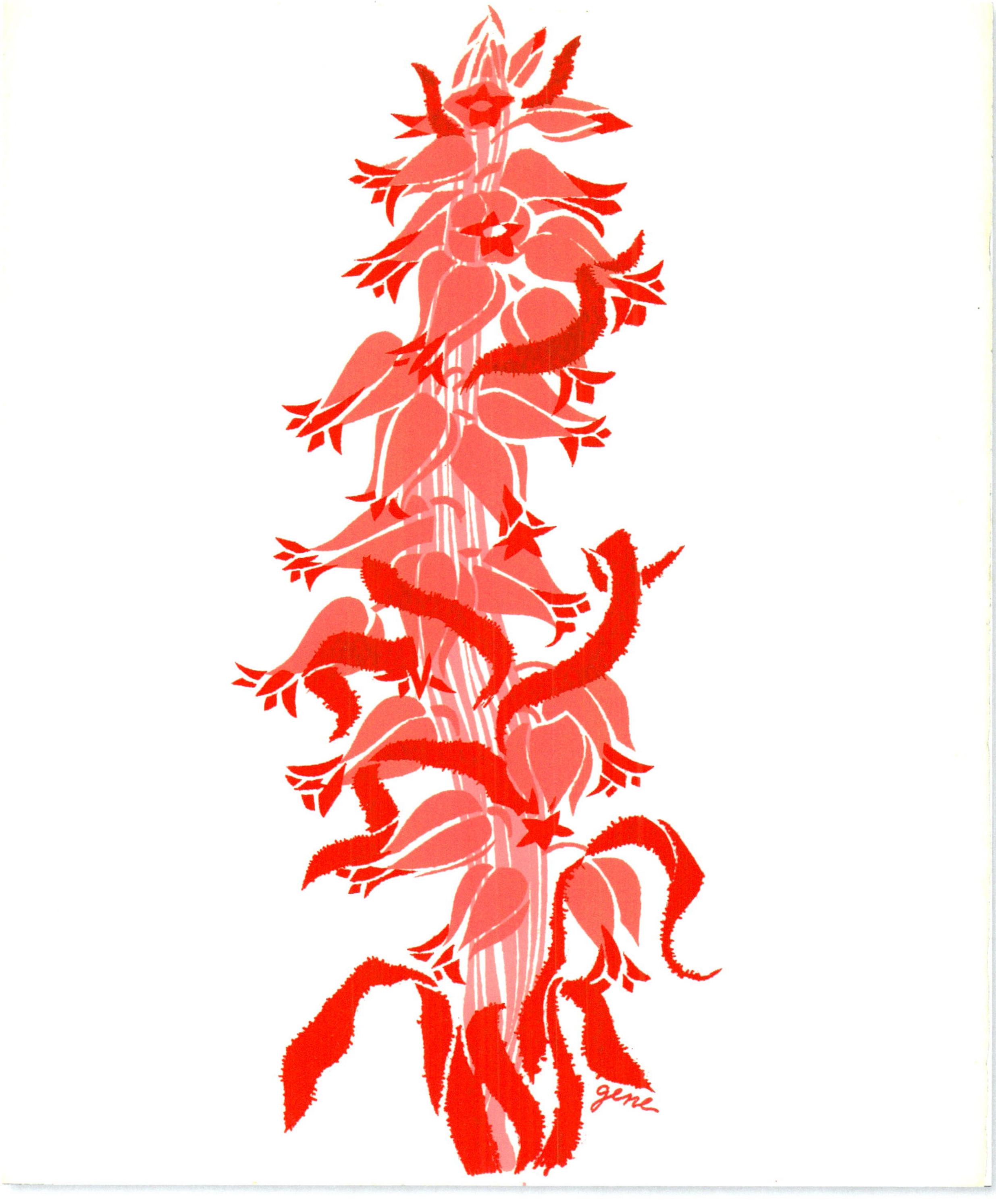
gene

Early July 1973

Of all plants growing in this world, possibly none are more captivating than those of the group known as saprophytes. A saprophyte is a plant living on dead or decaying organic matter and therefore usually lacking chlorophyll. It is both startling and exciting to suddenly happen upon one of these alien-looking plants thrusting up from the soil completely devoid of any green. It is almost as if it wasn't a living plant at all but some peculiar eruption on the surface of the earth.

California is fortunate in having many unusual versions of these already strange saprophytes. *Allotropa virgata* or sugarstick, a striking combination of red and white, does look much like a candy cane. The waxy white *Monotropa uniflora* (Indian pipe or ghost pipe) is ghost-like and rare. People think of *Hemitomes congestum,* a very lovely waxy pink, as false pink asparagus. These three saprophytes are usually found in the cool, dark, redwood forests of coastal Northern California. *Sarcodes sanguinea* is the saprophyte most familiar to us, purportedly because of its wide distribution, but perhaps also because, as John Muir described it as early as 1901, it is "the most attractive of all the flowers of the forest."

Sarcodes sanguinea

From two Greek words *sarx,* meaning "flesh," and *oides,* meaning "like," the genus *Sarcodes* has but one species and it is native to California, *sanguinea* (from the Latin, meaning "blood"). Though its scientific name aptly describes the fleshlike character of the stalk and flowers and the color of both, the plant is better known by a name that doesn't fit it at all: snow plant. *S. sanguinea* is found in the months of May, June, and July—at lower elevations, earliest around 4,000 feet, then progressively later at the higher elevations of 8,000 to 9,000 feet—and it really doesn't care for snow. But you can imagine someone thinking it does if all they've seen is the plant's intense red against the white background of melting snow or a late snowfall.

This extraordinary plant is composed of one fleshy stalk, straight and rigid and usually from ten to fifteen inches high. Sometimes it reaches only five or eight inches in height, but it can manage that within just about a week after its dome-shaped head pops up from the forest floor like a sudden, fiery gleam. From the stalk grow bell-shaped flowers with five lobes, and crowded among them, long, curling, graceful bracts. The whole plant looks as if it is brilliant red, but upon close examination, you see a great variation in the intensities of red. The stalk seems to be the flesh color of rosy cheeks; the exterior of the flowers are a deep rose color; and the interior and lobes, a glowing crimson. The bracts that grow between the flowers are a fiery blood red. Even the thickened scaly portion below ground level is red, a blood red.

In order to feature this plant, I had to find one. I looked in all the places where I had seen it growing in years past. A November 1970 fire had burned much of the area of my search, destroying the forest humus. Of course, no humus meant no snow plant, or so I thought. Leaving my immediate vicinity for a higher elevation, however, I found ten clumps, some single, some multiple, in an area smaller than an acre with nary a shred of visible humus, since the fire had swept here also. Two were growing in vertical cuts in the road, seemingly in nothing but decomposed granite. (Incidentally, two days later these were both gone—picked by two-legged beings.)

Obviously the roots go very deep in search of any decaying vegetable matter, and what a marvel that this is their only requirement for growing up so beautiful. Now that I had one close in front of me, I could see that my assumptions about its having such a complicated form were incorrect. The plant's parts had always seemed so assorted, mixed up, with the long-fringed bracts curling and spreading, that I had not realized the simplicity at the heart. In form it seemed complicated only because of the multiple intensities and close values of the single red hue composing it.

In fact, the special appeal of this golden native is its brilliance of color, which is multiplied many times when a ray of sunshine spills on it. Then for a brief but spectacular moment, there is a translucent quality about the whole plant, and it glows as if lighted from within.

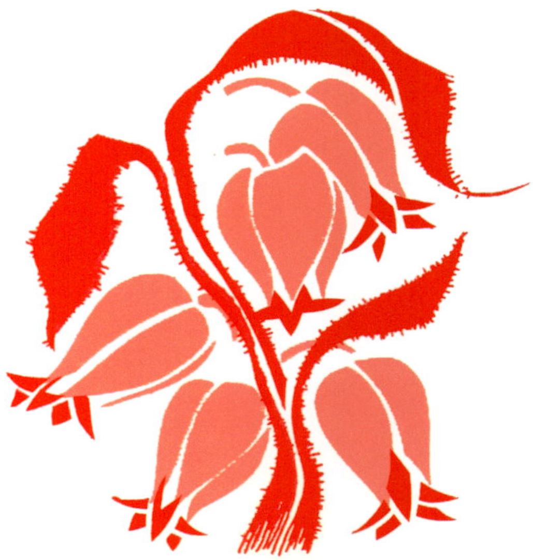

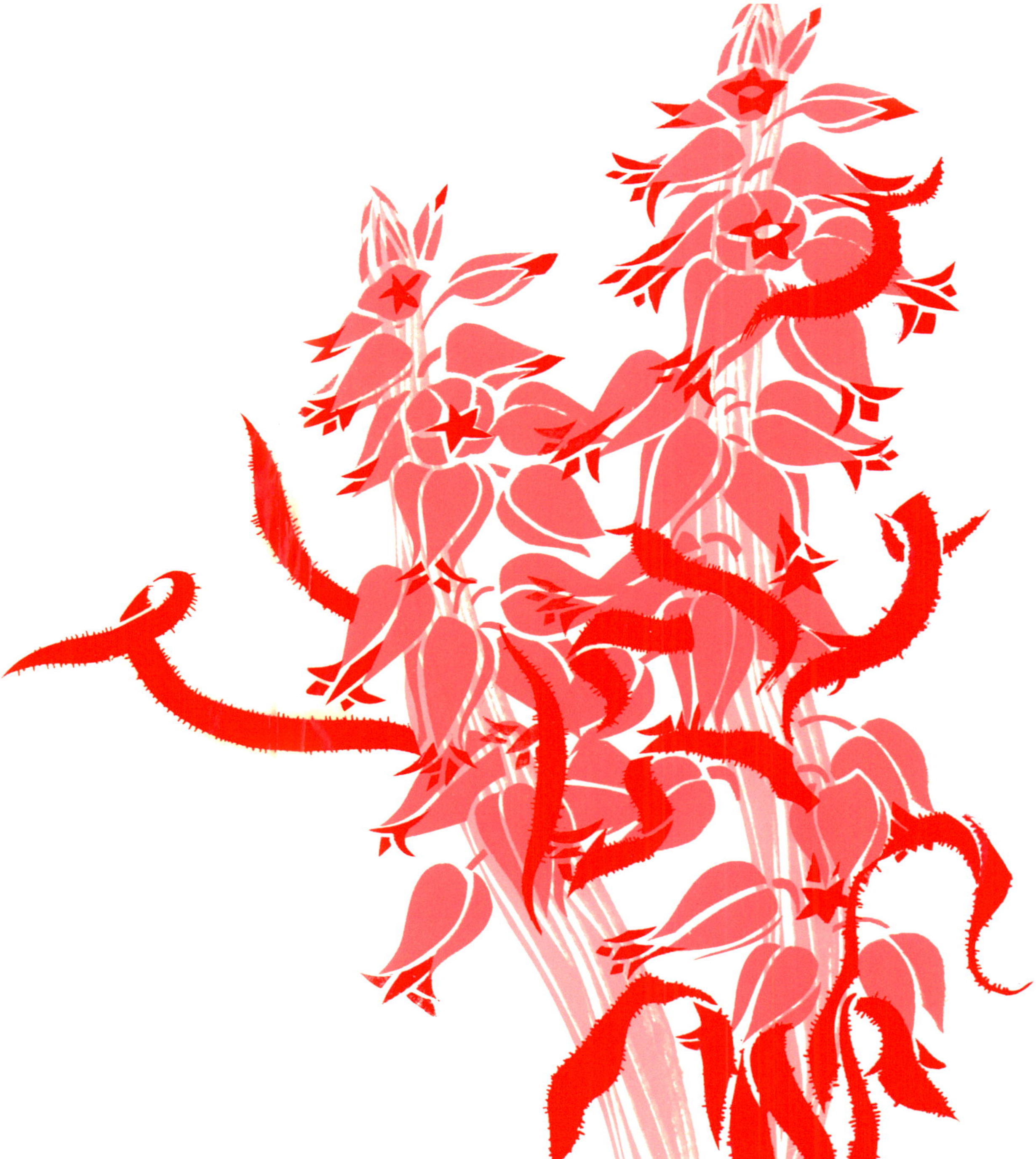

Matilija poppy *(Romneya coulteri)*

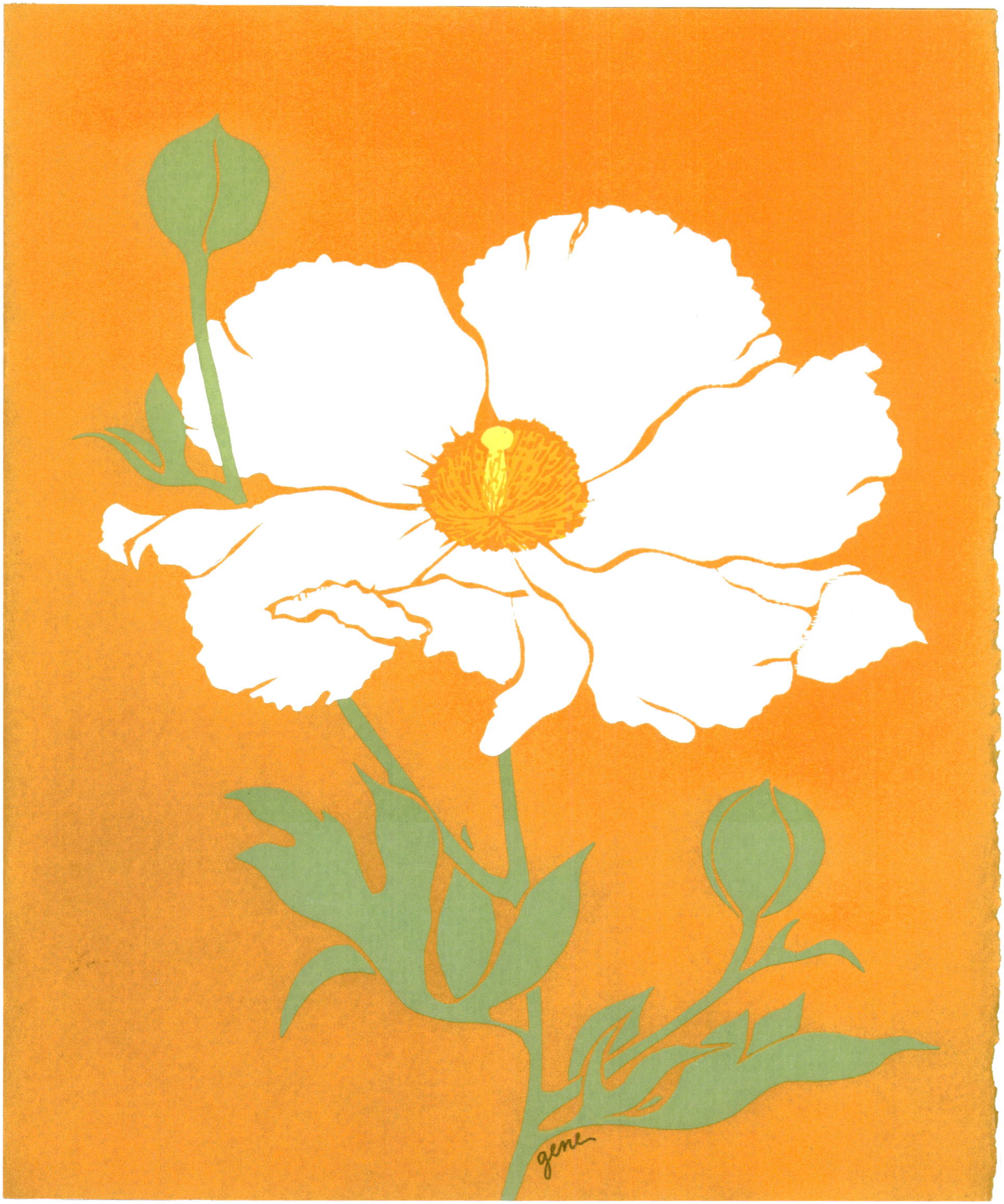
gene

Late July 1973

Romneya coulteri is sometimes referred to as the "queen of California wildflowers." This native plant's "commoner" name is the very lovely Matilija poppy (pronounced *ma-TI-la-ha*). Because it is not quite hardy at my mountain elevation (5,500 feet), I have never grown it. But I have always admired this poppy and its abundant flowering in Matilija Canyon in Ventura County. In addition to growing wild in the canyons of Riverside County as well, it is cultivated in many gardens in Redlands, Riverside, and San Bernardino, the latter of which was the source of the flowers from which I did my initial drawings.

The actual meaning of the word "Matilija" is shrouded in mystery, but it may have been the Chumash Indian name for a plant highly valued for its medicinal qualities. The name Matilija was also given to a Chumash rancheria in this area that was under the jurisdiction of Mission San Buenaventura. In 1898 the name Matilija was applied to a Southern Pacific station when the railroad linked Ventura and Ojai. Not as poetic but more descriptive is another name, the fried egg flower. It is also called mission poppy and giant California white poppy.

Often the scientific or botanic name of a plant is descriptive as well, chosen because of some outstanding quality or characteristic of the plant. The generic name especially is often chosen this way. The specific name sometimes honors a person, either the discoverer of the species or by him or her in honor of a friend. It is unusual when both the generic and specific names honor actual people, as in *Romneya coulteri.* The generic name honors T. Romney Robinson, an Irish astronomer who was a friend of Dr. Thomas Coulter, the Irish botanist who discovered the genus. First introduced to Great Britain by Dr. Coulter, the flower was blooming in Dublin by 1877. Many Europeans regard this regal beauty as the best looking of all ornamental plants from California.

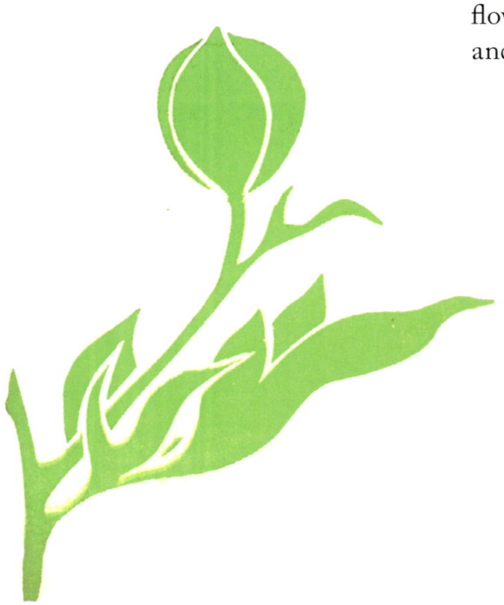

Romneya coulteri

Within the genus *Romneya* are but two species and both are native to California. *R. coulteri* is sometimes regarded as a perennial herb, sometimes as a shrub. Many rather woody stems do rise from the ground level and may grow from five to seven feet tall. The foliage is smooth, parted or divided, about five inches long—and handsome. It is that striking blue-green color, so prevalent among drought-resistant plants, that gets you. The stunning flowers, usually composed of six petals, are enormous, ranging from five to nine inches across, and produced in great quantities.

The petals are pure stark white and have an especially diaphanous quality about them, appearing crinkled as if made of crepe paper. At the center is a glorious tassel of golden stamens. The flower exudes a delicate fragrance not unlike that of a ripe orange.

R. coulteri is definitely a plant for large gardens. Once it is established, it spreads rapidly by underground roots and may crowd out less vigorous plants. It is difficult to grow from seeds, and even well-rooted sucker growth is somewhat iffy to establish. Best to purchase a plant in a container from a nursery. Very drought-resistant, the Matilija poppy requires but a minimum amount of summer watering (keep in mind that those in the wild receive moisture only from winter rains). It should be planted in full sun and is an excellent soil binder on hillsides. The plant becomes a bit scraggly looking in late summer. In early fall the entire plant should be cut to within six inches of ground level. New growth will pop up following the first rains. The plant will grow rapidly and the flowers will be borne on this new growth. Not completely hardy, it may survive temperatures down to 22 degrees.

I had never before attempted to make a drawing of this particular flower, and it was near the end of its blooming season after an extremely hot summer when I started. So the late bloomers were small—five inches across—and did not last long when cut, but I was very enamored of them. It was a delightful experience just to study the structure of the flower. Each petal was filled with what I like to think of as character, and I was excited to depict it in several different poses.

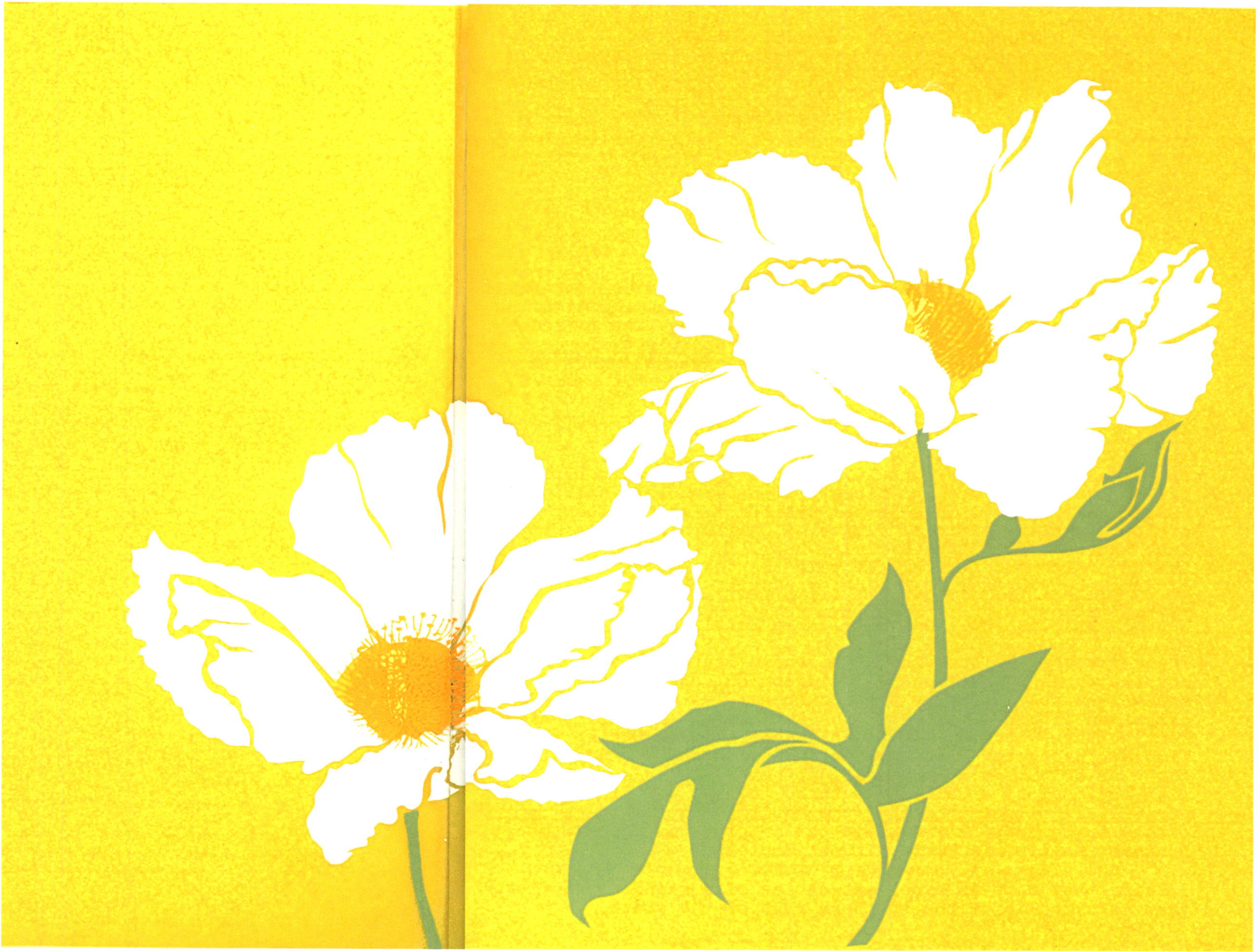

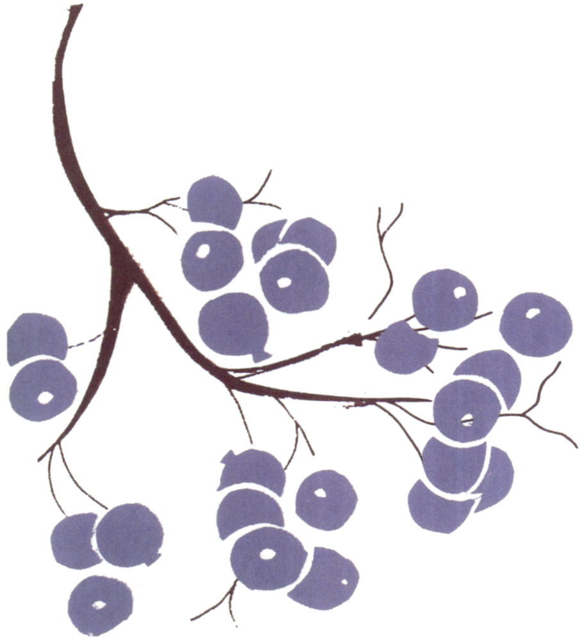

Blue elderberry *(Sambucus caerulea)*

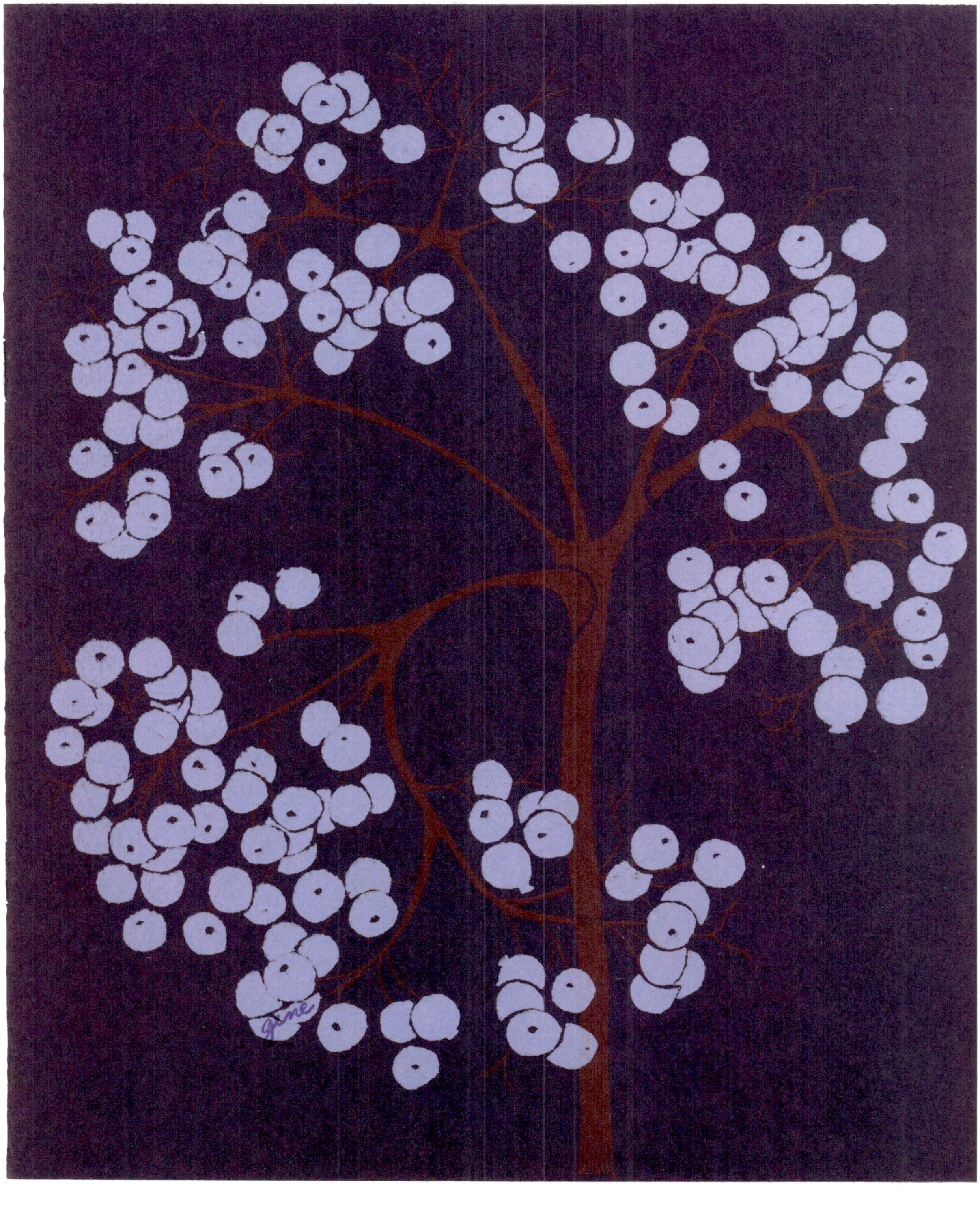

Early August 1973

Here in California summer opens with the appearance of elder flowers and the season ends with maturing elderberries. The plant's name, *Sambucus,* comes from the Greek word for a musical wind instrument made of elder wood *(sambuke),* indicating only one of its many uses. The twenty or so species of the genus, growing mainly in the world's temperate and subtropical regions, include five native to California, three of which can attain small tree size *(S. caerulea, S. mexicana,* and *S. callicarpa).* The other two California species are low shrubs *(S. microbotrys* and *S. melanocarpa).* The common name is usually just elder but sometimes elderberry. This plant, often found growing along streams or in damp areas, is universally known for its many good qualities, but if it had been better known earlier, I wonder if it might have changed history.

At least that was my whimsical thought as I chose the postage stamp for the booklet I sent on the elderberry. As I've mentioned, I give a lot of thought to selecting the stamps to be used with each serigraph, usually chosen for their appropriateness of color or subject matter or some other minor subtlety. The stamps I chose for this booklet were blue, yes, but they also commemorated the Boston Tea Party. Consider this possibility: the American revolution would still have been fought, but had the colonists been satisfied with elderberry tea (produced by the very abundant *Sambucus canadensis* of the eastern United States), would the high price of British tea necessarily have been the catalyst for it?

Sambucus caerulea

This is the species that is known as blue elder, blue elderberry, blue-berried elder, and mountain elder. Sometimes a tall, large bush or shrub and sometimes a small tree at twenty to thirty feet, under very favorable conditions it may grow up to forty feet high. Usually it has multiple stems from the ground level, and grows in washes and flats and along watercourses. Its size depends upon its location. It is often tree-like at low elevations and more shrub-like at high elevations. The largest specimen of *S. caerulea* previously grew near San Rafael in Marin County, but this tree has become the victim of urbanization.

The elder is not commanding of attention until early summer when enormous flat-topped clusters of cream-colored blossoms emerge to cover it. The individual flowers are very small but the clusters can be from eight to ten inches across, and they are so fragrant that it's no surprise to discover the plant belongs to the honeysuckle family. The best specimens grow in moist locations, and if the plant receives an adequate amount of moisture, the flowers develop into berries that ripen in August and September. This is the moment of elderberry glory when it is nearly overburdened by the heavy masses of juicy, sweet fruit. On fine specimens, the berries may be nearly a quarter of an inch in diameter and are a deep rich blue, appearing somewhat lighter because of the white bloom or the chalkiness covering each one.

Sambucus caerulea is a very persistent plant. Four years ago a bird deposited a seed in front of our house about three feet from the structure. I had always wanted an elderberry but not that close. Rather than take the time to dig out the roots, I simply kept cutting it off at ground level. As soon as I would cut it off, up popped a new shoot. Finally this year I allowed one stem to remain. It is now about five feet tall, and inspecting it a month ago I was thrilled to discover it was going to bloom. The cluster of flowers is fully developed right now and very sweet smelling. I am hopeful that these flowers will develop into berries. Where elders have been destroyed by fire or other causes, they do stump-sprout freely and grow very rapidly, probably good traits since they are so often damaged. The wood of the tree is very weak due to its pithiness (the trait that makes it easy to hollow out for wind instruments), and the branches break very easily under the great weight of the ponderous masses of berries.

Providing many necessities of life, *S. caerulea* was held in high esteem by the California Indians. They called it "the tree of music," making flutes and shafts for their arrows from the wood. The berries of this western species are the only ones that are edible, and they are good for making jams, jellies, pies, and syrup as well as tea. The flowers are not only beautiful and fragrant but also edible and drinkable. A favorite way to prepare the blossoms is to dip the whole cluster in egg batter and fry it. The flowers are often added to pancakes and muffins and also used to make a tea, a cough syrup, and a pale yellow, delicate wine. Not to be left out, elder leaves are used in making hot poultices to relieve the pain and promote the healing of sprains and bruises.

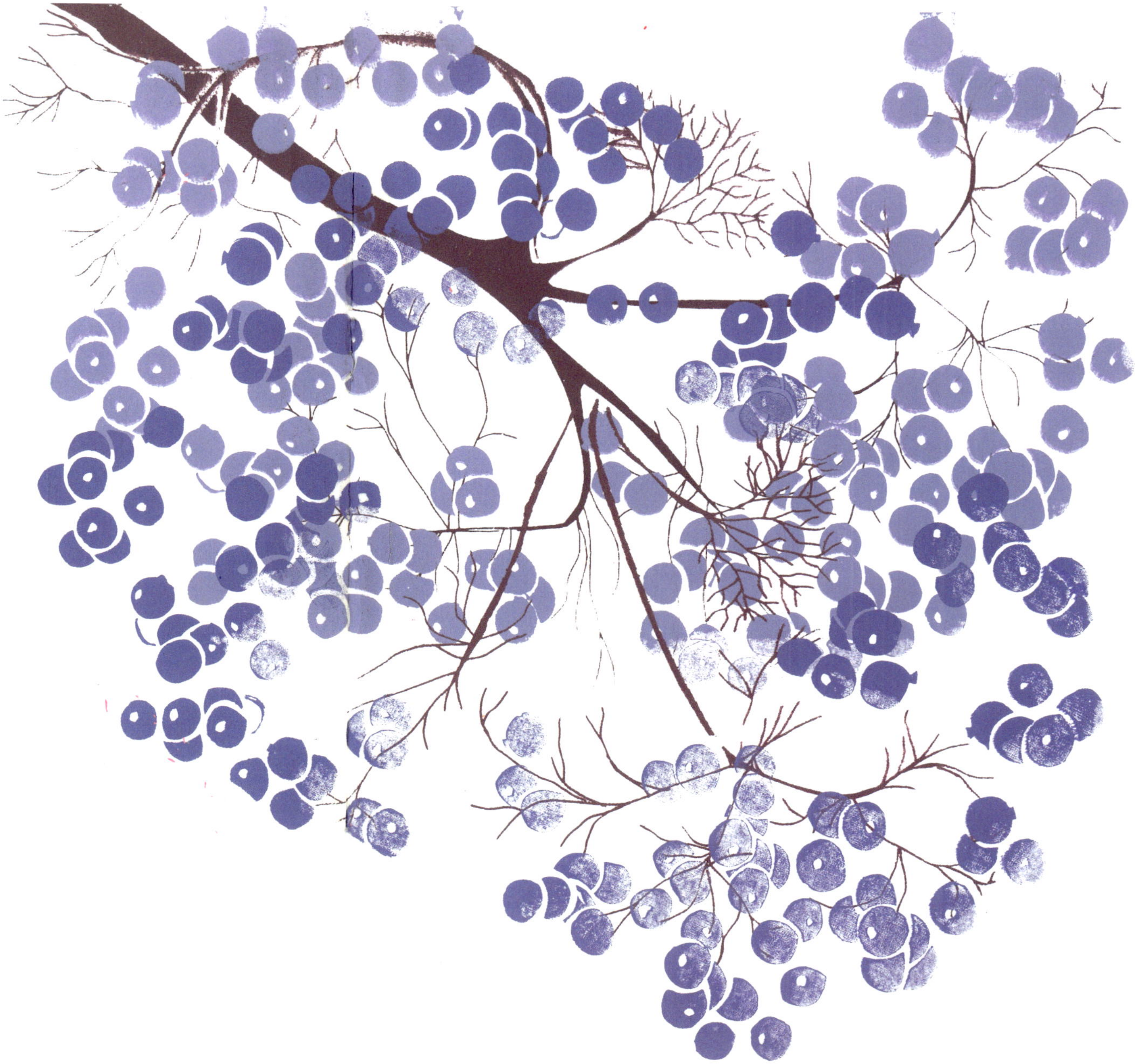

Cord grass *(Spartina foliosa)*

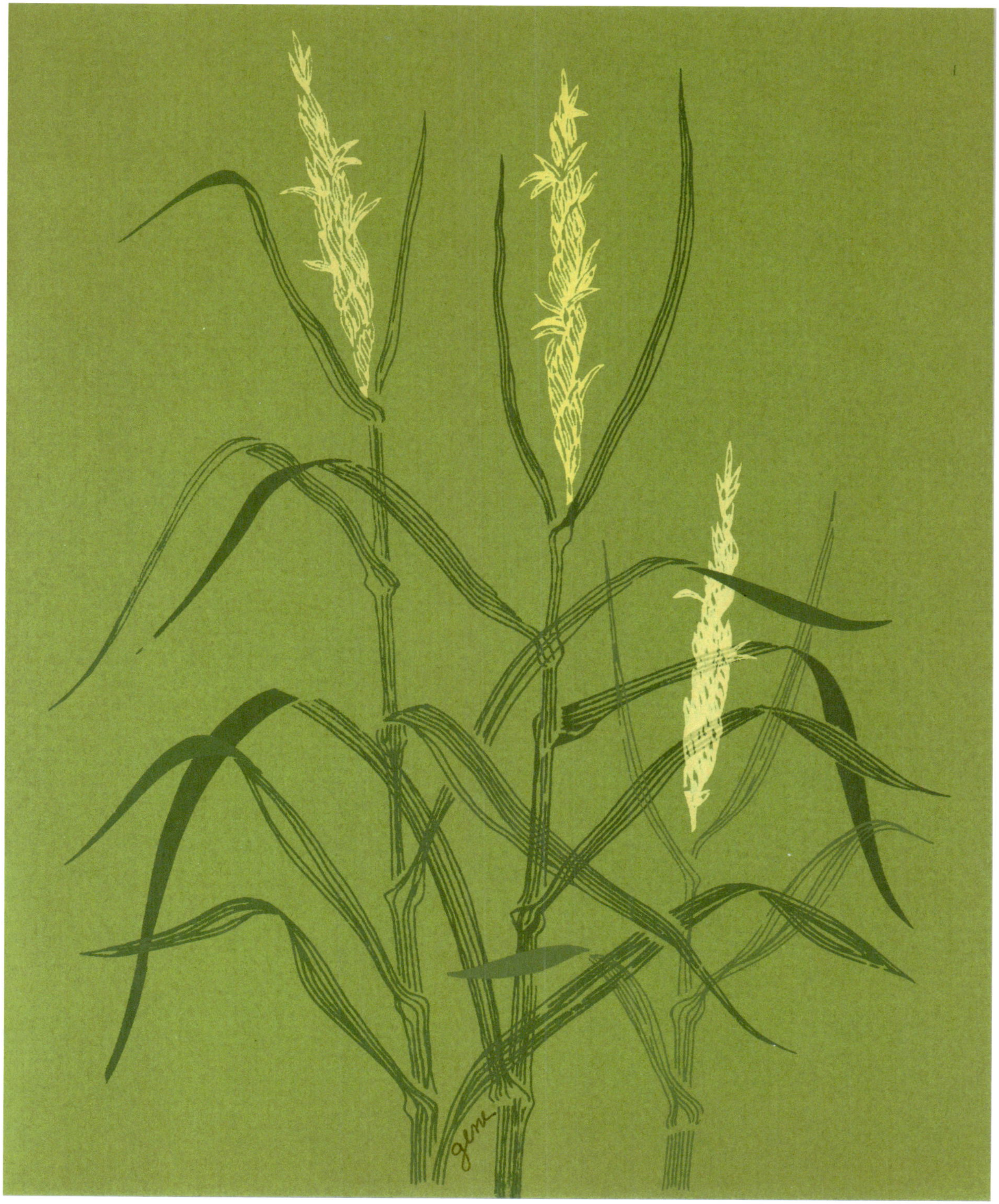

Late August 1973

Some California native plants live out their lives barely noticed even when they are of tremendous ecological importance. A region in Northern California, bordered on the east by San Francisco Bay and on the west by the Pacific Ocean, harbors a plant species as underappreciated as it is vital. It is called *Spartina foliosa* or cord grass. University studies have shown that, while manufacturing food for its own growth, cord grass removes more carbon dioxide from the air and gives off more oxygen than any other known plant. And that's not even the only valuable function it performs, quietly and all without rave notices.

Spartina foliosa

Marshlands are the most productive places in the world for exchanging carbon dioxide for oxygen. One region that includes such vast expanses of flat salt marshes is San Mateo County, well-endowed also with mild, Mediterranean-like weather and gently rolling hills. The salt marshes stretch great distances along San Francisco Bay, yet seeing in them only dismal-looking wastelands, people rarely take the time to investigate them. This is unfortunate because baylands are as important to the whole of California as are coastal forests, inland valleys, foothills, high mountains, and deserts. Within the salt marshes may be found quiet beauty, space, and solitude. The peaceful community flourishing here—plants, birds, and land and sea animals well-adapted to this way of life—would all vanish if the salt marshes were allowed to disappear. The most dominant bayland plant among those growing in these salt marshes literally holds them all together, like a cord might. In fact, the plants genus name of *Spartina* means "cord" in Latin.

Spartina foliosa is a true grass with hollow stems, swollen nodes, and dark green leaves that are long and narrow. Cord grass produces strong creeping rhizomes or roots that help the plant to spread and also help to anchor it in shallow soil. Because of this, cord grass is an extremely important dune binder and without it many of the dunes and marshy areas would be washed away. In turn, if the salt marshes were to disappear many small land and sea animals and water and shore birds would disappear as well. These creatures, many of which are considered delicacies for human consumption, depend on the bay salt marshes for food, shelter, and resting places. Cord grass not only binds this community with its roots but also provides food for birds and animals with its very small flowers and seed. Then when the cord grass dies and falls into the water, it decomposes into proteins, carbohydrates, and vitamins that are consumed by very tiny animals which then become even more nourishing morsels for larger animals.

Cord grass comes into its own as a plant with great value as an air purifier and oxygenator beginning with the promise of low tide. In early spring the young shoots are exposed to air and sunlight only during low tide. The plant grows steadily throughout the summer and by fall may reach heights in excess of four feet. After reaching this height it is rarely totally submerged. A stout perennial plant, cord grass continues to grow for several seasons. In early fall the tall stalks are topped with slender cylindrical plumes of golden grass flowers sometimes a foot long.

Very few plants are able to exist under the trying conditions that exist in salt marshes. Saltwater produces dehydration in most plants. But cord grass actually absorbs the saltwater, and because it is equipped with special cells, is able to excrete the salt then utilize the freshwater it has created. Its genus includes fourteen species that are widely distributed throughout the world. Two are native to California: *Spartina foliosa* itself is strictly a low-elevation species growing in coastal salt marshes; the other native, *S. gracilis,* grows in interior alkaline places at a high elevation of from 6,000 to 7,000 feet.

If you've ever thought of salt marshes as drab and dreary, I hope now you will reconsider their worth. These shallow waters are breeding grounds for many birds and spawning grounds for fish and home base for the give-and-take that one underappreciated golden native orchestrates. Though cord grass might not be glamorous, its reason for existing is vital to our very breath and to the life of all those creatures.

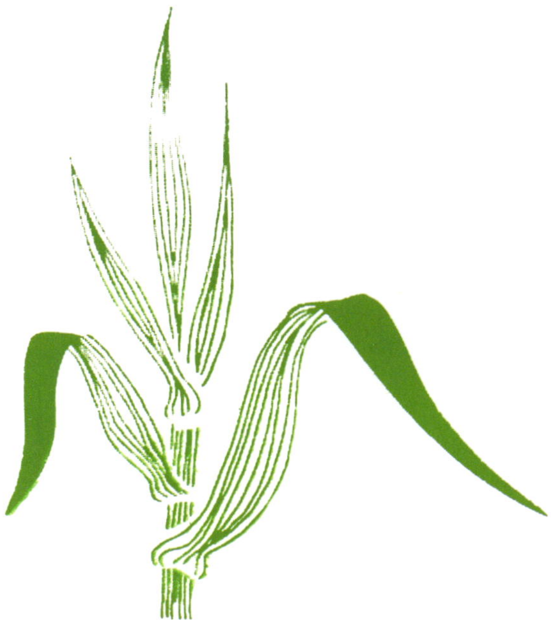

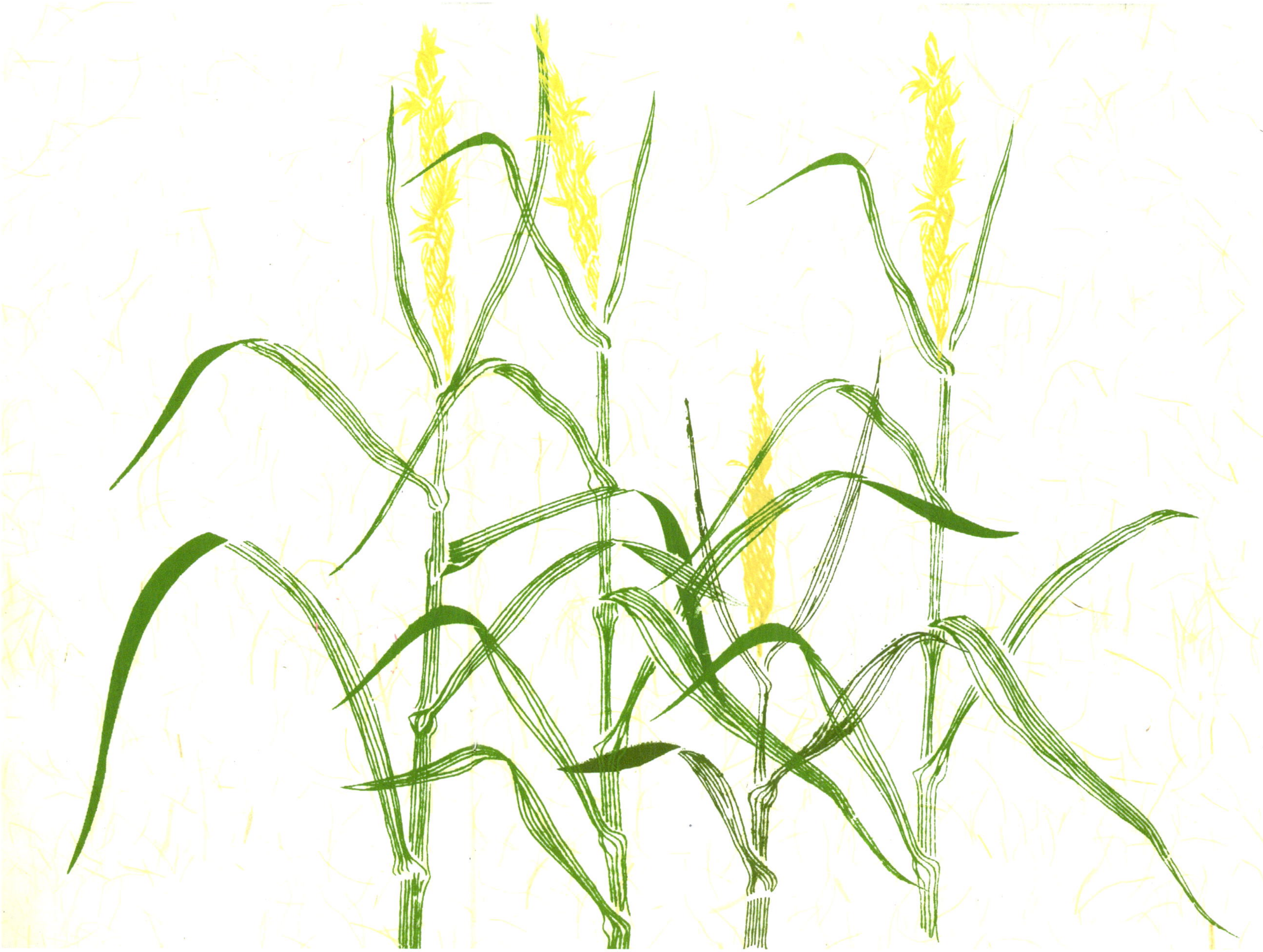

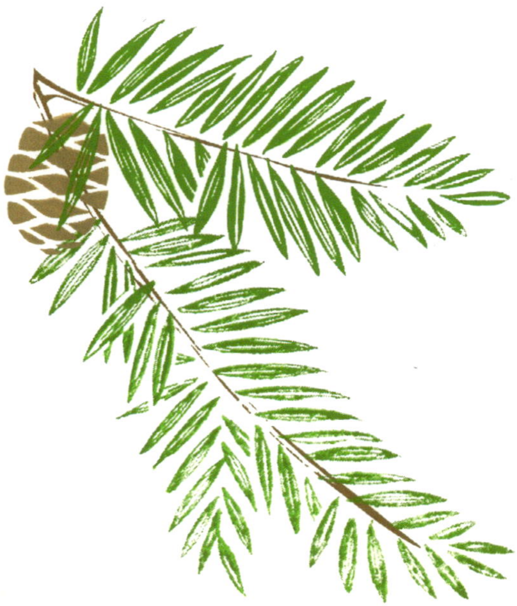

California redwood (*Sequoia sempervirens*)

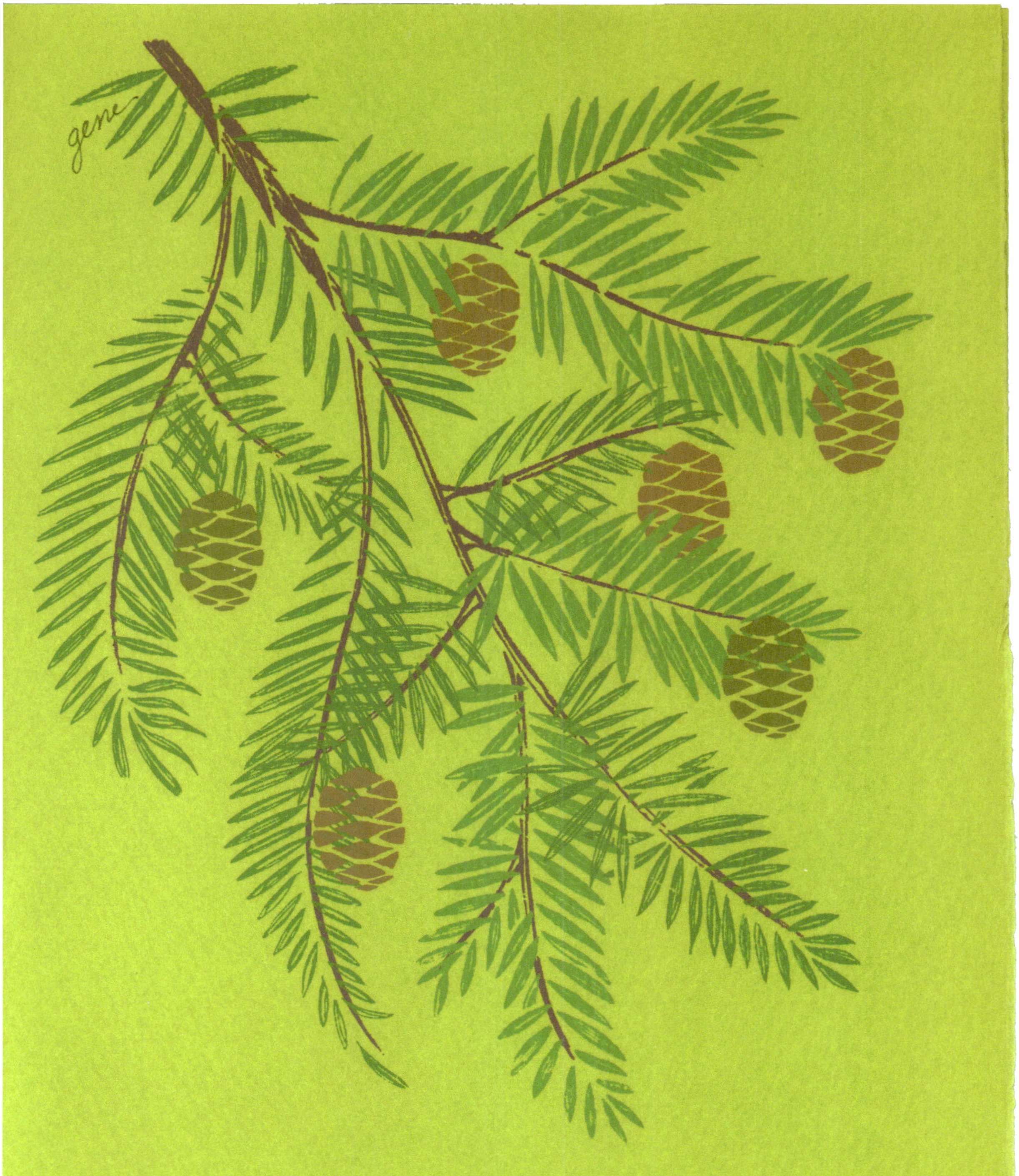
gene

Early September 1973

Millions of years ago, great coniferous forests covered the earth. Fossil remains indicate there were at least forty different species of the genus now known as *Sequoia* growing in all parts of the world. All that remain of the original forty are three species, each indigenous to only one part of the world. These three are closely related, somewhat like cousins, so despite the fact that their distinctions merit the status of genus, each of the three generic names bears a word in common, in honor of such close kinship. The three generic names are *Sequoia* (coastal Northern California), *Sequoiadendron* (inland California's Sierra Nevada mountain range), and *Metasequoia* (China's Shui-Sha Valley).

Sometimes understanding a nonnative illuminates characteristics of the native, and such is the case with the dawn redwood *(Metasequoia glyptostroboides).* From fossil remains it is known to have inhabited the earth, possibly even western North America, millions of years ago. It was believed to be extinct until a living tree was found in a remote province of China in 1941. Because of world conditions at the time, seeds were not available until 1946. Growing readily from seeds and cuttings, the dawn redwood has been widely planted all over the United States since then and in nearly all botanic gardens in California. The foliage and general appearance of this tree is almost identical to our *Sequoia sempervirens* except for the softer feel of the dawn redwood's foliage. The characteristic that makes this tree distinct from the two native Californians is that it is deciduous.

Sequoia sempervirens

The generic name, given by Austrian botanist Stephan Endlicher in 1847, honors the Cherokee Sequoyah. (The botanist was also a linguist, so perhaps he was impressed with how Sequoyah devoted his life to inventing an alphabet for his people.) The only species in this genus is native to California—it exists in a native state nowhere else on earth but in coastal Northern California and a small corner of southwestern Oregon. But that's not the only reason this tree is phenomenal. Its physical attributes are awe-inspiring. The specific name *sempervirens* means "ever-green"; however, some writers use the word to mean "ever-living," and with this particular tree I find that quite appropriate.

Its beautiful cinnamon-red bark inspired the early Spaniards to call it *palo colorado* (red tree). A very tall version of it was found in 1776 and given the name *palo alto* (tall tree), and a town was named after it. As of 1963 the original tree still stood. *Sequoia sempervirens* is known as redwood, California redwood, and coastal sequoia, and its natural range is a narrow strip of land averaging twenty-five miles wide that stretches five hundred miles, from southwestern Oregon to Monterey County south of San Francisco. Within this small area are the tallest trees in the world. Not only are they the tallest as single specimens but also as a race of trees. Many of these wonders exceed three hundred feet in height, with trunk diameters of twelve to twenty feet. Is it any wonder that a tree of such grandeur was chosen as the state tree? These monarchs of sublime beauty have been planted throughout the state, though this is not a tree for the deserts nor for elevations much higher than 4,000 feet where temperatures drop below 15 degrees.

A redwood is a graceful beauty from its earliest youth, as if already aware of its potential. It is pyramidal shaped with dark green leaves that are narrow, flattened, and arranged like feathers on slender branchlets. The world's tallest trees produce some of the smallest cones in the world, less than one inch long and egg-shaped with unusual shield-shaped scales. Within each scale are borne five or six tiny seeds—it is incredible that such enormous, long-lived trees are produced from such wee seeds. The wood of this tree is so magnificent that unlike most coniferous trees it is capable of sending up new shoots when the tree itself has been destroyed. Fallen logs left in the forest can sprout anew: Often after a tree is cut, a "fairy ring" of trees shoots up all around the stump. This is not like producing offspring or a new generation from seed but more like a continuation of life. I like to think that immortality comes closest to describing the lifespan of a redwood.

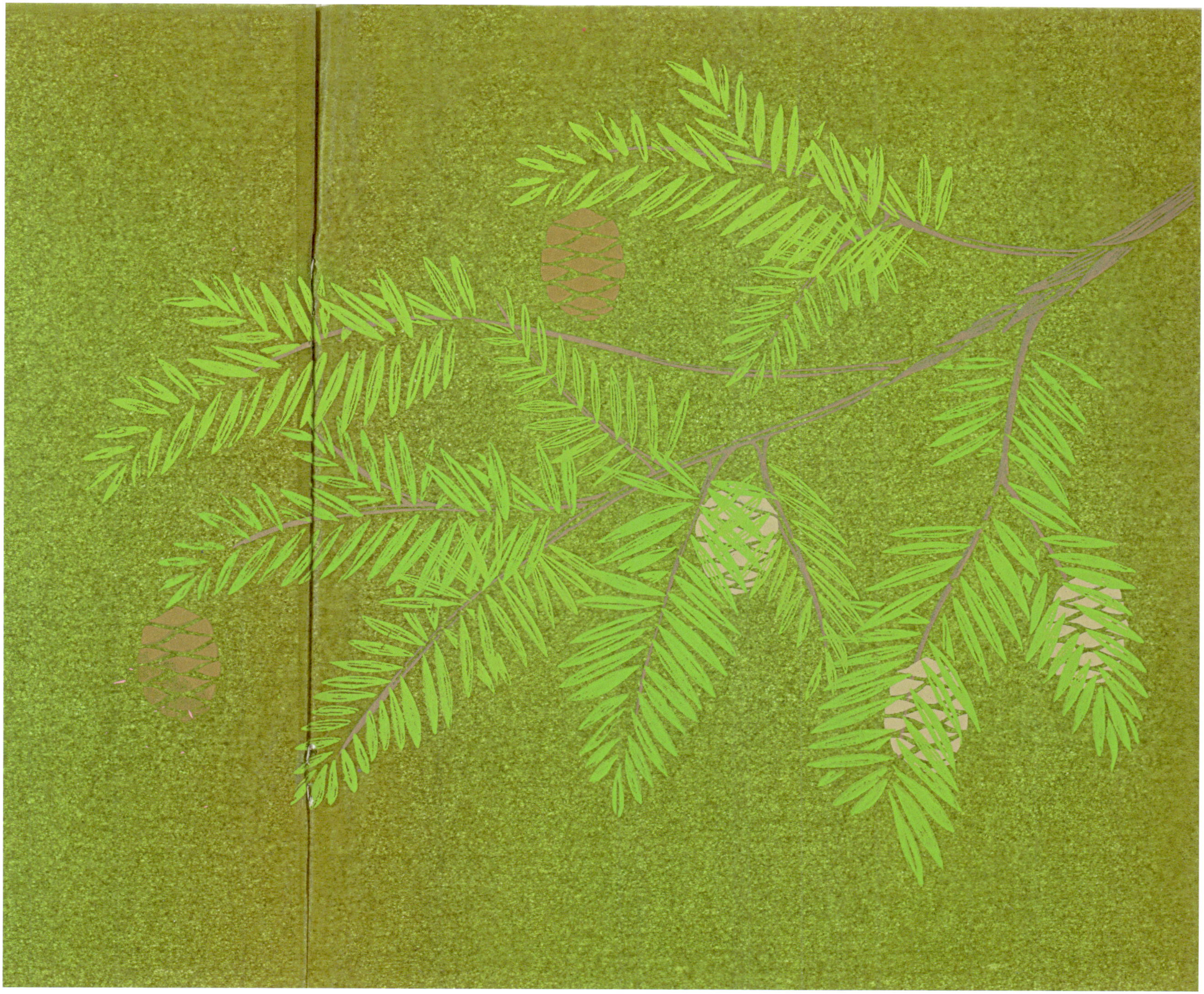

Giant sequoia *(Sequoiadendron giganteum)*

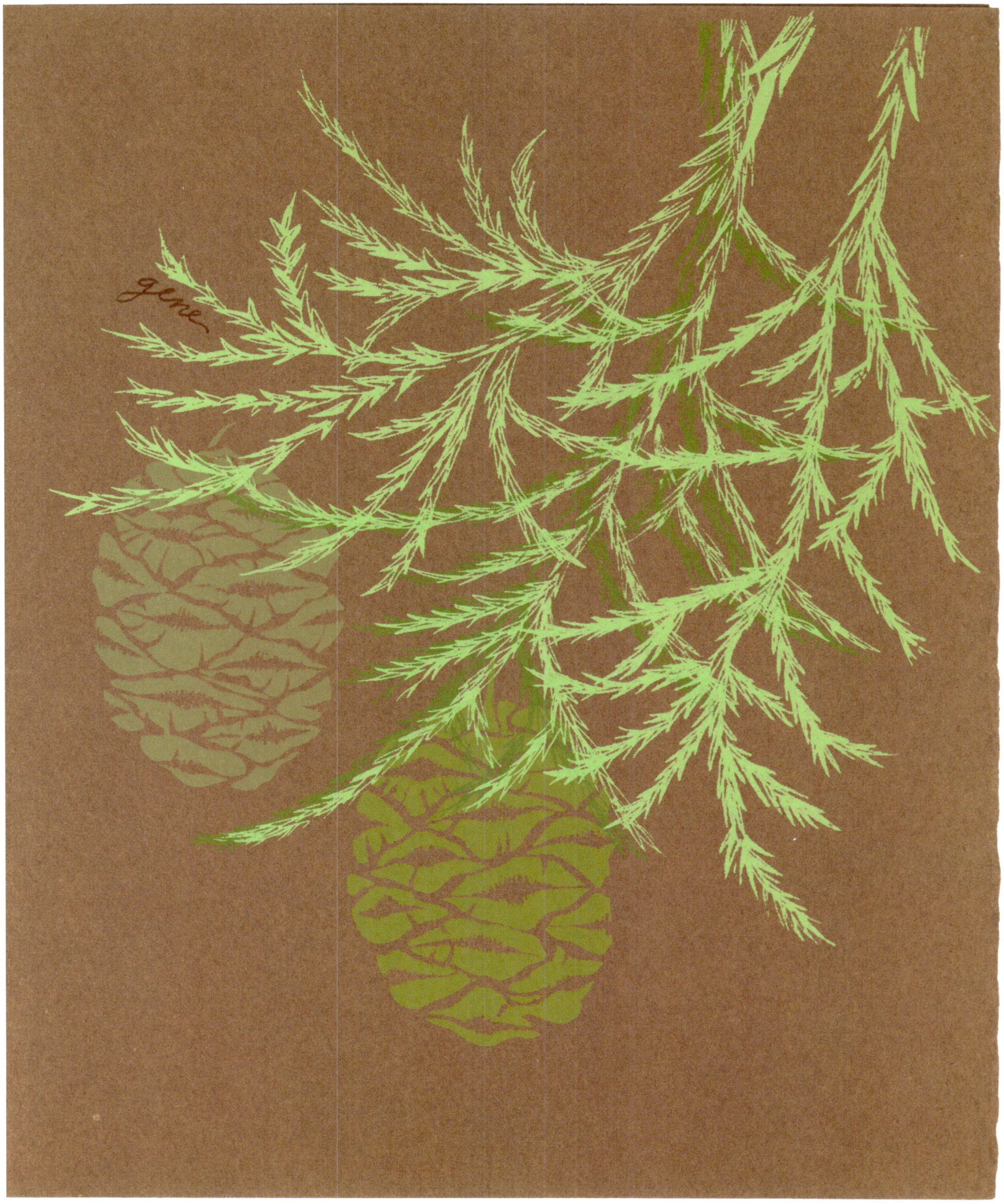
gene

Late September 1973

This is the second of the two very famous California trees, the big tree, the giant sequoia, aka *Sequoia gigantea.* But its real name is none of those. After much research, the American taxonomist John T. Buchholz decided in 1939 that this tree is sufficiently different from *Sequoia sempervirens* (page 71) to be elevated to generic rank. Because so much had been written about this tree under the name *Sequoia gigantea* he tried to retain some kind of continuity by renaming it *Sequoiadendron giganteum.* Some botanists refused to accept this name, but most authorities have by now and I will, too, because I feel the tree is deserving of its own special niche.

Sequoiadendron giganteum

The *dendron* part of its generic name means "tree"; *giganteum* speaks for itself. English botanists were the first to identify this tree and named it for the Duke of Wellington, *Wellingtonia gigantea.* American botanists believed it to be a species of bald cypress and renamed it *Taxodium gigantea.* In 1854 a French botanist claimed it was of the same genus as the coast redwood and named it *Sequoia gigantea.* At least everyone agreed it was gigantic.

Sequoiadendron giganteum is by far the most massive tree in the world. It is not as tall as its coastal cousin, *Sequoia sempervirens* (the redwood), but for sheer bulk it has no equal. The trees of other genera growing beside it—pines, firs, cedars—are magnificent specimens themselves, but they are dwarfed by this giant monarch. *Sequoia sempervirens* has often been called the "queen of trees," and just as frequently *Sequoiadendron giganteum* has been the "king of trees." California is the only state in the United States that can boast of two different trees of equal distinction as state trees. I'd go so far as to say that these two are the most magnificent trees in the world.

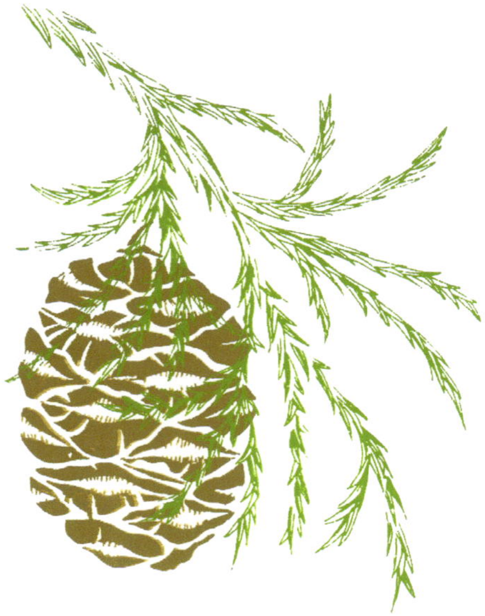

Each is singularly distinctive and spectacular. The natural range of *S. giganteum* is smaller than that of *S. sempervirens,* only a narrow strip of land about eighteen miles wide and extending for two hundred fifty miles or so along the western slopes of the Sierra Nevada Range at elevations ranging from 3,000 to 8,900 feet. As a naturally high-elevation tree of the interior, the giant is hardier than its low-elevation coastal relative and accustomed to snow packs ten to twenty feet deep. It is also somewhat slower growing and a bit more drought resistant.

This tree does not have the remarkable ability to stump-sprout as does its coastal cousin. New trees are produced only from seed. The cones are similar to the redwood, though a trifle larger at two to three inches long. The foliage of the two are completely different. *S. giganteum's* awl-shaped leaves are spirally attached and a bit sharp feeling. The tree as a whole is more silvery blue, with a metallic sheen sometimes.

Perhaps because these trees are not as abundant as *S. sempervirens,* they were the first to be protected. John Muir was the prime defender of them, writing about *S. giganteum* in 1911 as "the king of all conifers, not only in size but in sublime majesty of behavior ... seedlings and saplings growing up here and there, thrifty and hopeful, giving no hint of the dying out of the species." Muir insisted that national parks be created so that these trees might be protected. Today nearly all the giant sequoia still existing in their natural habitat are in protected areas.

Sequoiadendron giganteum was among the first trees that I ever planted, and it grows beautifully at 5,500-foot elevation. With summer water it will gain one to three feet a year; without it, three to six inches per year. Yet even when its growth is minimal it maintains its exquisite pyramidal shape and extremely dense foliage. Under my conditions for survival it has already withstood heat, cold, drought, a forest fire, browsing deer, perching pigeons, squirrely squirrels, and it even seems to be immune to that pesky pest known as the California pocket gopher. Studies by both state and federal government agencies reveal it to be especially resistant to the effects of smog and insect attacks.

Since I do not care to have a one-of-a-kind specimen type planting, I have planted these trees in quantity. Hopefully, I may live long enough to enjoy my very own giant forest.

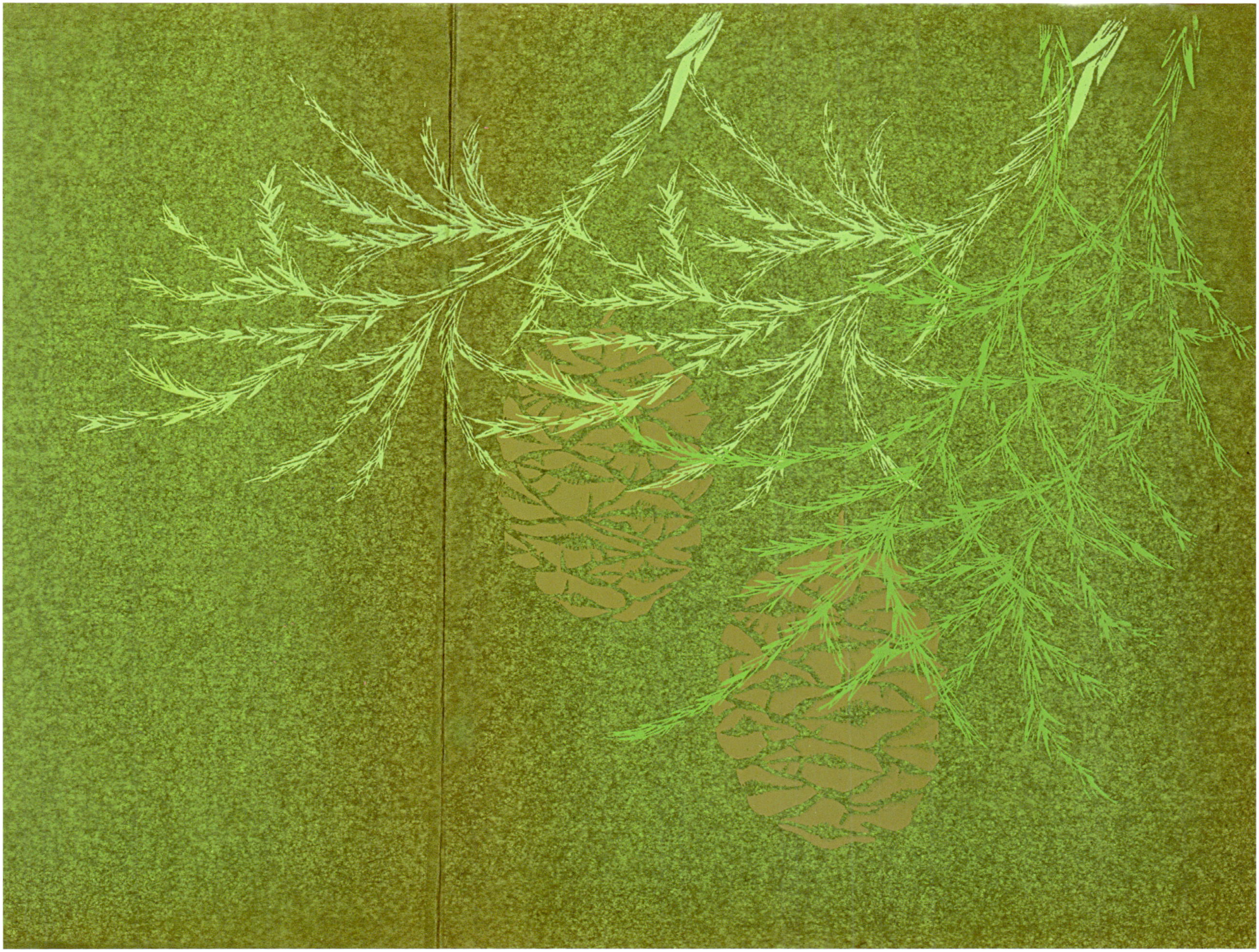

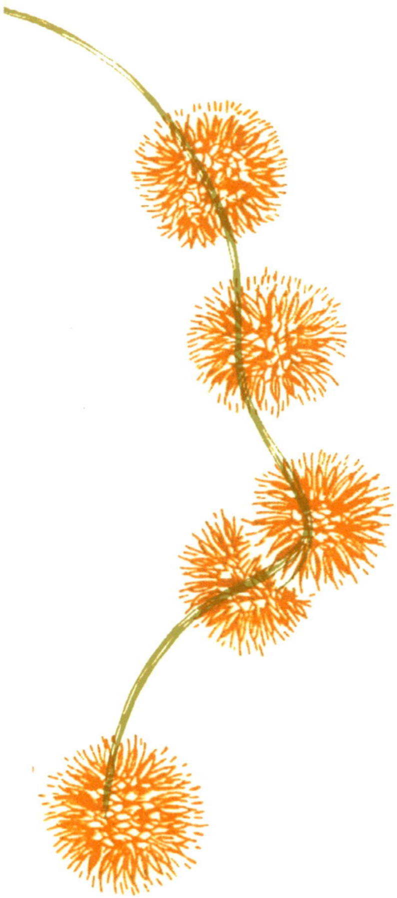

Sycamore *(Platanus racemosa)*

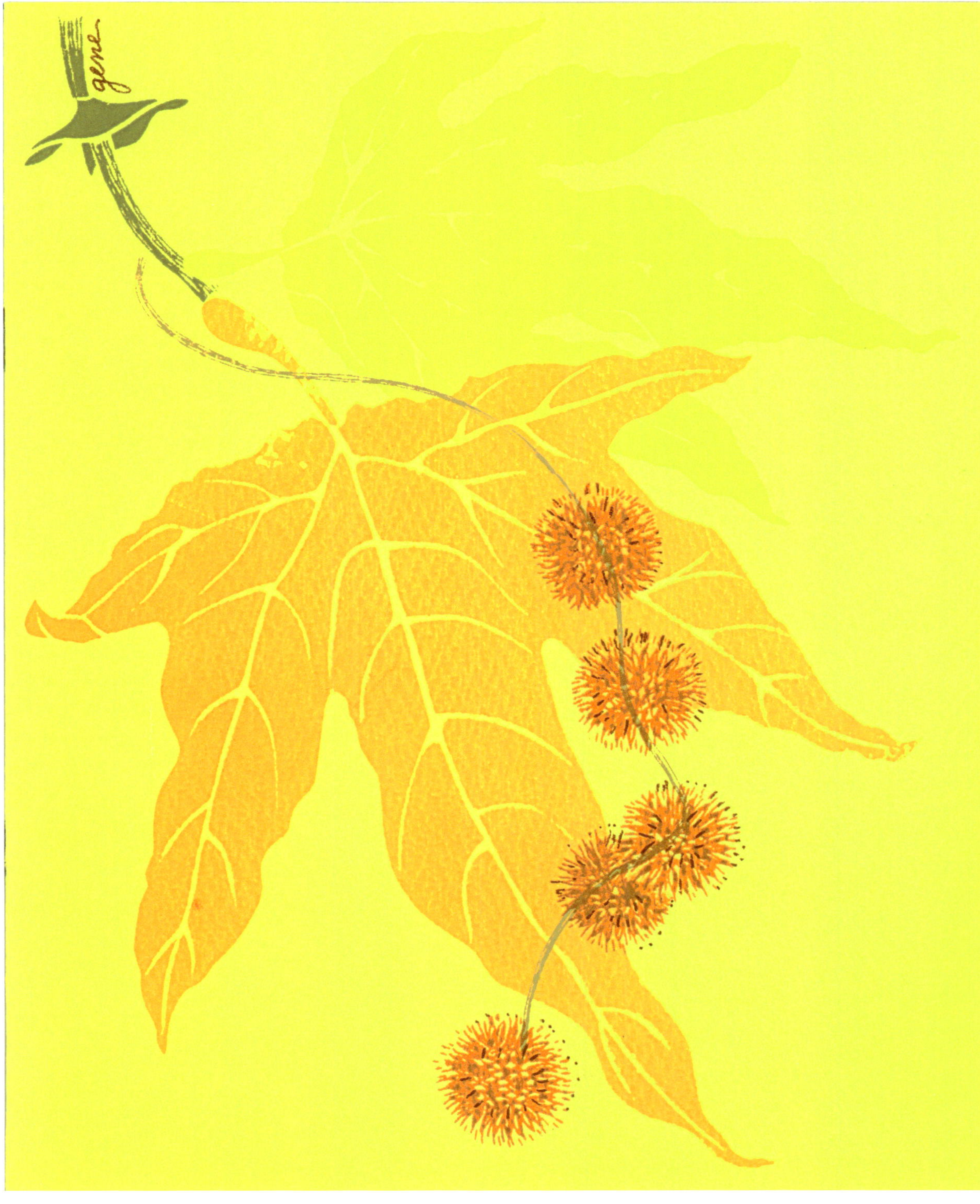
gene

October 1973

Even when a particular tree exists in abundance, there is one tree that likes to be the only one in its locality, the picturesque sycamore. Also known as western sycamore, buttonwood, buttonball-tree, and planetree, *Platanus racemosa* grows singly or in little groves in sandy washes and river bottoms of interior valleys and canyons.

The generic name is from the Greek *platys,* meaning "broad," and usually considered to be in reference to the leaves of the tree. Yet it could just as well be descriptive of the appearance of the tree itself, which is never dense or stuffy, but welcoming with open arms. As the only shade over vast open landscapes, it was a happy sight for early Spanish Californians trekking through the heat and often became a landmark to guide them on their way.

Platanus racemosa

Of the nine species in this genus, only *Platanus racemosa* is native to California. The flowers of this tree grow on long slender stalks in the form of a raceme. The flowers appear in the spring and by fall have developed into fruit or seed balls that are sometimes called buttonballs. The deciduous *P. racemosa* is considered typical of the state and especially characteristic of Southern California. It is usually found at low elevations, but I have found isolated trees in deep mountain canyons at a 5,000-foot elevation.

One of the easiest ways to identify it is by its exquisitely smooth bark. As the tree grows the inner bark is green, but upon exposure to the sun turns a chalky white. The outer bark or that of previous years does not grow but becomes flaky—children are tempted to peel it off in large pieces like scales. The bark turns blue gray, gold, and brown at different times, thus the trunk always has an interesting mottled look made up of unusual and beautiful colors.

Sycamore trees usually grow from forty to fifty feet tall but may even achieve eighty or ninety feet. Their strong, massive trunks support large, irregular crowns that are composed of gracefully sprawling branches.

The leaves are from six to ten inches long and as broad; they are thick-textured, light green above with a yellowish fuzz on the underside. Before falling in October they turn a golden bronze, a warm color that looks stunning against the creamy-white bark. The leaves are unusual in the manner in which they attach to the branch. Rather than being flat, the stem is hollow and forms little cone structures under which are found the buds for next season. This way the newly formed buds are protected, not exposed to the elements. When the leaf breaks away, the little circular clown's cap remains until the following spring when the bud bursts open. Sometimes the dried, cone-shaped cap remains on the tree all summer. These are fascinating little cones or hoods and well worth looking for in autumn and spring.

The handsome little string of four to six bristly fruit balls, each ¾- to 1¼- inch in diameter, develop during the summer and at maturity are a tawny brown.

The sycamore, very possibly the finest shade tree in the southern part of the state, is an open tree that permits breezes to move through it. It always seems to grow in warm areas where its cool shade is needed and appreciated. I drew the leaves and buttonballs you see here from trees growing in the bleak, sparse Santa Ana River wash on the outskirts of the town of Redlands. This is a dry river during the summer months but periodically wet during the winter with the runoff from melted snow of the high mountains above. Eventually this water makes its way down through the Santa Ana Canyon and through Orange County on its way to the Pacific Ocean.

P. racemosa adapts well to garden culture but should be planted only in an area large enough for its full development—it really likes to sprawl. Even as a young tree it is graceful and engaging, growing rapidly and requiring little attention. It is able to withstand abuse and neglect, but anything that can grow in the lean, poor, rocky soil where it is found would certainly enjoy being in someone's garden where it might receive a little love and attention. It certainly deserves it. With its interesting branching habit, beautiful peeling bark, and handsome leaves, *Platanus racemosa* has earned its status as one of the most vigorous, rugged, and charming of our golden natives.

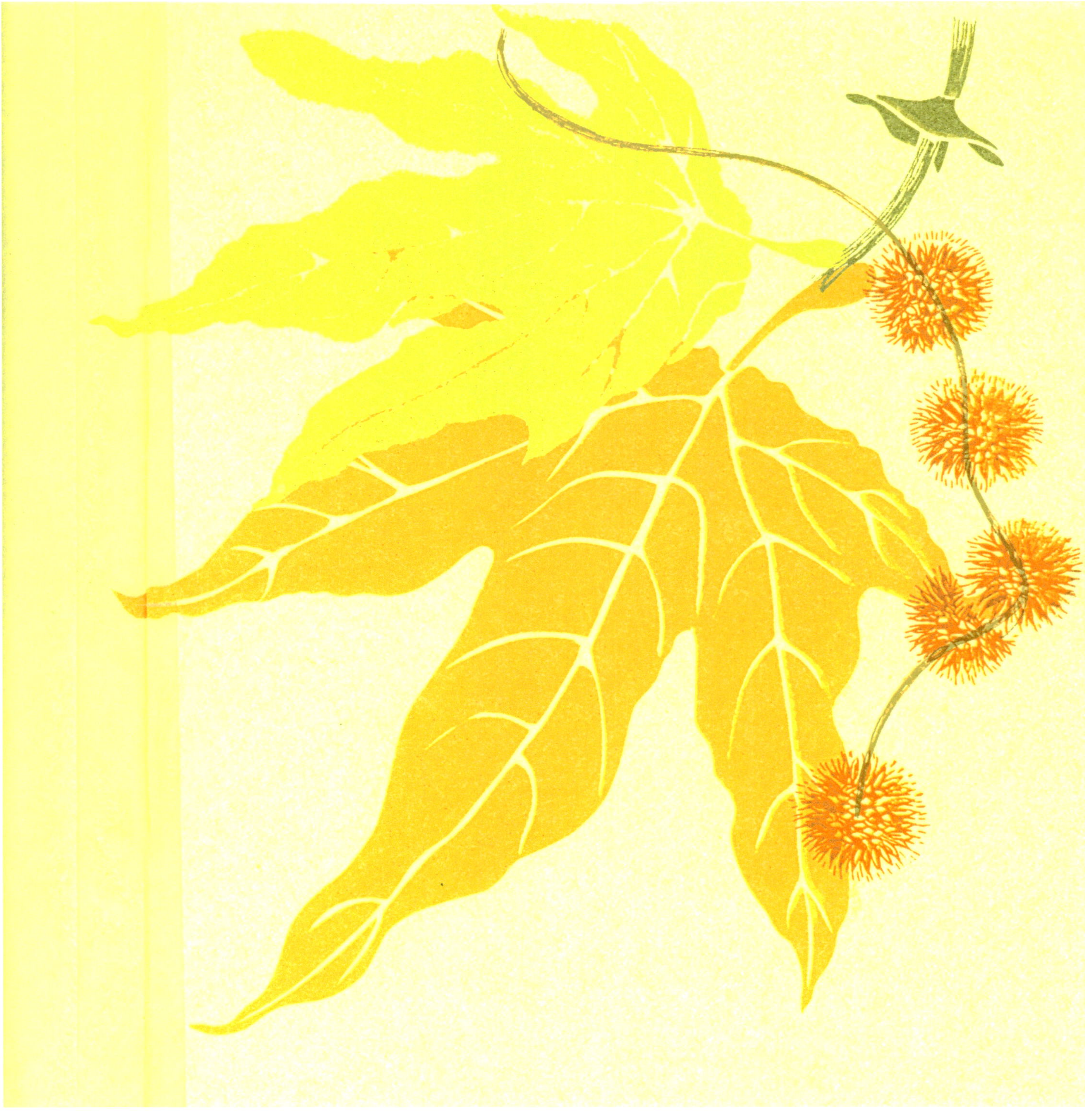

Digger pine *(Pinus sabiniana)*

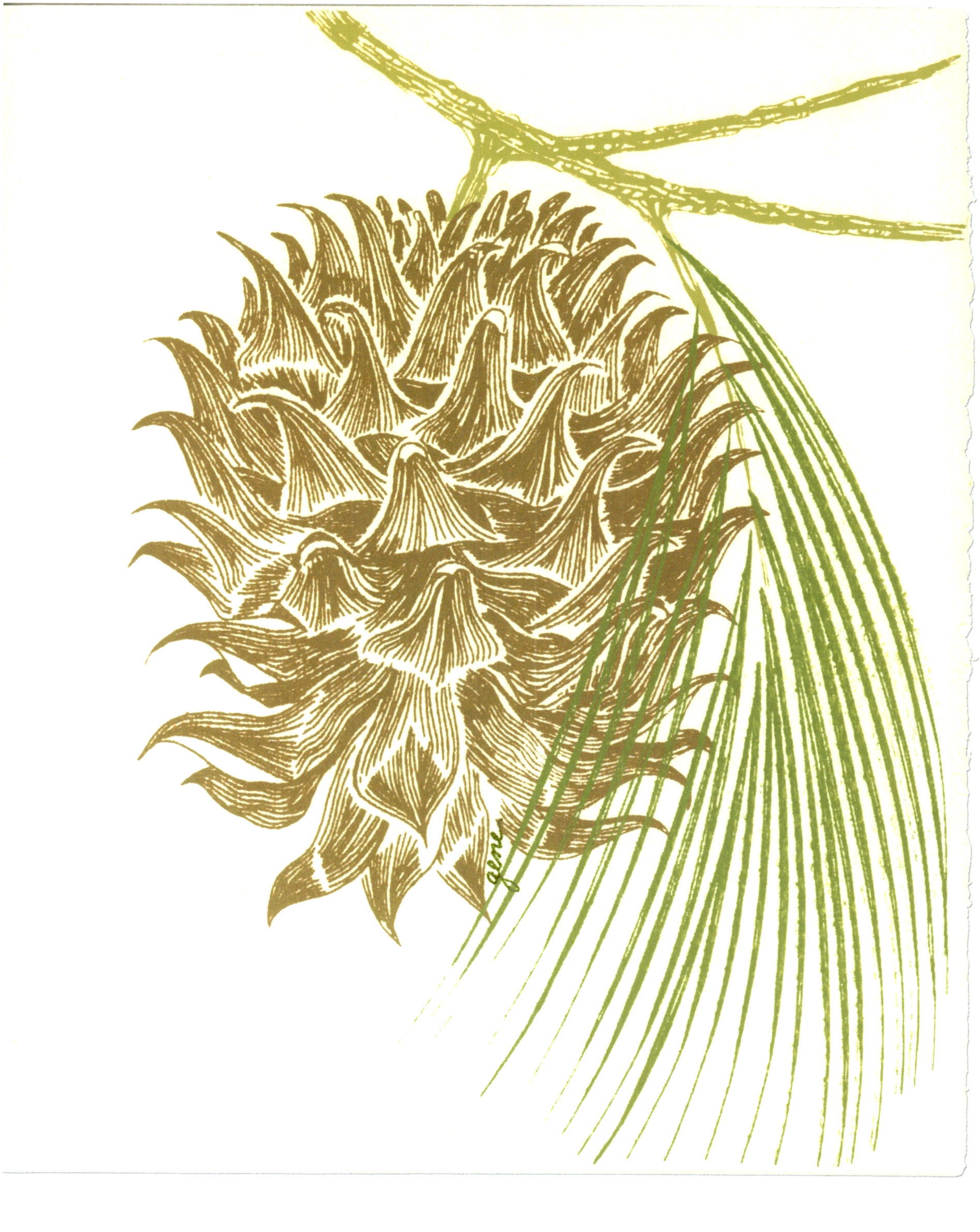

November 1973

The state of California abounds in magnificent evergreen trees, perhaps none more numerous, widely distributed, and diversified than the genus *Pinus*. Botanists may not agree about how many true species of pine there are, but it is safe to say that thirty-four are native to the United States and seventeen of these are native to California.

In general, pines are identified by their foliage (needles) and their fruit (cones). Pine needles are neatly packaged on the tree in bundles, and the first thing to do in attempting to identify a species is to determine the number of needles in a bundle. In only one species, *P. monophylla* (singleleaf pinon), are the needles attached singly—this is the only one-needle pine in the world. In all other species the needles are in bundles of two, three, four, or five. Also needle length varies greatly with each species, ranging anywhere from 1¼ inch to 10 inches. Another way to identify a pine is to know the cones, which also vary greatly—in size, shape, and viciousness—ranging (roughly again) from one to twenty-two inches long.

Long and slender or short and thick and armed or not, the cones of most species mature in two years. During the fall months of the second year the cone opens its scales and seeds are released. But some species do not open their scales and their cones remain on the tree for an indefinite period of time. The cones of these closed-cone pines usually do not open until the intense heat of a forest fire destroys the tree, opens the cones, thus releasing and dispersing the seeds; nature's marvelous way of reforesting the land after a fire.

Pinus sabiniana

Some of the California pines are very well known whether due to their rarity (torrey), adaptability to cultivation (Monterey), food value (pinon), age (bristlecone), aesthetic qualities (sugar), or economic value (ponderosa, Jeffrey, sugar). Some are little known due to their inaccessibility (foxtail and whitebark). Others are known but overlooked either because people are unaware of their true virtues or because these virtues are no longer important now. *Pinus sabiniana* seems to fit into this last category. Best known as the Digger pine (see page 134), it has other common names (Sabine's pine, bull pine, gray pine, grayleaf pine, blue pine, and squaw pine). By whatever name, *Pinus sabiniana* is indigenous only to this state and no other tree is so strongly associated with the romance of early California.

P. sabiniana grows abundantly in the dry, hot foothills that ring the Great Central Valley of California. A three-needle pine with needles from seven to fourteen inches in length, it is an open, airy tree with such sparse foliage as to project an almost lacy appearance. Like so many plants native to hot, dry areas, it has blue or blue-gray foliage. Yet, perhaps through some combination of color and lacy quality, it is like no other tree in its area and even from a distance its identity is unmistakable. As a young tree it grows straight and slender, but with age its strong tendency to branch into a cluster of secondary stems emerges. These second stems further branch into wispy sprays, so much so that from a great distance the tree appears to be a thin, misty cloud of gray.

Possibly the tree's most attractive feature, its cone is heavy and as robust as it is vigorous appearing. From six to eight inches in length and five to six inches in diameter, the cones hold within their scales the large, sweet nuts that were so highly prized by California's earliest inhabitants. But beware, the tip of each scale narrows into a really fierce, up-curving spine or hook that is very sharp. Great numbers of Indians settled where this tree grew because they knew the value of the food that it produced. Not only did they eat the nuts from the cones but they also ate the cone's soft core when young and green. The young buds and the inner bark also provided nourishment.

Not only was it the most prevalent tree along parts of El Camino Real where the Franciscan missions were established, but also it was and still is the most dominant pine of the golden foothills of the Mother Lode country. This golden native is typically Californian all year long but never more so than in spring when at its feet the hills are aglow with lupines the color of the sky and poppies the color of gold.

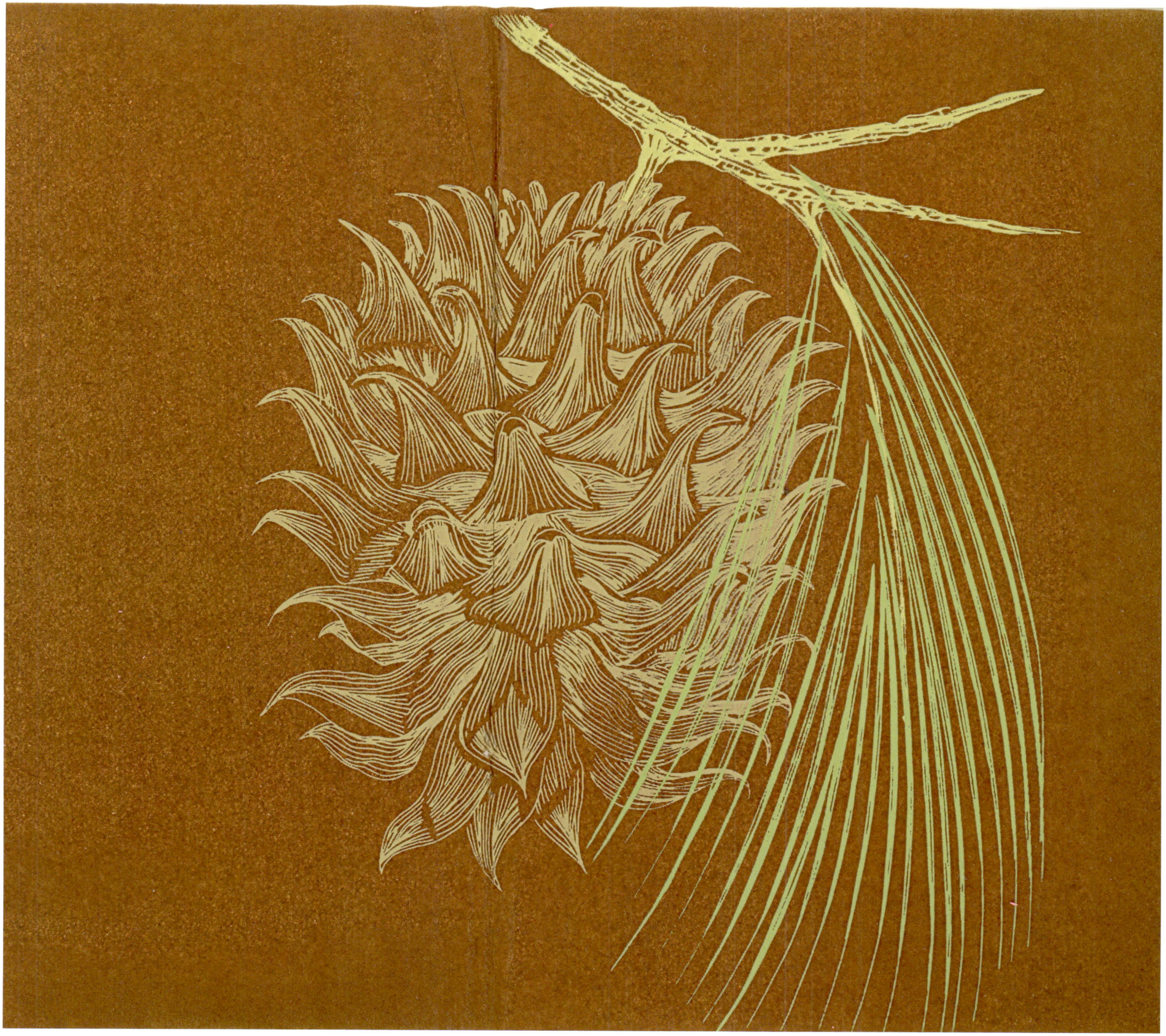

Toyon *(Heteromeles arbutifolia)*

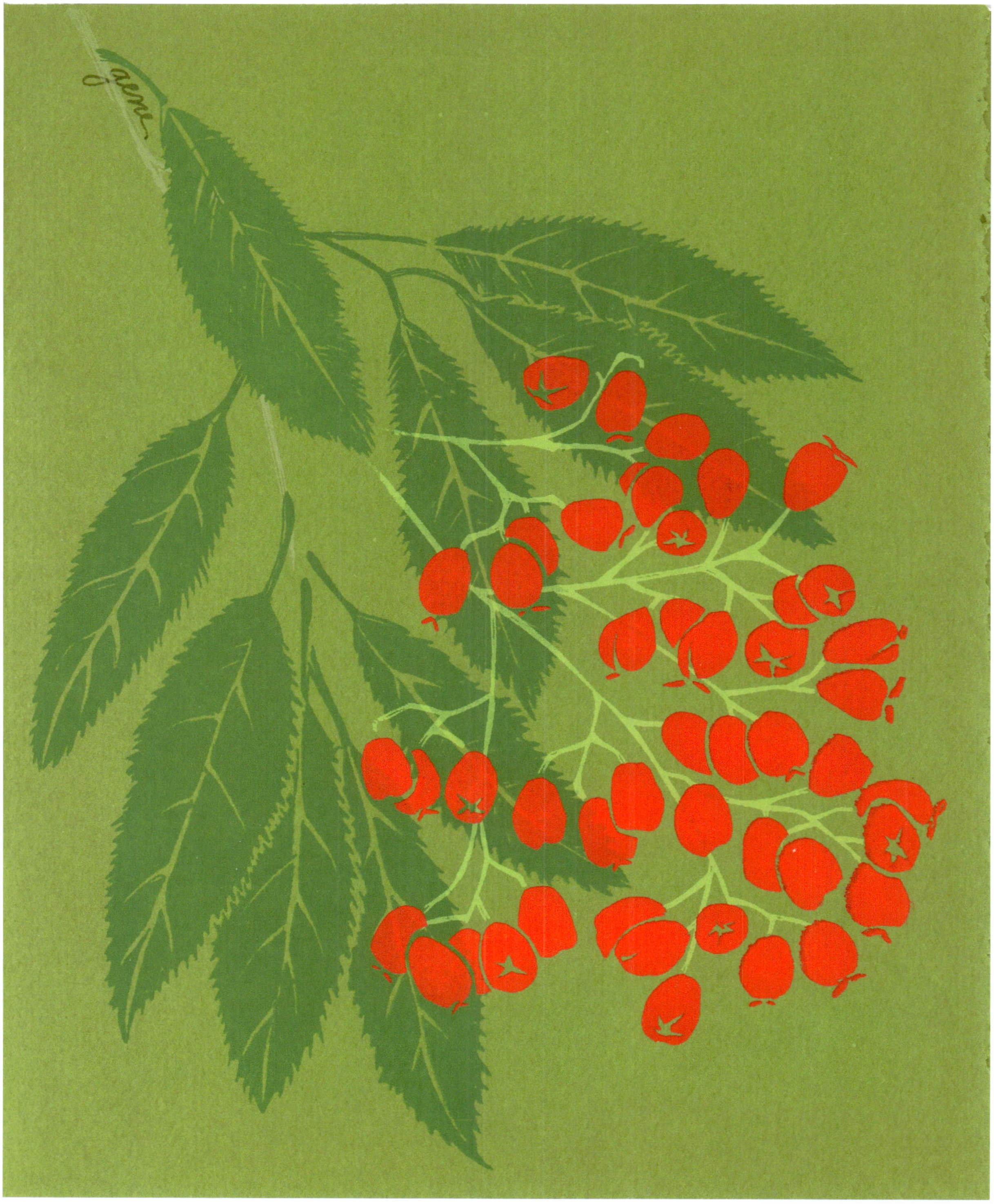

December 1973

In a location as densely populated as the Greater Los Angeles area, you'd be hard-pressed to find any plant still existing in its native form. Yet in some of the hilly sections of this area *Heteromeles* might yet be discovered in a native state. Though in no way related to holly, one of its common names is California holly. It is said that because this shrub was so plentiful on the hills surrounding the original subdivision, the town that evolved was named Hollywood. This popular plant adapts so well to cultivation that it has been widely used in landscaping freeways, parks, school grounds, and other public spaces as well as in many private gardens. This shrub was introduced to the horticulture of the British Isles more than one hundred years ago, given the name Maybush by the British because of its similarity to the Hawthorn or May-blossom. In the genus *Heteromeles* is but one species, *arbutifolia,* and this one species is native to California. The species name reflects how much its leaves resemble another of the state's natives, *Arbutus* or madrono. This plant's other common names are toyon and Christmas berry.

Heteromeles arbutifolia

I first became acquainted with this native about fifteen years ago right here in the San Bernardino Mountains. Shortly after Thanksgiving Day, some magnificent red berries glistening in the sunlight caught my eye. Upon closer examination of these enormous clusters of red berries against their rich, deep green background, I found its leaves to be like so many plants accustomed to the harsh conditions of the chaparral region where it often lives. The evergreen leaves are thick and leathery—glossy on top and lighter underneath—with a serrated edge. A rounded, tall shrub or a small tree, *Heteromeles arbutifolia* usually grows on semidry, brushy slopes and canyons below 4,000 feet, but it may be found almost down to sea level. The species has one variety, *macrocarpa,* which means large fruit. *H. arbutifolia* variety *macrocarpa,* an especially large-berried form, is abundant on Santa Catalina and San Clemente islands.

The plant's large flower panicles had developed into these big clusters of berries possibly as early as mid-October. The individual berry is about ¼ to ½ inch long and formed exactly like a wee, tiny apple (thus the genus name from the Greek *heter,* meaning "different," and *malus,* meaning "apple"). Birds love them, but the bees are happy in late June and July when the small, creamy white flowers appear and form into the large panicles. With a somewhat spicy fragrance, the flowers are particularly attractive to the honey makers, but it was coming upon the berry-laden shrub that made me exclaim out loud, "What a beauty!" The berries are usually a vivid scarlet red, though they can vary from lighter to darker. California Indians used the berries as a food source, usually boiling or roasting them to take away the bitter taste of the raw fruit. The berries also make a pleasant cider.

Nearly all the plants that I discovered on that day so many years ago were completely burned in the Bear Fire of November 1970. Since then I have watched them closely. That first spring the *H. arbutifolia* plants sprouted anew at ground level, and almost like magic they grew to three feet by December. That first year there were no flowers and consequently no berries, but just three years later all are solid, compact, fresh-looking plants festooned with berries of indescribable beauty. I was quite startled at how rapidly they came up after the fire, all due to their magnificent underground roots.

It makes sense that a plant with such versatile utility would have such roots. I am reminded of my favorite description of *Heteromeles arbutifolia,* which was written by the first professional landscape architect to locate in Los Angeles, Ralph D. Cornell, in his 1938 book *Conspicuous California Plants:*

> "Any plant that encourages bird life, supplies the bees with an unexcelled source of honey, gives food to man, furnishes tannin from its bark, protects arid slopes from erosion, paints the landscape with vivid colors and carries joy into the home at Christmas time, when no other native berries are available to most Californians, surely deserves the protection of man, whom it serves so well."

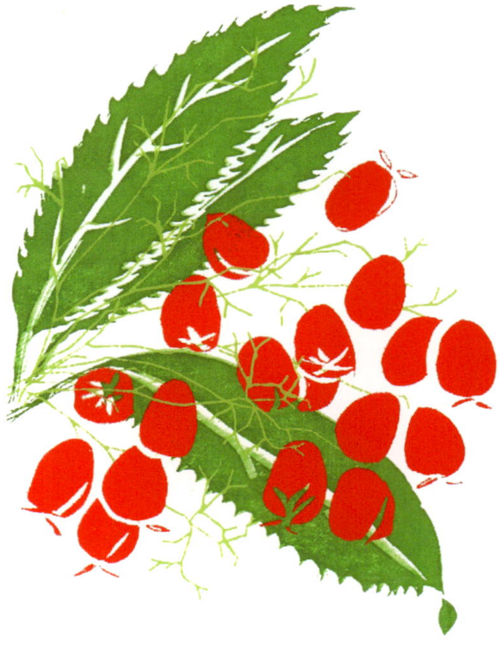

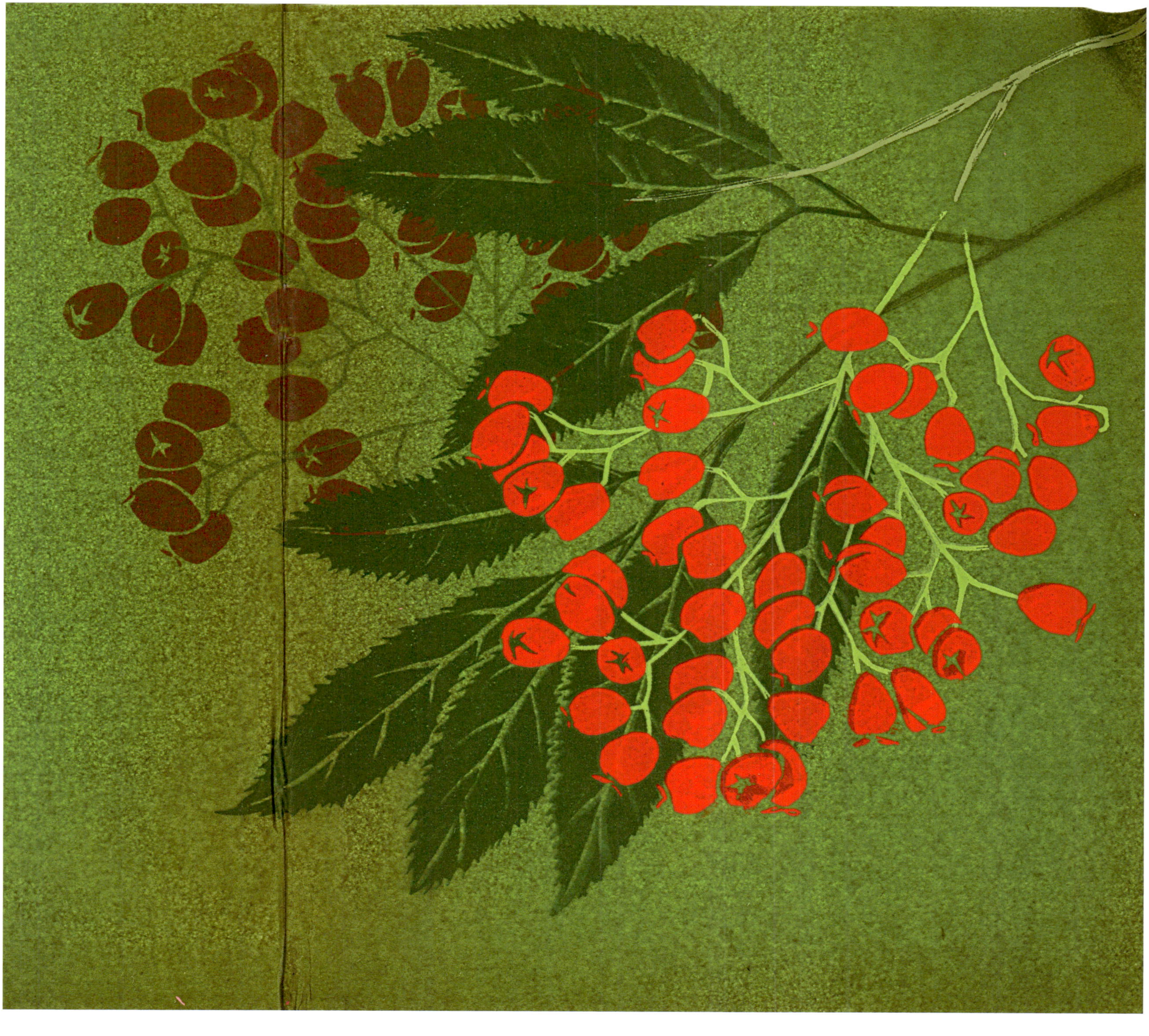

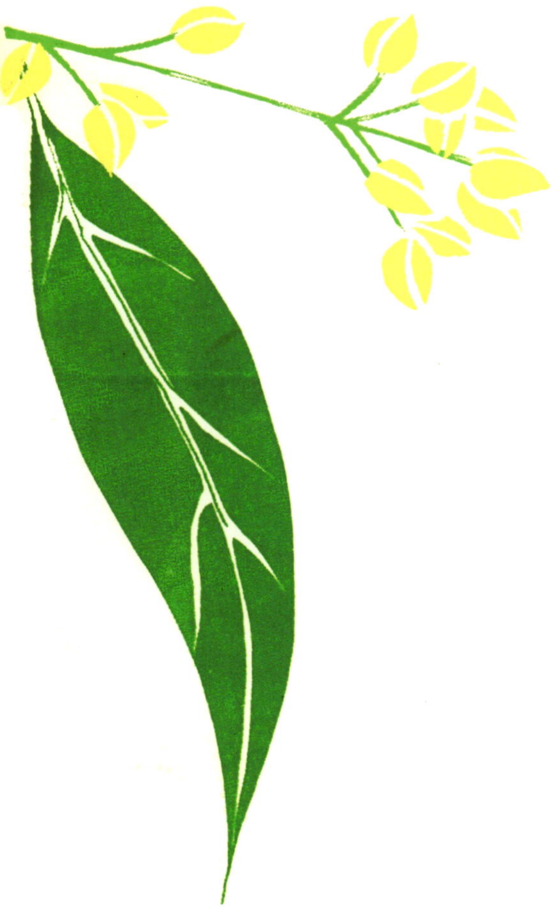

California bay *(Umbellularia californica)*

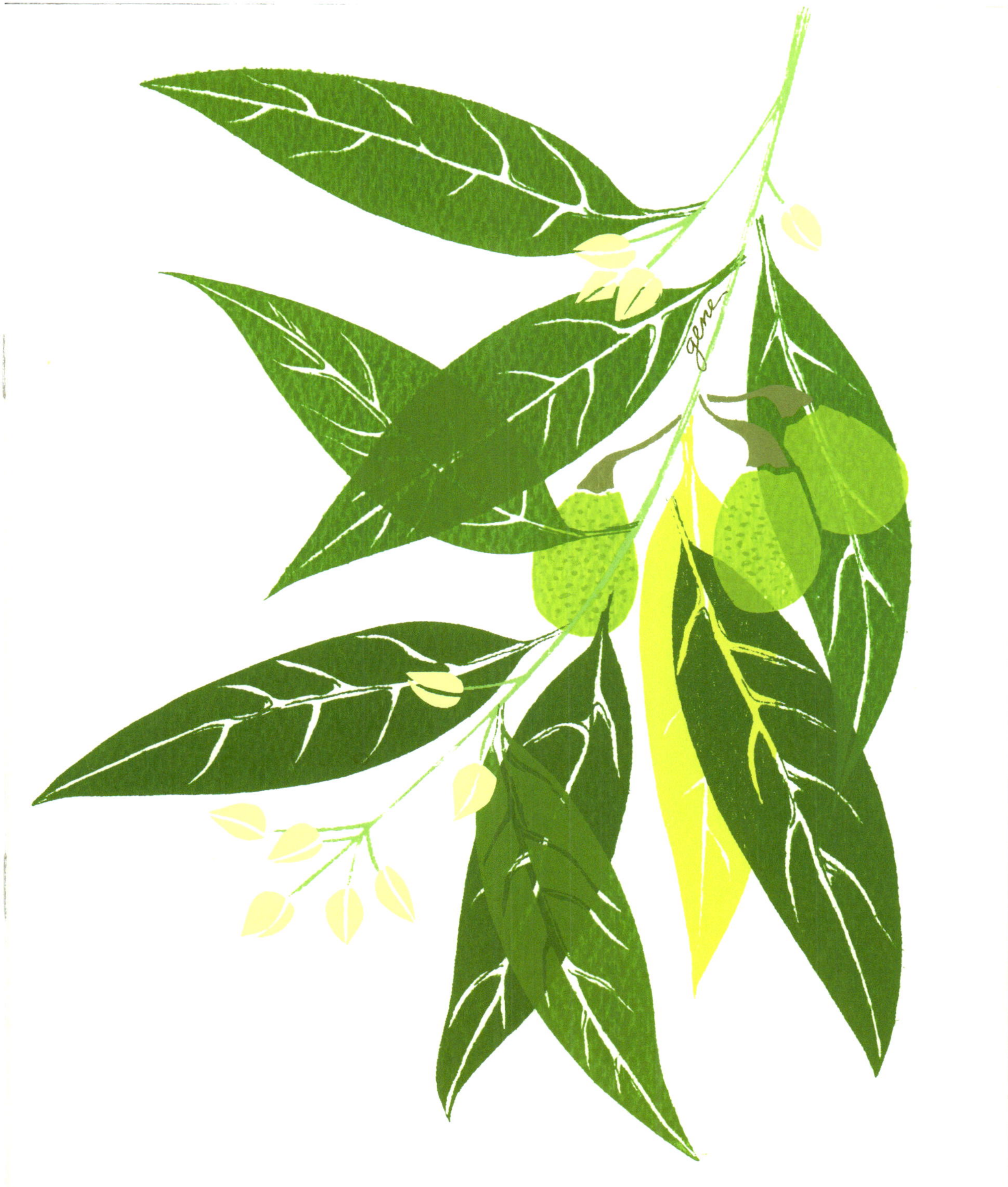

January 1974

So many of the trees in California that are the best known—by residents and visitors alike—are exotics, introduced from a foreign country. In this category are the orange, lemon, grapefruit, loquat, olive, avocado, eucalyptus, pepper, magnolia, and ornamental fig trees. Not one of them is indigenous to California and all are broadleaf evergreen trees. It seems a pity then that all of these are so celebrated while our very own broadleaf evergreen trees are often overlooked. One of our most handsome broadleaf evergreens possesses many fascinating qualities, provides the finest hardwood to be found on the Pacific Coast, and is even related to the laurel of antiquity from which heroes' wreaths were made. It is *Umbellularia californica,* the only species in its genus and a native of our state.

Umbellularia californica

Any Oregonian will tell you it is Oregon myrtle. Northern Californians call it California laurel or pepperwood. People in the southern part of the state refer to it as the California bay tree. It is also known as mountain laurel, bay laurel, and spice tree. Where it is growing frequently determines by what name a tree is known. With this tree the leaf is one sure way to identify it. The smooth, shiny leaves of *U. californica* are from three to six inches long, 1 to 1½ inches wide, and when crushed emit a pungent, camphor-like aroma. This is not the true bay leaf for cooking, though one may use it that way; if so, expect it to be a bit stronger in flavor.

Perhaps more than any other native tree, the bay tree is capable of adjusting and adapting itself to many different conditions. Its size, shape, and character can differ greatly depending on where it grows. It flourishes most luxuriantly in cool, moist canyons where it may reach heights of sixty to one hundred feet. Here where it seems happiest the bay tree has a single trunk, its crown is dense and broad, and it provides deep and constant shade. In contrast, when growing on high, dry, exposed hills, it takes on a more shrub-like appearance. Along wind-swept sea bluffs it assumes a low, almost prostrate form, giving the appearance of a rich, green ground cover. It is sometimes found growing in dense thickets of trees of the same height, with slender straight trunks topped by a flattened canopy of dense green foliage. They look pruned, but actually constant winds from the sea shape them this way. Again adapting, this time in a rather dwarf form to the hot, dry conditions of the chaparral, this shape-shifter becomes many-stemmed and as twiggy and tough as other plants found there.

As a hardwood provider, *U. californica* is highly valued for cabinet work and furniture making. The wood is heavy—very firm and very strong—and when finished becomes one of the most beautifully grained woods to be found anywhere. This tree even has famous relatives. As a member of the laurel family, it is a distant relative of the true laurel of antiquity, the classical sweet bay tree *Laurus nobilis.* A European tree native to the Mediterranean region, the latter is the laurel of the poets, the laurel linked to mythology whose leaves were made into wreaths for the Olympians of ancient Greece—and are still used for cooking.

Yet renowned family aside, *U. californica* shuns the spotlight. Even its flowers though plentiful are quite inconspicuous, a creamy yellowish green. Borne in small clusters or umbels (from which the tree derives its generic name), the flower buds appear in late summer, most opening in very early spring. The fruits, which ripen in October, are about one inch long, globular in shape, and yellowish green as well, looking much like a plump green olive.

Come to think of it, perhaps it's understandable why this tree goes unrecognized so much of the time. Consider how it appears in those many various forms ensuing from its ability to conform to its environment. Consider that its fruit looks like an olive. Consider that I myself, in creating my representation of it, was constantly reminded of how hauntingly it is like the avocado tree. In fact, as a member of the laurel family, *Umbellularia californica* is related to the avocado, the camphor tree, and others that grow almost entirely in tropical or semitropical regions. As the lone member of this family indigenous to California, it does somehow seem like a spillover from the tropics—almost as if it doesn't really belong with other vegetation found here.

And yet it is ours, and long recognized as a stately tree by experts such as Charles Sprague Sargent in his 1923 *Manual of the Trees of North America,* who wrote of it, "…no evergreen tree of the temperate regions surpasses it in the beauty of its dark dense crown of lustrous foliage … [it is] one of the most striking features of the California landscape and fit to stand in any park or garden."

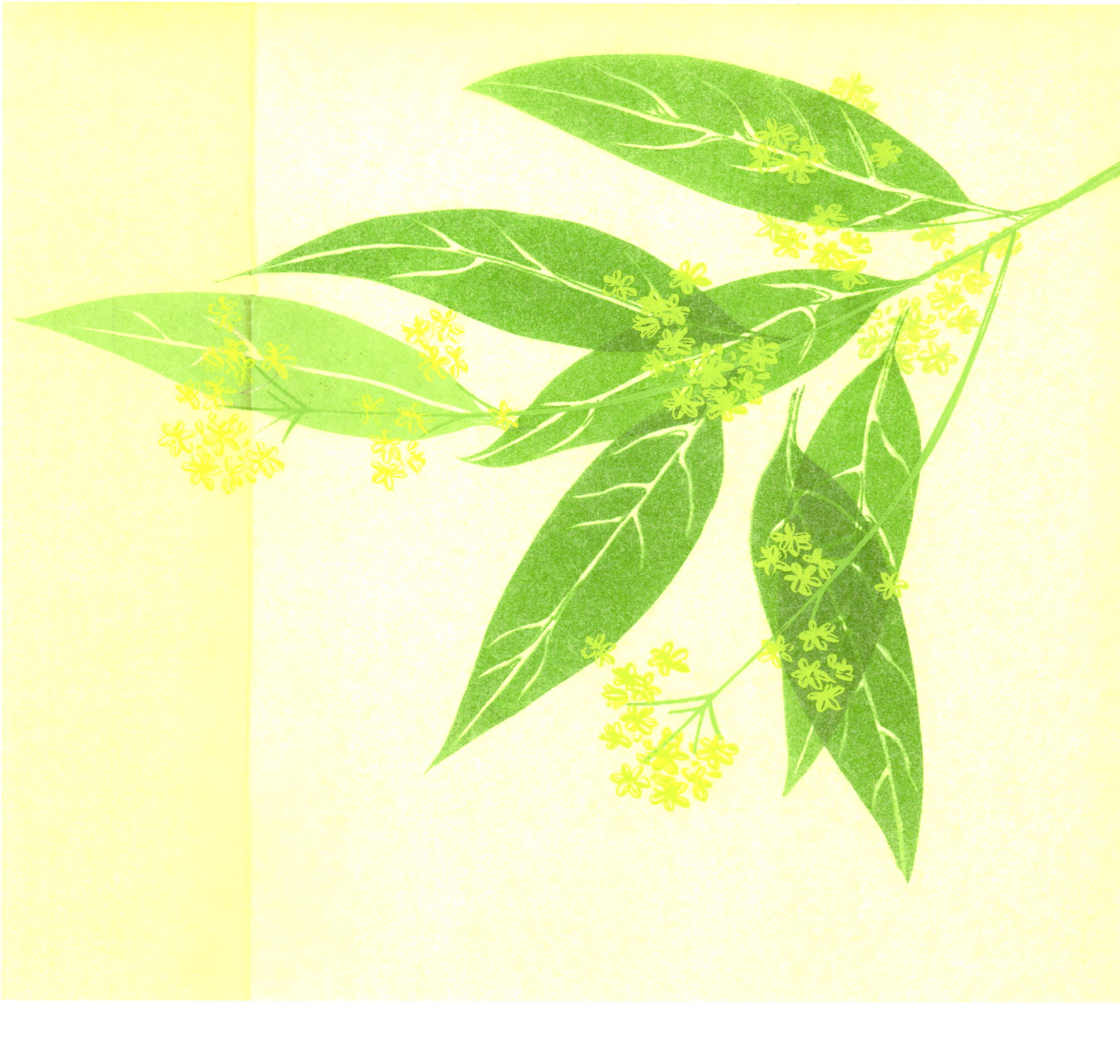

Pink sand verbena *(Abronia umbellata)*

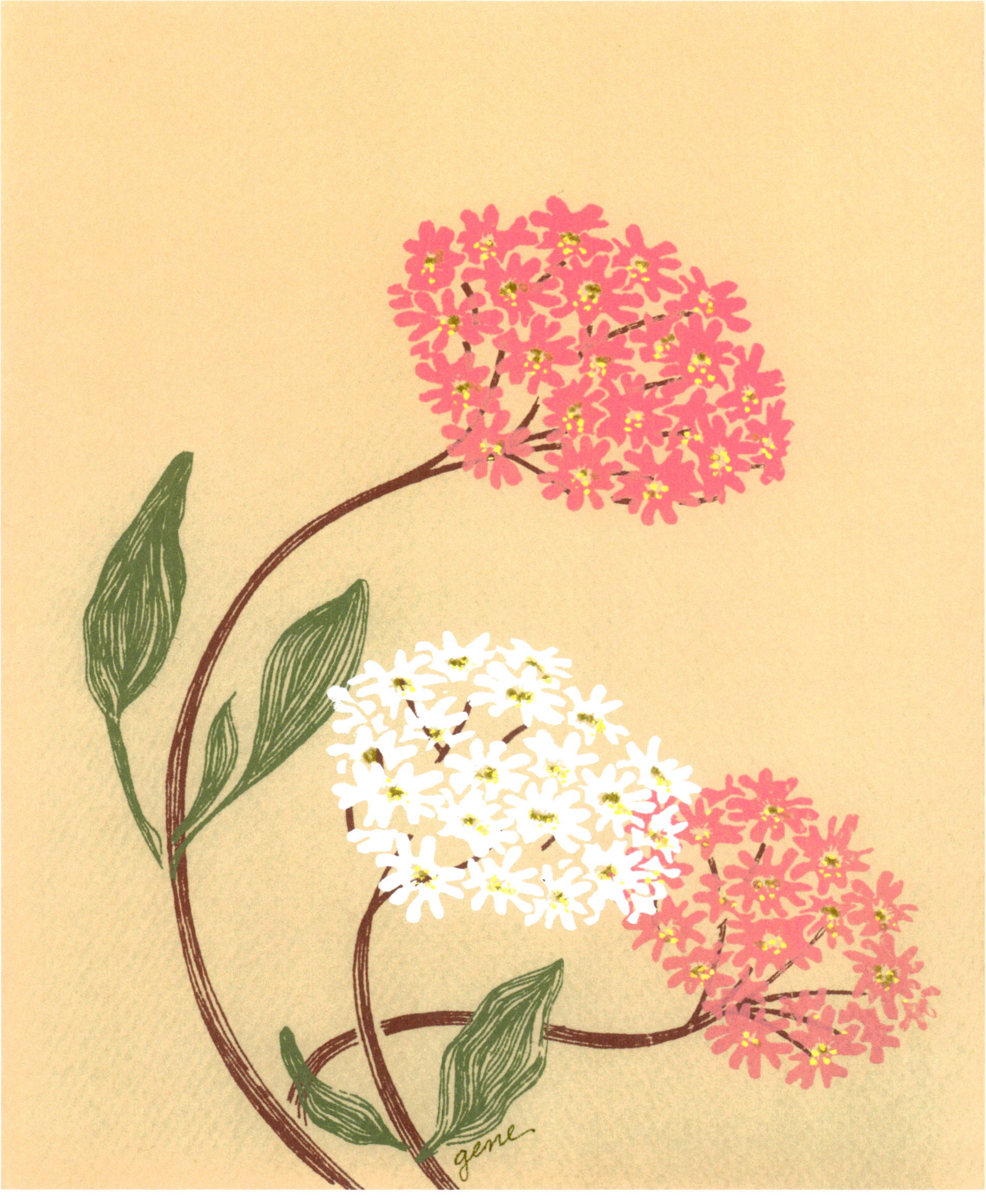
gene

February 1974

Drifting sand, salt spray, mist, and fog are ever-present in the area known as the coastal strand, elements natural to the slightly elevated terrestrial borders of the sea but hard on plant life. Regions like this, such as the south coast of Los Angeles County including the Port of Los Angeles, are rarely inundated with seawater, but vegetation along this higher strip of land is greatly affected by its proximity and therefore considered among the most fragile to be found within the state. Already bleak and sparse due to the constantly shifting sands, the coastal strand is becoming thinner and thinner due to the continuous invasions of human beings. People who make their homes on the bluffs and highlands overlooking the sea could counteract this somewhat by reestablishing *Abronia umbellata,* the lovely plant that once grew so abundantly along the coastal strand.

The generic name *Abronia* is from the Greek *abros,* meaning "graceful" or "delicate"; common names include wild lantana, sand puffs, and sand verbenas (although they are not true verbenas). The twenty-five species in the genus *Abronia* are widely distributed over western North America. The ten species that are indigenous to California are very interesting insofar as they are found growing under such diverse conditions: three species are found only on the coastal strand; three others grow only in the desert; and four are limited to high elevations (4,000–9,000 feet) and rare even then because of the severe conditions (found usually in sandy meadows or on the desert-facing slopes). *Abronia villosa,* smaller and more slender than the coastal plant, is the species that blooms so abundantly in the Palm Springs area. It is a striking beauty that covers the desert for miles during favorable winters.

Abronia umbellata

Pictured here is the best known of the coastal species, the pink sand verbena, but it is so prevalent that people often call it the common sand verbena. *Umbellata,* the specific name refers to the fact that the flowers are borne in an umbel or cluster. The flowers are very fragrant, especially in the evening. They are usually a lovely rosy pink but may sometimes vary to a soft pink and even to a pinky white. The plant's unusual reddish stem coloring against the flower's pink is a particularly handsome combination against the soft beige beach sand. This species of *Abronia* provides the most striking color on the coastal strand from spring until autumn. Some isolated blooms may be found throughout the entire year.

A. umbellata grows from six to eighteen inches high and can spread out from three to five feet. A prostrate, many branched, trailing plant, it roots at the joints becoming an excellent dune binder, despite being constantly buffeted by sea breezes that at times become fierce winds. Though a tender plant, it possesses the extraordinary ability to adapt both to a very high salt content in the atmosphere and to shifting sands. It may even hold a place in history. According to some authorities, *Abronia umbellata* may have been the first plant from California to be scientifically described and named. In 1786 a French expedition visited the Spanish mission in Monterey. A member of the expedition, French botanist Jean-Nicole Collignon, collected seeds of an attractive rose-colored flower blossoming in abundance along the beach. He sent the seeds to Europe and they were sown in the Jardin des Plantes in Paris. The plant that grew from the seed was new to Europe and the French botanist Jean-Baptiste Lamarck gave it the name *Abronia umbellata,* no doubt in honor of its delicate grace.

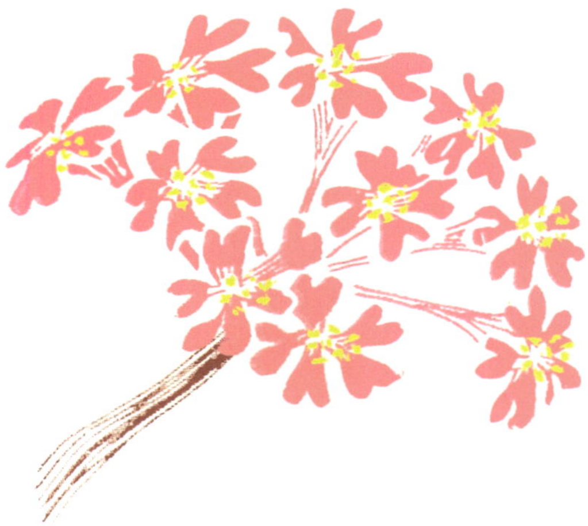

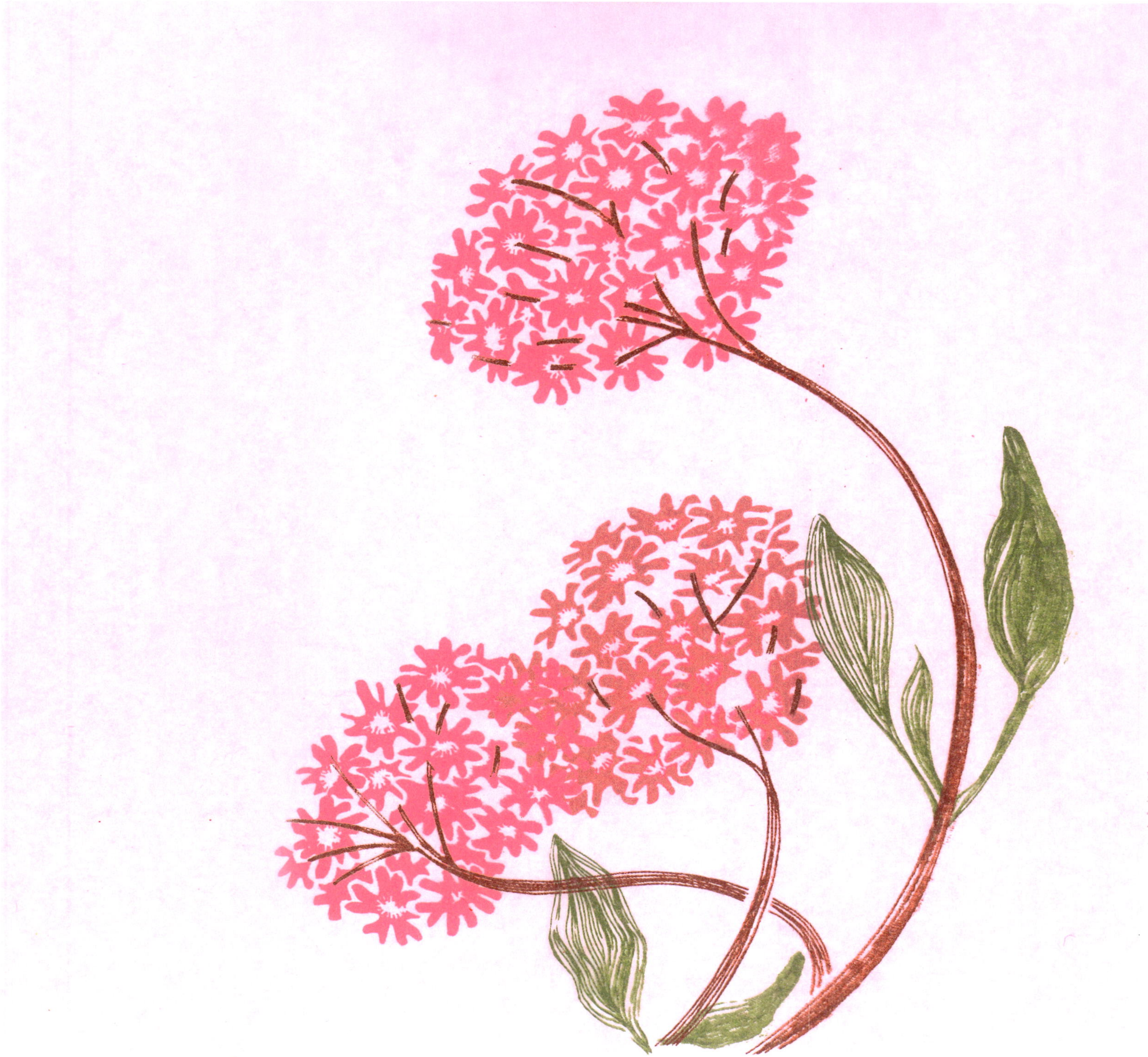

Mendocino mission bells *(Fritillaria lanceolata)*

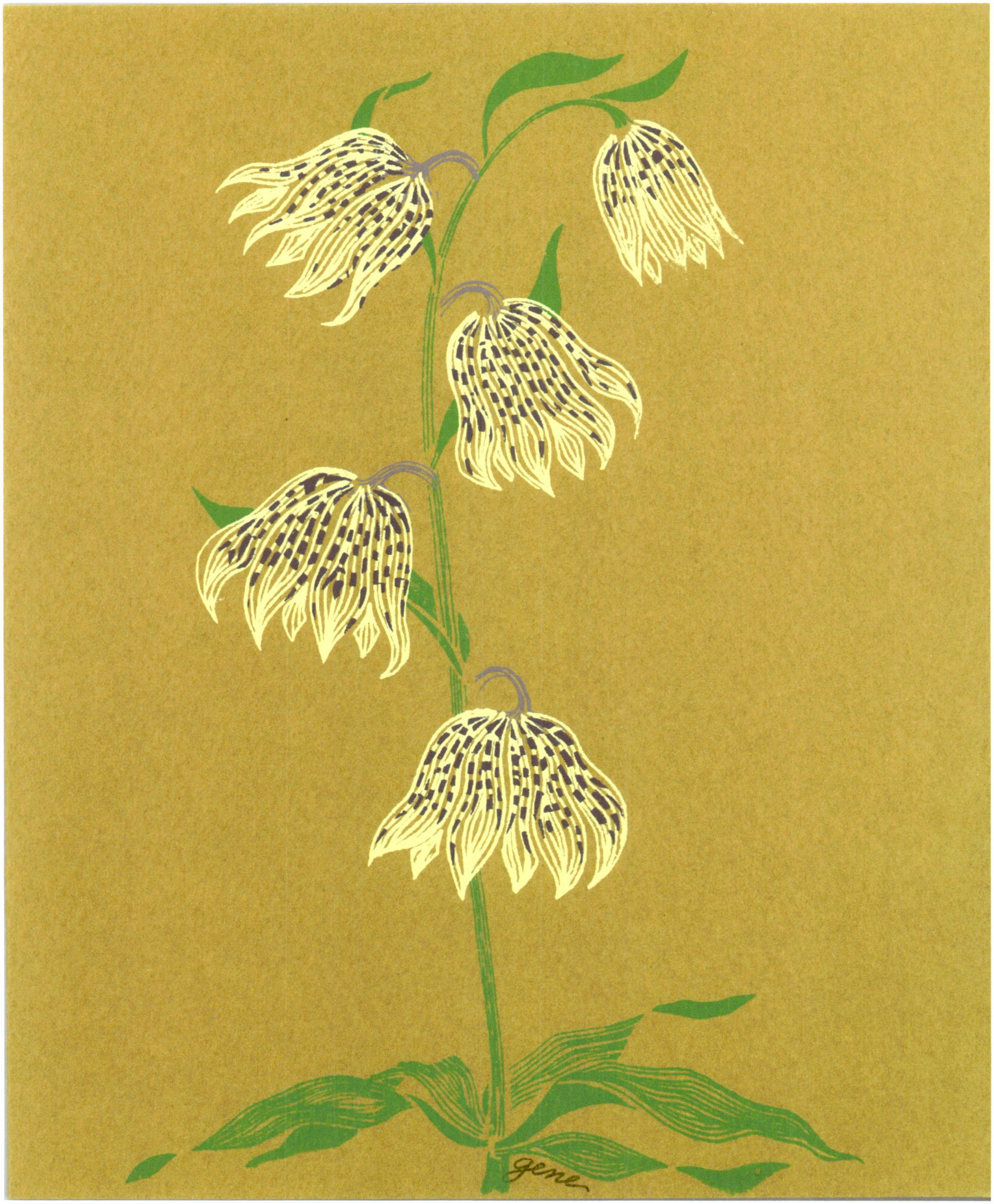
gene

March 1974

Bulbous plants—bulbs, corms, tubers, and rhizomes—differ both in their structural aspects and method of reproduction, but they all share two functions in common: the bulb of each acts as a storehouse for carrying the plant through its dormancy into its next growing period and serves as a means of propagation. A few of the most interesting bulbs found in California are species from the following genera: *Allium, Bloomeria, Brodiaea, Calochortus, Camassia, Erythronium,* and *Lilium.* One of the least known but most interesting is the *Fritillaria,* a species of which is depicted here.

Flowers produced from bulbs hold a special fascination for me. They all carry an indescribable, almost pristine, purity. They make a statement as if aware they are made of great substance, each petal holding its presence as if it had been dipped in wax. Usually the flower colors of bulbous plants are clear, deep, and rich, but the genus *Fritillaria* is an exception. You find a clue to its unusual flower coloring in the Latin word from which the genus name derives: *fritillus* means a dice box or checkerboard, here referring to the mottled effect on the perianth of many of the one hundred or so species in this genus that are widely distributed throughout the North Temperate Zone. Seventeen of these species are native to California, known by various common names such as chocolate lily, adobe lily, yellow bell, brown bells, checker lily, and my favorite name, Mendocino mission bells.

Fritillaria lanceolata

Searching for the *Fritillaria* at the University of California at Berkeley Botanic Garden, I meandered through the native section noting that many of this genus were barely out of the ground, and then suddenly I saw a whole colony—with just one flower in full bloom—the *lanceolata* you see here. The name refers to its lance-shaped foliage, and the species as a whole is in fact among the earliest to bloom. Most species in its genus are petite, growing only about one foot high with one to five flowers, but *lanceolata* can reach a height of four feet with eighteen nodding flowers. Fritillarias in general are related to tulips and lilies and like them, their flowers are composed of six petals.

The flowers of the genus as a whole are 1 to 1½ inch long, bell-shaped, and hanging or nodding. *Lanceolata* is also a fine example of the genus as a whole insofar as its most distinctive characteristic is the flower color. The colors are basically very subtle or somber, usually a combination of browns, purples, greens, and yellows. With very few exceptions, the California species are mottled, splotched, dappled, freckled, or checkered. Therefore also in terms of aspect or pattern as well as color, they are a challenge for eyes not trained to discern them, and exist quite effectively camouflaged in the wild.

Perhaps there is a reason for that. It takes from three to seven years for the new little bulb to become large enough to produce a flower. Within that time many things can become attracted to this fat storehouse of energy called a bulb. In earlier times, the abundance of available bulbs provided added natural food to the pine nut and acorn diet of the California Indians. Highly nutritious and palatable, many bulbs have a nutty flavor and were eaten raw, while others were cooked. Nowadays, only rodents appreciate it as food.

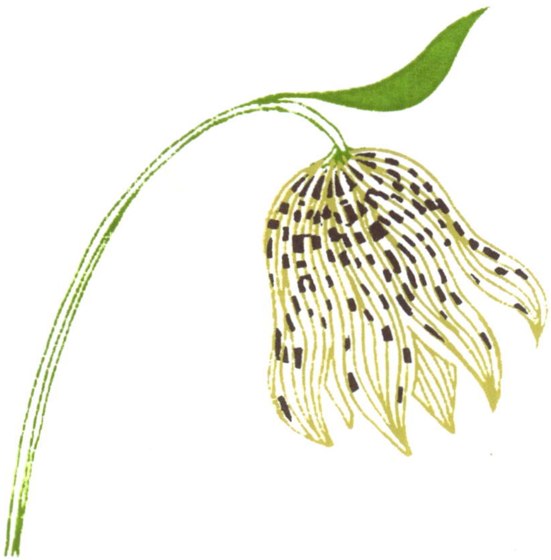

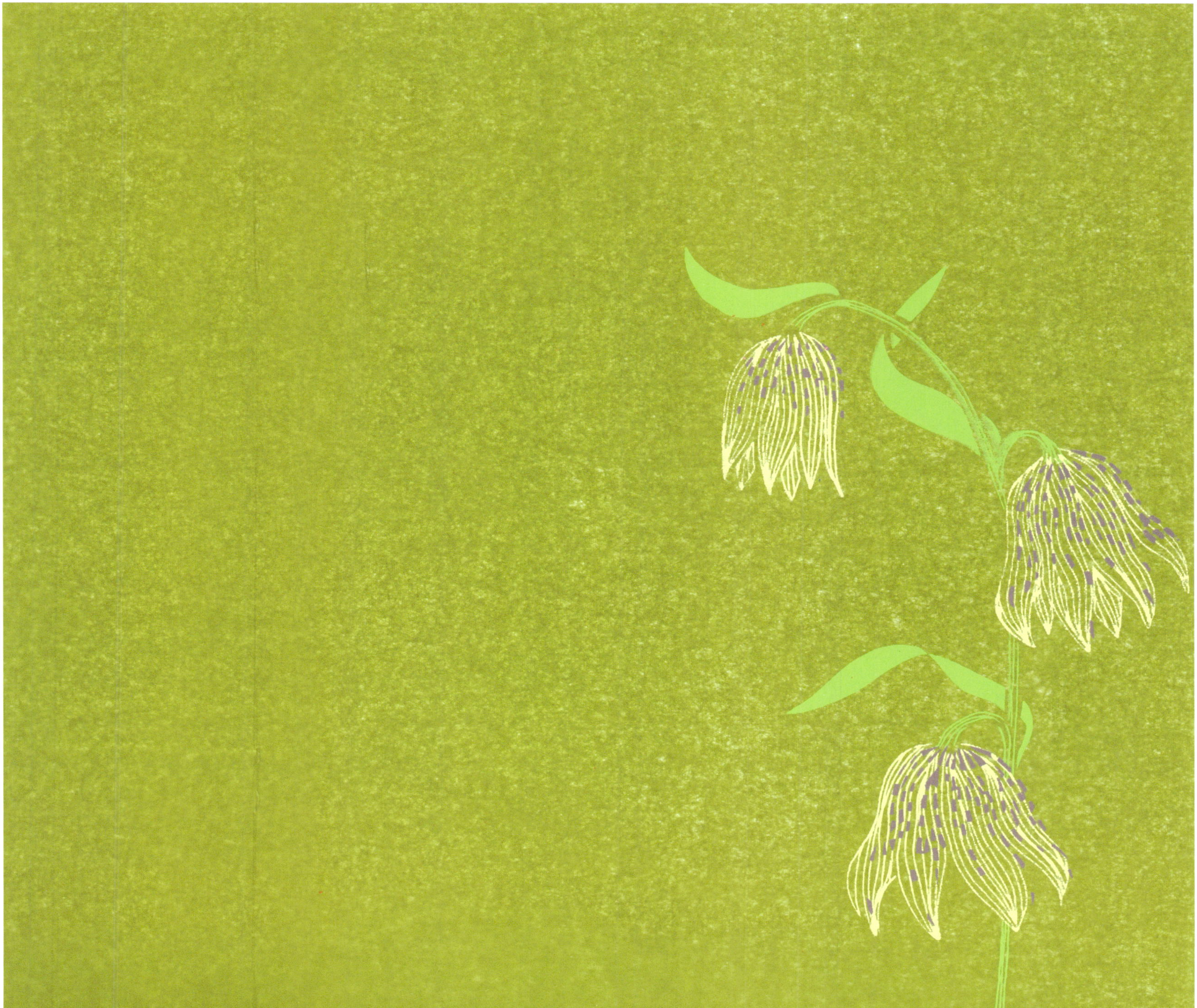

Columbine *(Aquilegia formosa)*

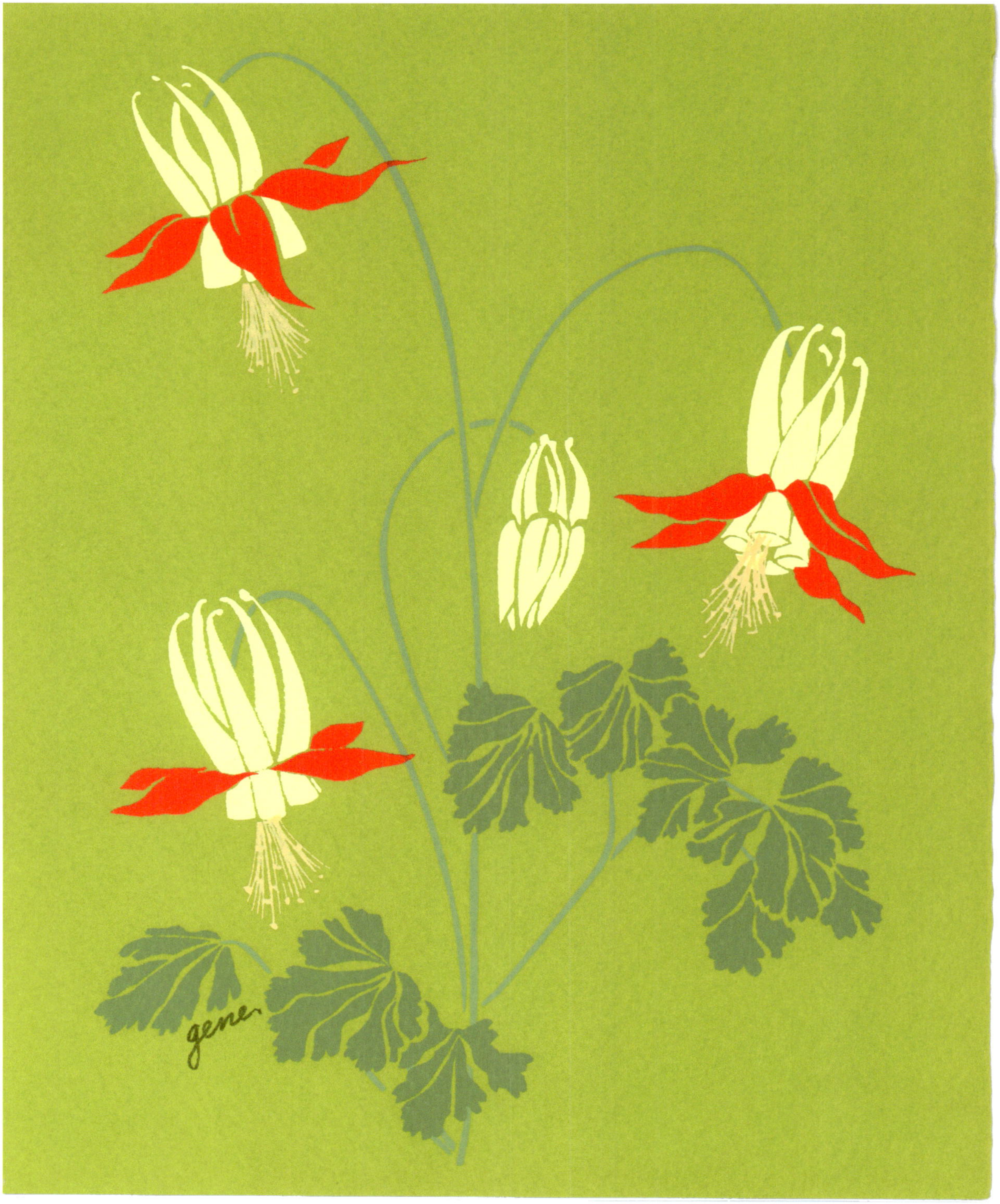
gene

April 1974

One flower genus is always near the top of the list when proposals are made for a national flower. Each of the fifty states can boast of at least one species of this genus as native to it, so perhaps that's why. Indigenous across the United States, the genus *Aquilegia* could very well be classified as all-American. Though several of its important species do occur in Europe, many are native to North America. Ranging in height from one to three feet tall, about seventy species in the genus *Aquilegia* are found in the North Temperate Zone. Four of them are native to California (including *A. formosa* depicted here). The flower is commonly known as columbine, and in 1930 became the official flower of the National Council of State Garden Clubs, Inc. From the Latin *columba,* meaning "dove," the common name refers to the shape of the flower.

It would be difficult to confuse the columbine flower with any other because its shape is so unusual. Each flower is composed of two basic parts—the sepals and the petals. In most flowers the sepals are the external, leafy part of the blossom as opposed to the inner, showy part. In the columbine it's the opposite. The five colorful, wavy parts that look like fluttery petals are actually the sepals. The five true petals are the smaller parts inside. Hollow and in the shape of a cornucopia, these inner parts are long and slender and end in a spur that contains nectar. Only long-tongued insects and hummingbirds can reach this sweet nectar easily; however, bees have learned that by puncturing the spurs they too can feast upon it. The flowers of some species are hanging or nodding, while the flowers of others in the genus are erect. Some species bear flowers of one lone color, while others combine two colors, one color for the sepals, another for the petals.

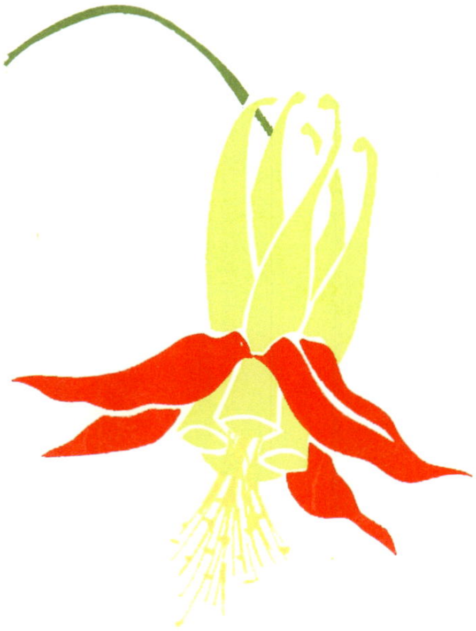

Aquilegia formosa

This is the species depicted here and found in many areas of California, from sea level to high mountains. You'll notice it when you see its dancing flowers held on thin, wiry stems high above exquisite foliage. Like all columbines, these are perennials and their foliage is amazing and beautiful. The leaves are divided into three segments and each segment is further divided, with edges that are delicately rounded and scalloped. The plant is reminiscent of the graceful maidenhair fern as well as *Thalictrum* or meadow rue. Even if the plant did not flower, the foliage alone is distinctive and desirable, and this is true of all the species in the genus. Flower and foliage together combine to give the columbine an airy, lacy affect that adds a light and lilting quality to many woodland areas.

Aquilegia formosa grows in several different forms, and at least three varieties are now known. Depending on its location it blooms anytime from April through July, and with its wispy presence, makes quite a statement when it does. The willowy plant likes to be in moist meadows or along stream banks tucked in among rocks. It's always surprising to find the ethereal foliage of the columbine popping up seemingly from nowhere between large boulders at the edge of a stream. A short time later it is crowned with the fragile and delicate flowers.

Keep in mind their natural habitat if you are going to cultivate these or other columbines. Basically a woodland flower, the plant should have only morning sun in warm regions, and it does require moisture. The California species are available from nurseries specializing in natives (like the Theodore Payne Foundation, see page 225). Remember that species are wild forms and will reproduce themselves from seeds. Columbines are a friendly lot and interbreed among themselves freely. If growing more than one species, keep them separated so that their seed will remain pure. New plants true to the parent may be easily grown.

Much work has been accomplished with the hybridization of the genus *Aquilegia.* Many hybrids are available and they are usually strong, vigorous plants with very large flowers in a wide assortment of colors. Hybrids do not reproduce themselves true from seed, though, so divide the original plant for more plants. After flowering, cut off the flowering stem to encourage new growth.

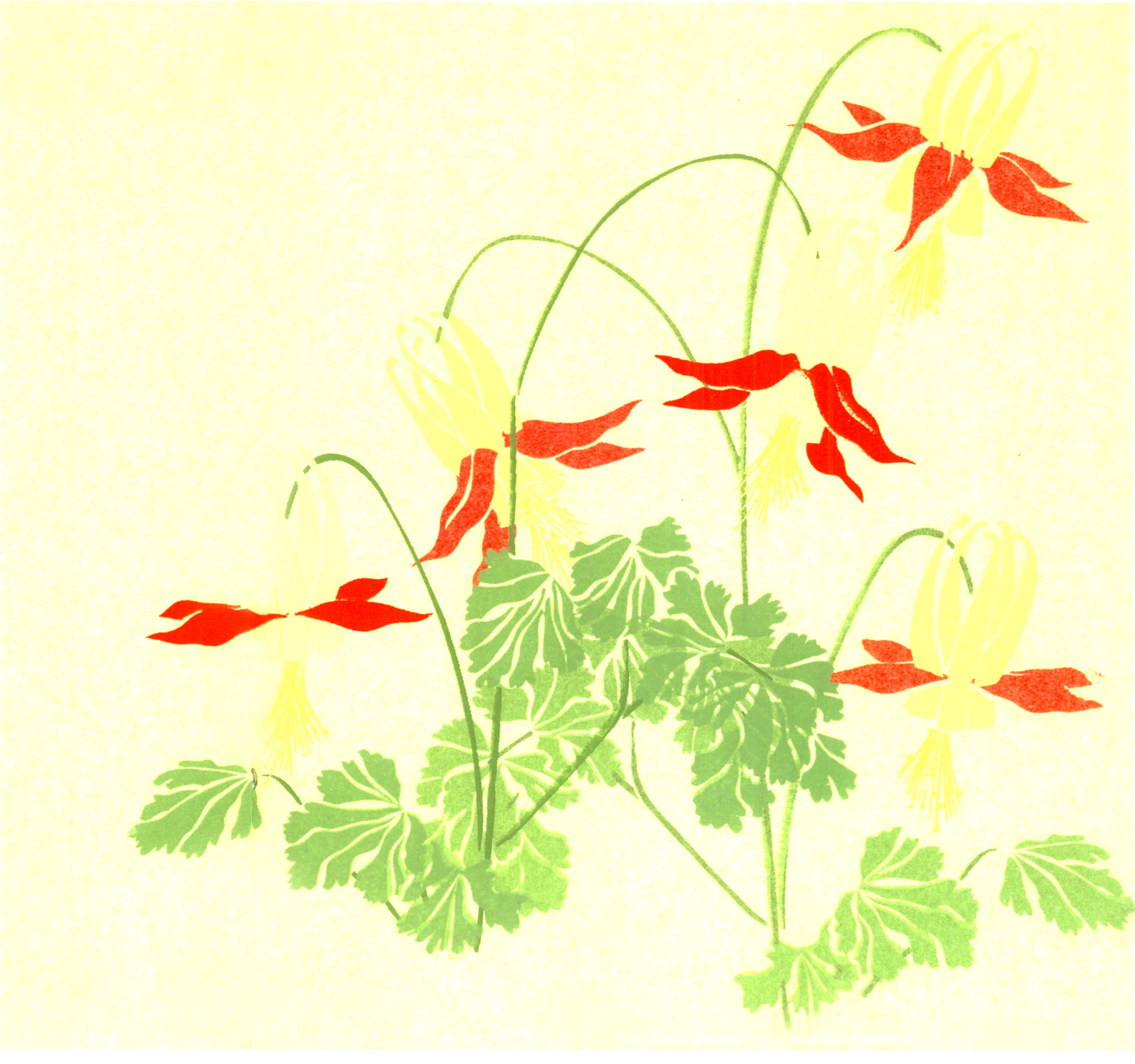

California poppy *(Eschscholzia californica)*

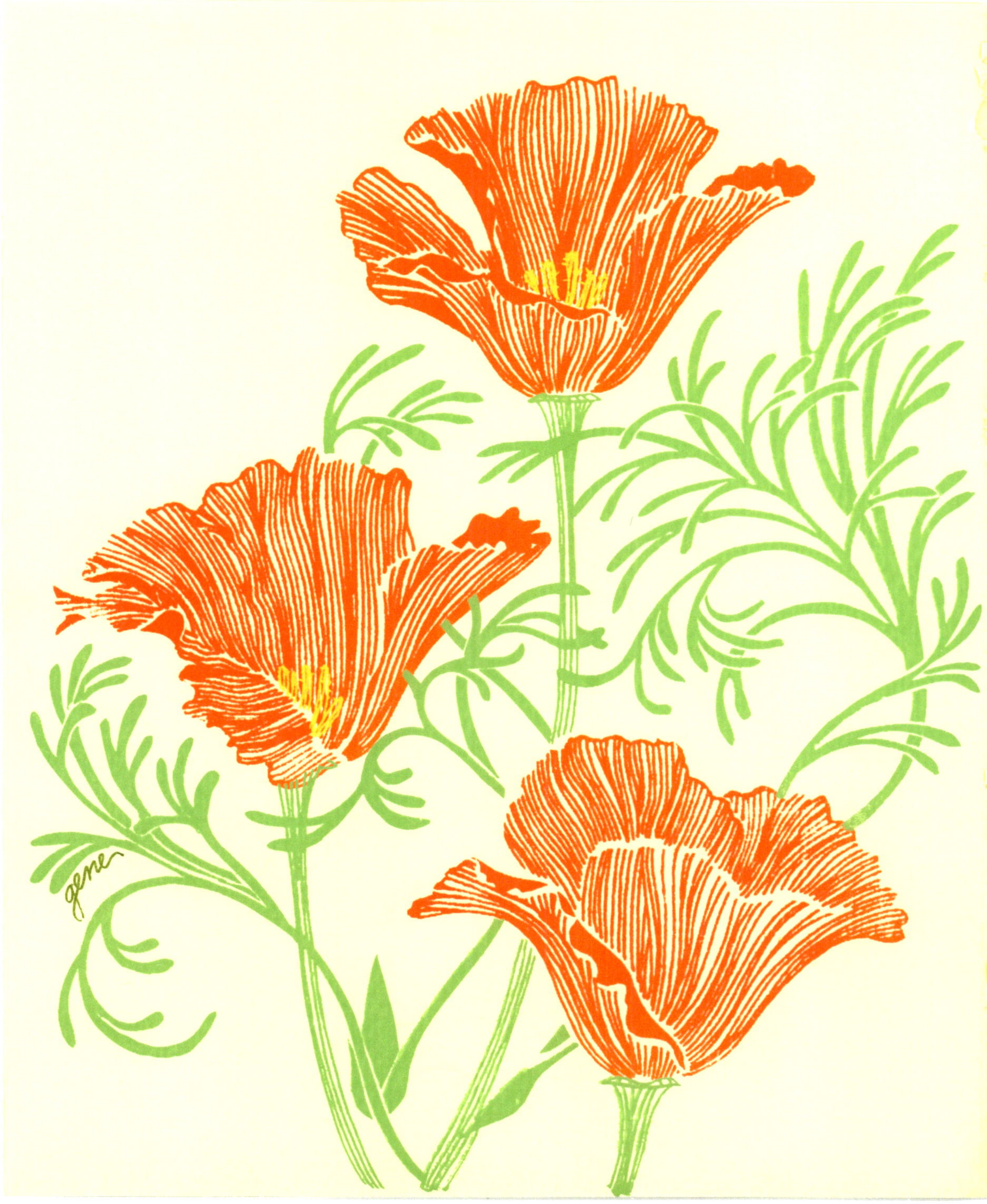
gene

May 1974

I have known the California poppy all my life. No other flower is as symbolic of the Golden State. Its cup is the symbol of plenty, and its glorious color also speaks of the wealth of the land: the gold of the mine, the orchards, the grain fields, and the gold of the radiant sunshine. The California poppy is truly the queen of all the golden natives in this, the Golden State. In fact, California was so moved by it that it became the first state to select an official state flower, in 1890. At that time it was abundant nearly everywhere: a flower of the Channel Islands, the coast, the inland valleys, the foothills, and in favored years and in favored places even in the high mountains.

The eight species in this genus, *Eschscholzia,* ranging from northern Oregon to northern Mexico, are all native to California. Commonly, this plant is known as California poppy, golden poppy, California gold flower, orange flower, and crocus poppy; some of the very charming Spanish names are *calce de ora, la amapola, torosa,* and *amaflores.* One of the state's most famous landmarks, the Golden Gate Bridge may well have derived its name from this flower. "Golden Gate" first appeared in print in 1848, and it is said that the name came from the abundance of golden poppies that grew in and around San Francisco.

The following year, gold was discovered in California. Among the legends associated with the poppy is that Native Americans believed the petals of the "Great Spirit Flower," falling year after year and sinking into the soil, gradually formed the shiny metal the strangers were rushing to find. In 1897 a great wave of commercialism swept the state. The golden poppy was chosen to adorn many labels and trademarks. Some of the popular items in the state were poppy brands of oranges, poppy buns, poppy butter, and poppy baking powder. In 1932 *Eschscholzia californica* was chosen to be the flower on the official seal of California Garden Clubs, Inc., the organization to which I am sending this series of serigraphs.

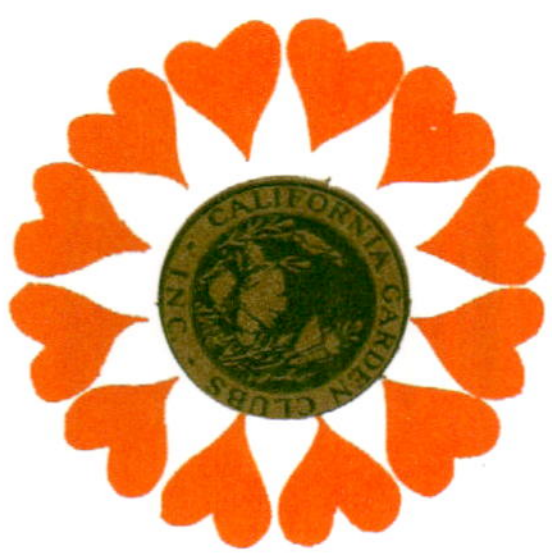

Eschscholzia californica

Eschscholzia californica, its specific name honoring the place where it was originally discovered, was once abundant the length and breadth of the state. (Even today the typical form is found in wild form only in California.) The generic name honors Dr. Johann Friedrich Eschscholz, a surgeon and naturalist and a member of a Russian expedition that visited San Francisco Bay in 1816. The plant still exists in a native state in some areas, but many large areas have now been seeded. It has proved itself to be a perfect plant for seeding on some of the bleak and barren landscapes that resulted from the harsh cuts of freeway construction. If seeds are sown in the fall of the year and rains are plentiful in the winter, a grand show of magnificent color will ensue the following spring.

New young plants should be thinned, not transplanted, because each plant has a very long taproot that does not take kindly to transplanting. This species is extremely variable and may be annual or perennial depending on its location. The one thing that distinguishes *E. californica* from the other fifty or more species in its genus is the little flat disk or rim just below each flower that is quite conspicuous after the petals have fallen away. The cotyledons or seed leaves are very distinctive, too—I always refer to them as "little forked tongues."

E. californica is usually twelve to eighteen inches high with delicately cut blue-green branching foliage, slender graceful stems and cup-shaped flowers two to three inches across whose four golden petals have a satiny sheen. The sepals are a very interesting part of this flower: they enclose the bud in the form of a tiny peaked, pale yellow-green cap that is pushed off and falls away as the bud expands. This feature is often noted in poetry about this flower. The flower opens in the warmth of morning sunshine but by late afternoon the petals have folded together just as they were in the bud but not as tightly. Because of this it was given one of its Spanish names, *dormidera,* meaning "the drowsy one."

Basically, *E. californica* is a spring and early summer flower, though sometimes it will bloom intermittently throughout the year. I found the flower that served as the model for my drawings in full bloom on December 12, 1973, in the town of Los Banos, almost the center of the state. One flower and one stem of foliage were cut, placed in water, and rode around in an automobile for the next five days. Each evening the flower closed up, but early each morning when it opened I made sketches of different poses. I became very well-acquainted with that one flower. It was a real beauty and it held up marvelously under very trying circumstances for four of those days.

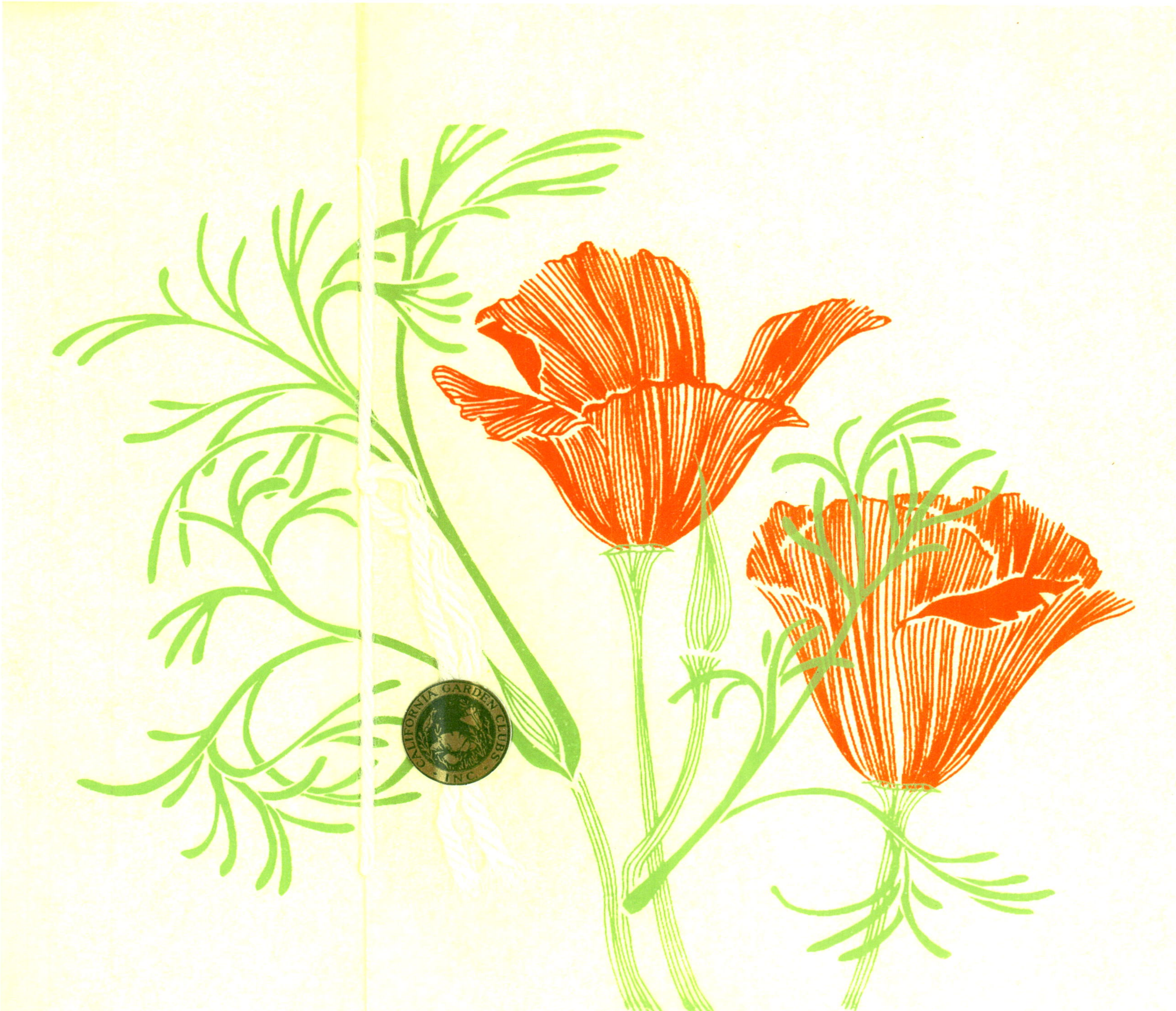
CALIFORNIA GARDEN CLUBS
INC.

Collection 2

Golden Botanical Gardens 1976–1978

June 1976
While attending the Spring Extravaganza at the Los Angeles State and County Arboretum on May 22 in Arcadia, I was enchanted by a specimen of the tulip tree in full bloom. This noble tree stands adjacent to the tram entrance located near the exit gate of the arboretum. I doubt that it is fully appreciated in this particular location, but patiently it stands there anyway in its dignified and serene glory waiting to be admired. No one could possibly have admired it more that day than did I. Standing there with it, I was reminded of the words of Professor John Merle Coulter in an address he gave at the opening of a botanic garden in 1917: "No one knows what may turn up in a garden like this one of yours. It is a gold mine of opportunity. See to it that it is cultivated."

Liriodendron tulipifera
I chose to draw a tree found in many arboreta and botanical gardens in California, known as the tulip tree, and a month later a magnificent coincidence occurred: I discovered that the Arnold Arboretum in Massachusetts had decided to commemorate the nation's 200th anniversary by distributing small tulip trees throughout the state. Apparently, during the Revolutionary War period, *Liriodendron tulipifera* came to be known as the "tree of liberty," such were its grand proportions, and the colonies planted it as a symbol of their quest for independence. It still is one of the most distinctive and elegant trees indigenous to the eastern United States, and so admired there in its native habitat that it was chosen as the state tree of no less than three states, Indiana, Kentucky, and Tennessee.

The generic name is a combination of two Greek words meaning "lily" and "tree." The specific name refers to the tulip-like flowers. Among the most attractive features of the tulip tree are its unusual leaves. It would be difficult to confuse these leaves with those of any other tree. With a breadth of about four to six inches and a notched or inverted tip, the leaves are often referred to as saddle-shaped. In addition to these handsome leaves, the tree produces extraordinarily beautiful flowers, in abundance during the month of May. Few people ever see them because most people never look up. But even if they did, the flower color is so subtle that they blend in with the fully developed leaves making them difficult to discern. The flowers are cup-shaped, a greenish yellow with a blotch of orange on each of the six petals, and are enhanced by a ring of golden filaments inside. At the center of each flower is a cone composed of many pistils that eventually develop into a cone-shaped cluster of dry fruits in which seeds are contained.

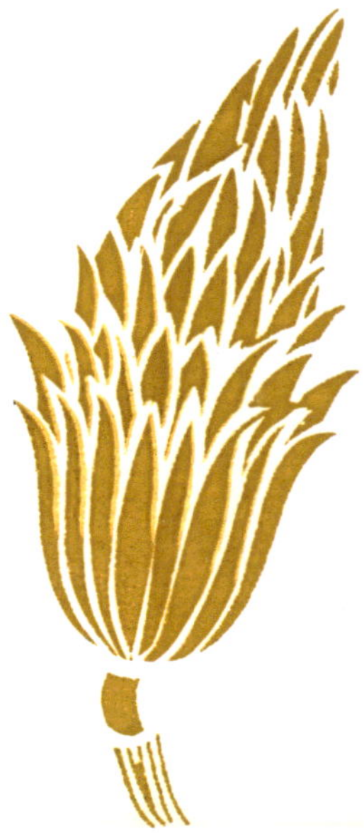

Tulip tree *(Liriodendron tulipifera)*

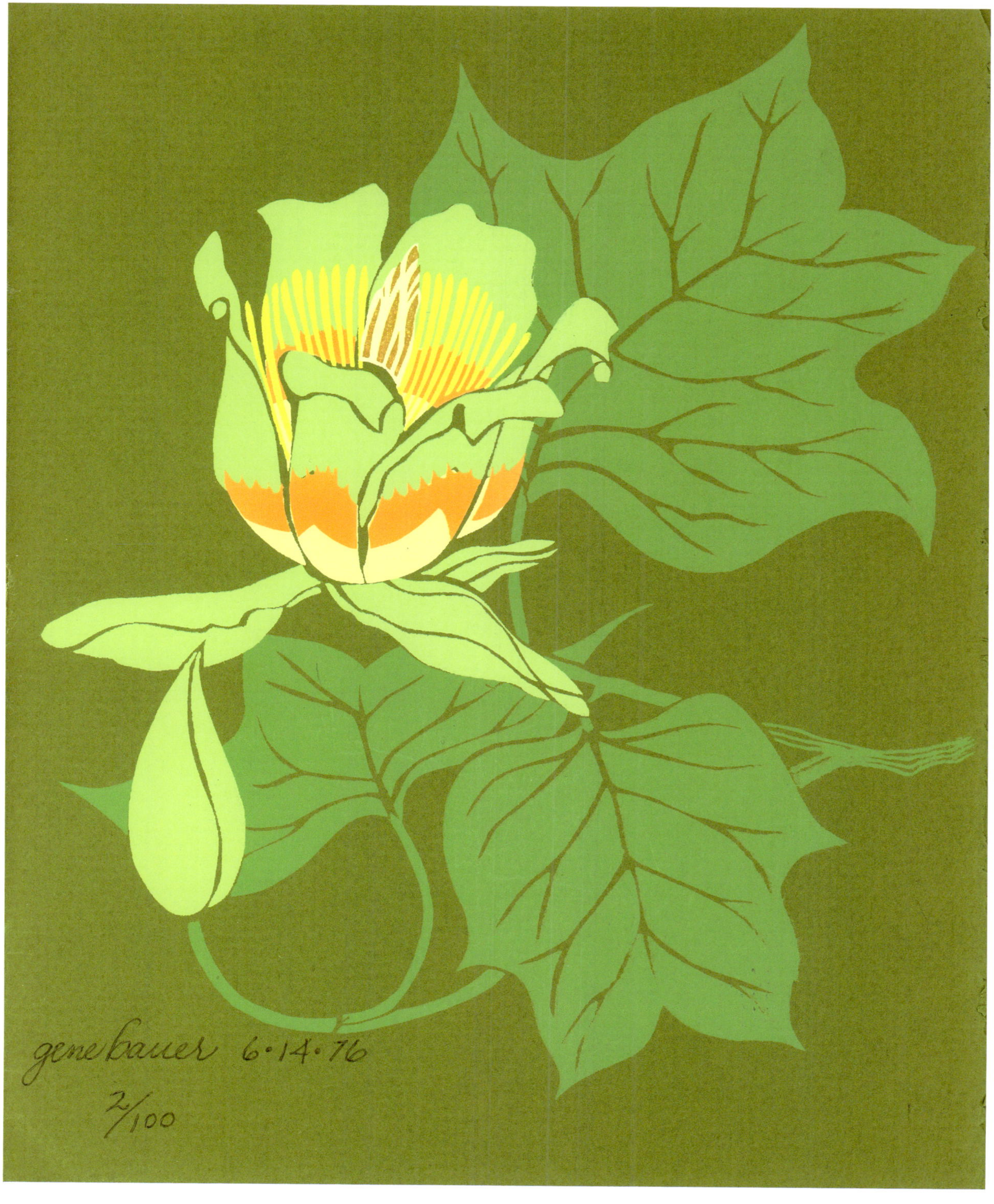
gene bauer 6·14·76
2/100

July 1976
I spent some time considering which plants to choose for the first few booklets in this Golden Botanical Gardens series. Strolling through the many gardens and arboretums in late spring and early summer, I saw that the *Cassias* were quite remarkable; the *Agapanthus* were cool and refreshing; the water lilies—and especially the Egyptian lotus—were ethereal. However, I was not completely satisfied with any of these lovelies. And then, suddenly, right in front of me was the most stunning tree imaginable. Glistening in the full sunshine and enhanced by a rich evergreen background. *Eureka! I have found it!*

Brachychiton discolor
What an amazing beauty! A thirty-foot high specimen and every inch from tip to ground festooned with exquisite pink flowers. The tree was incredibly gorgeous. Immediately, I took out my pencil, paper, and eraser. I spent about forty-five minutes drawing, and the entire time the lovely pink flowers kept falling from the tree, gigantic pink raindrops falling on and all around me. They fall while still in excellent condition, so the ground under the tree was as stunning as the tree itself.

Brachychiton discolor, a native of Australia, is commonly known as pink flame tree (I now know why) and sometimes as Queensland lace bark tree. This is a tree for warm inland areas and does not do well in damp coastal regions, but when everything is right it grows to be from forty to ninety feet high. The leaves are from four to six inches long and resemble those of the maple. The tree may as well be evergreen since its leaves fall just before the flowers open and then appear again, fresh and new, toward the end of the flowering period. The flowers are between 1½ and 2 inches long, bell-shaped, and a remarkable shade of rose pink. The outside of the flower is covered with short, brownish-colored wool; the same brownish wool covers the handsome six-inch seedpods.

Flowering is usually during the months of June and early July and not to be missed. I found the tree from which I did my drawing near the Coach Barn at Los Angeles State and County Arboretum in Arcadia, California. A truly lavish display of these trees may be seen in the Australian section of the Huntington Botanical Gardens in Pasadena, where there are also outstanding specimens growing near the gate to the Japanese Garden as well as at the main entrance and for a block on a road alongside it. I think someone at the Huntington Gardens likes this tree and I certainly agree, so make that two.

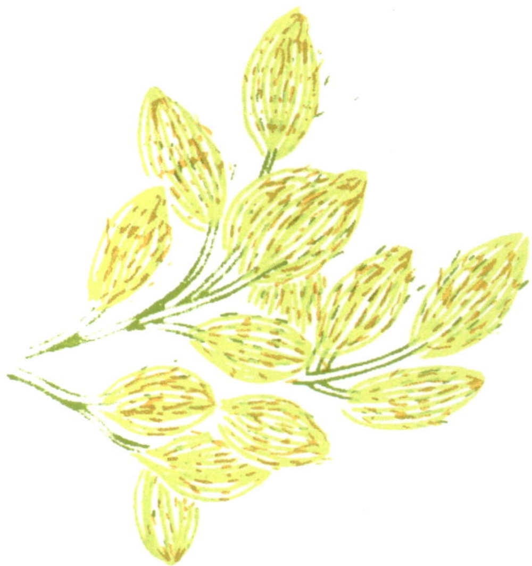

Pink flame tree *(Brachychiton discolor)*

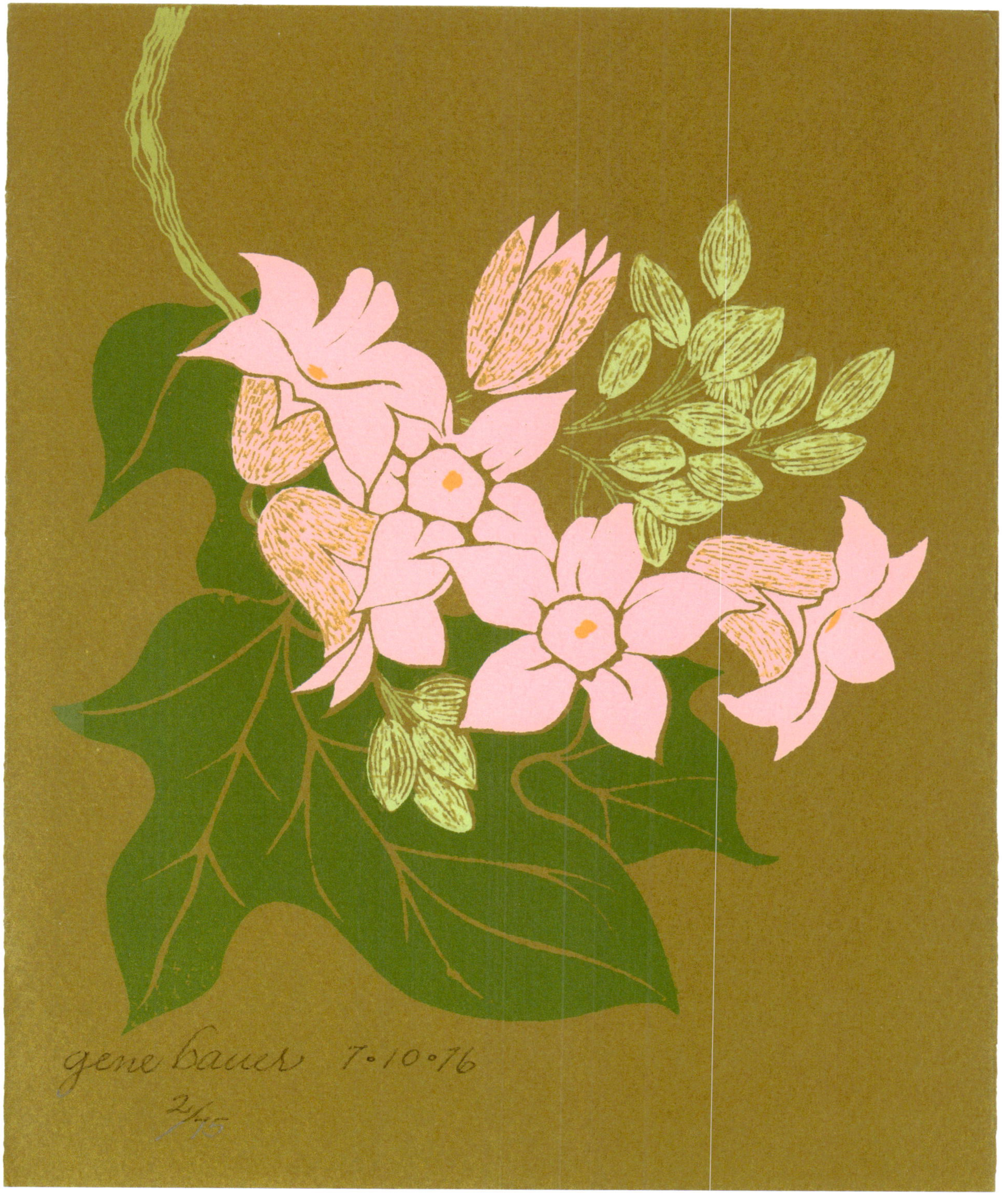
gene bauer 7·10·76
2/15

Fuchsia (*Fuchsia* 'Trumpeter')

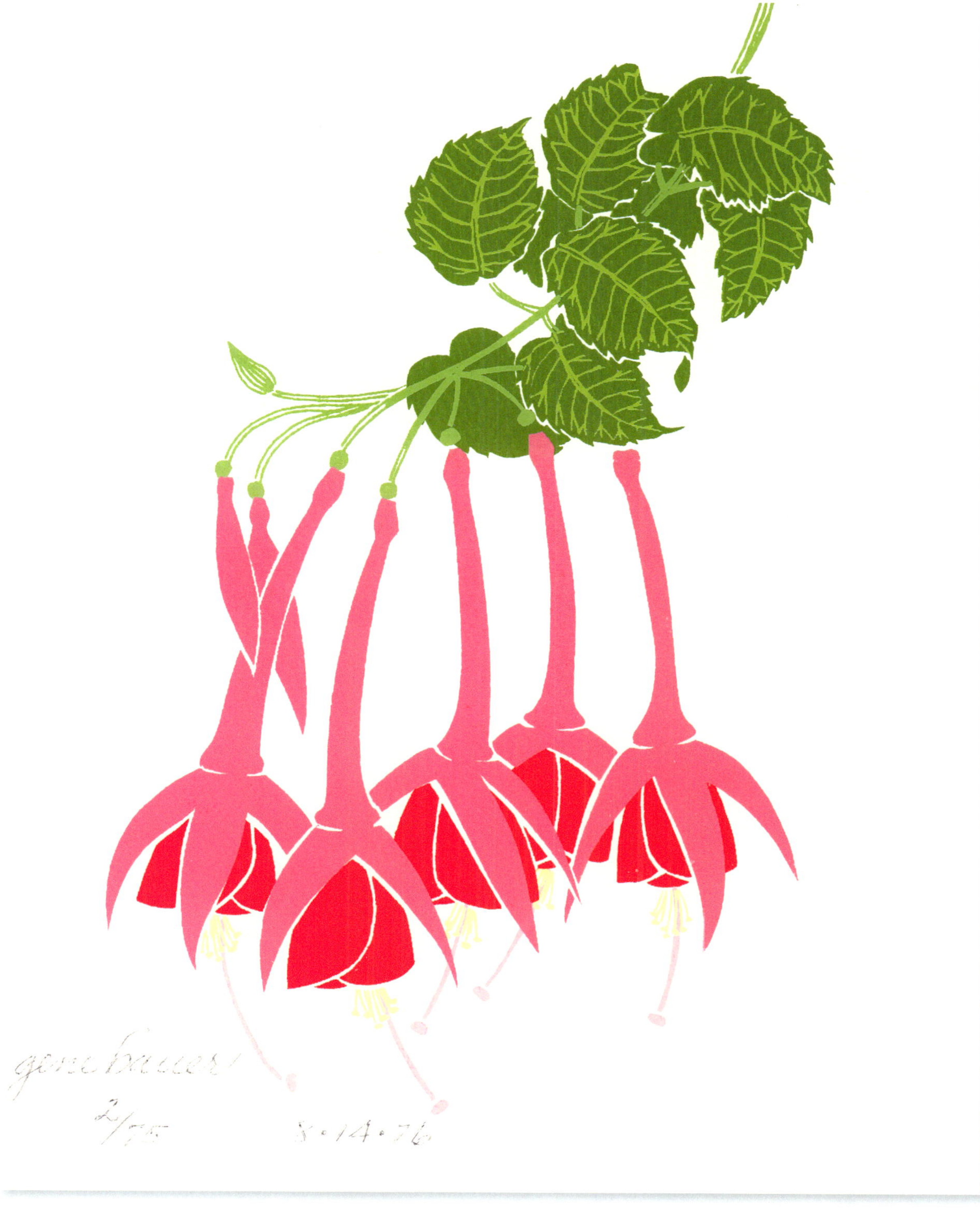
gene bauer
2/75
8·14·76

August 1976

The consistently top-notch appearance of a garden is accomplished by a program of digging up the flower beds as soon as they have passed their prime and replanting with other plants just as their colorful buds begin to open. This is what the Sherman Foundation Center [now the Sherman Library and Gardens in Corona del Mar] made possible for its small garden's beginnings in the 1960s, by maintaining a nursery elsewhere on a larger piece of land. In this way, it has served as a model for other towns striving to establish a small garden for the enjoyment of its citizens.

Apparently an abundant variety of flowering plants is maintained all year long, as I have visited at different seasons of the year and have always found the flower beds in full bloom and the hanging baskets well-grown, well-groomed, and bountiful. Large flower beds and containers throughout this botanical garden are changed seasonally in order to feature the colorful annuals for which the mild coastal climate here is ideal. Seldom have I encountered a two-acre piece of land so beautifully planned and meticulously maintained. Once inside the Sherman Gardens entrance gate there is a marvelous feeling of peacefulness and calm in sharp contrast to the fast-paced, hurly-burly world outside. I think of this joyful place as a very small, intimate edition of the Henry Huntington Botanical Gardens, with love and sheer perfection evident in both.

Fuchsia 'Trumpeter'

Container gardening is the answer when space is limited. Pots, tubs, and jardinieres literally spilling over with both the rare and the commonplace can be found any day of the year throughout this garden, along with a collection of supremely beautiful hanging baskets: baskets in the sun; baskets in the shade; baskets with curiously textured and colored leaves; baskets with plain green foliage that just keeps growing longer and longer and longer. The most sensational are the baskets of color; difficult to surpass the beauty I found there this month.

Magnificent baskets of tuberous begonias in all those exquisite colors of apricot, peach, salmon, sunset, orange, rose, yellow, and white cascaded from overhead, while in a nearby bed on ground level, growing upright, were additional tuberous begonias in the same yummy colors. In the same area an equally stunning collection of baskets of fuchsias, hanging from the trellis, caught my eye.

Even with such variety in this enchanting floral atmosphere, my attention was riveted by the fuchsias: single, semidouble, double, ruffly, and frilly petals, and a color range almost impossible to comprehend. How could I decide? Finally, I chose a long, sleek, slender, single flower extremely attractive in its simplicity. The tube and sepals were a strong, clear pink and did not recurve in the classic fuchsia manner. The corolla was a bright cerise. Quite appropriately, the name of this cultivar was 'Trumpeter'.

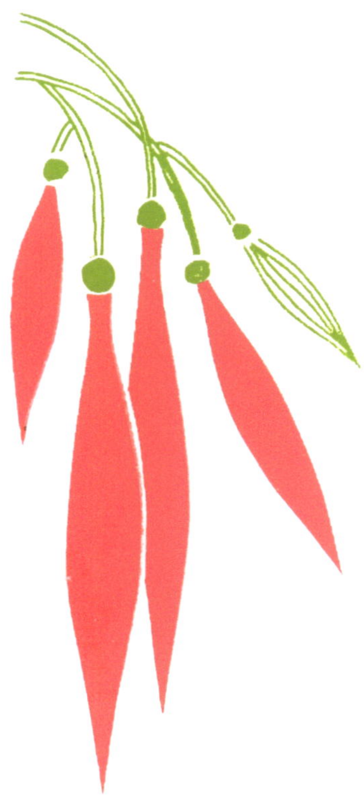

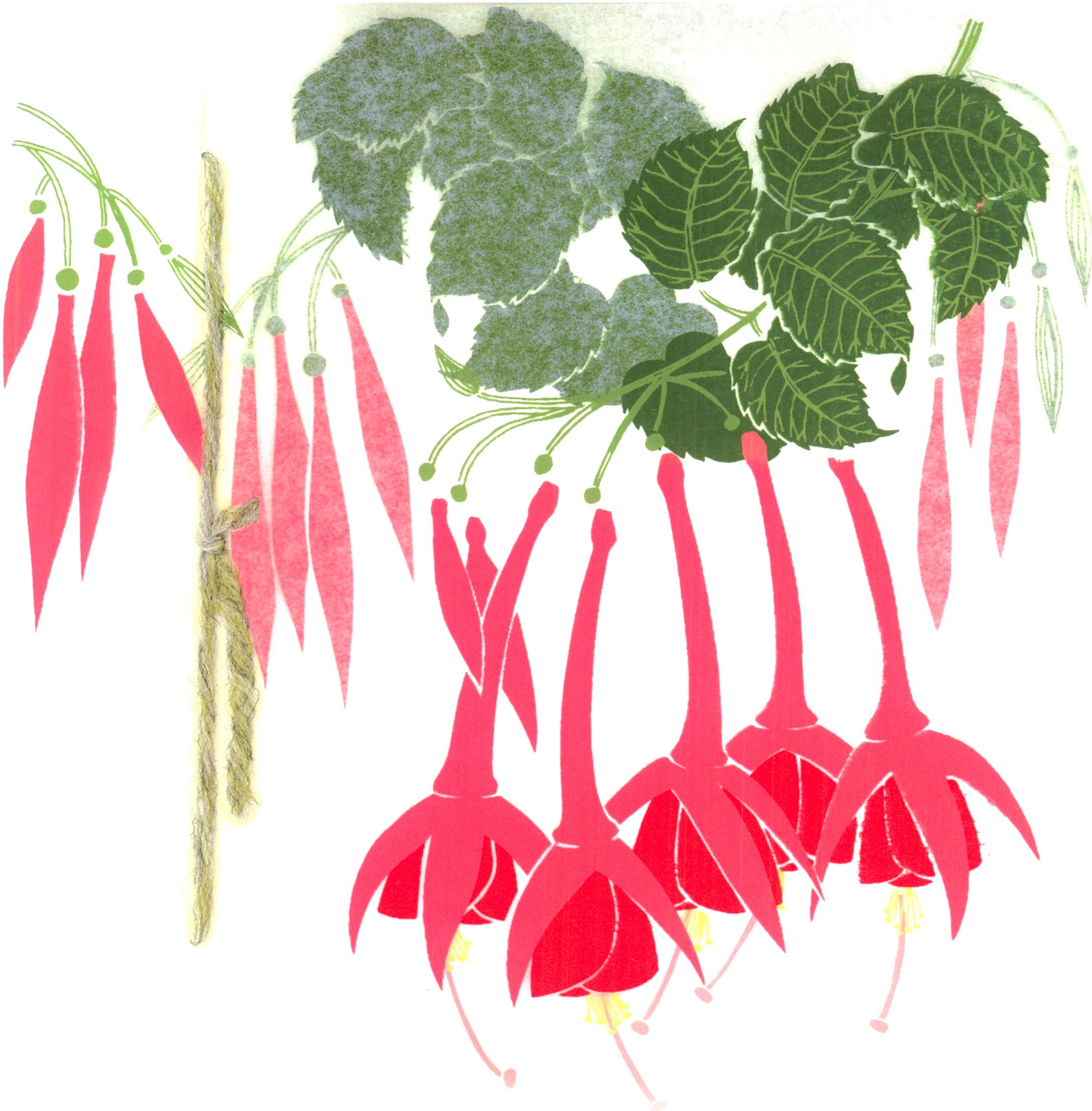

Ginger lily *(Hedychium gardnerianum)*

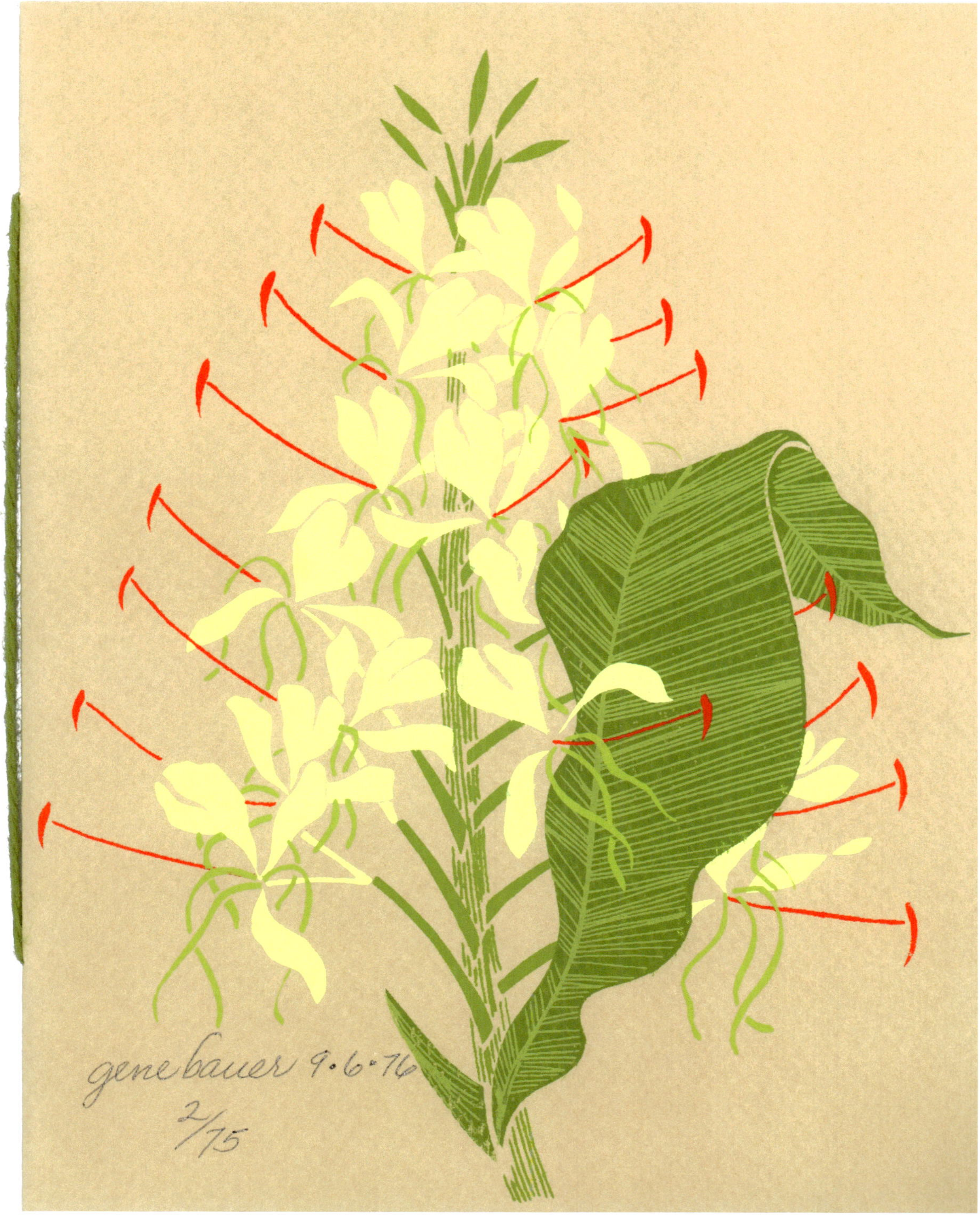
gene bauer 9·6·76
2/75

September 1976

It was more than twenty-five years ago that I first became aware of the University of California at Los Angeles Botanical Garden, primarily a university teaching and research facility covering eight acres and begun in 1929. As an art education major with a specialty in painting at UCLA, I not only lugged the necessary books for academic classes but also toted drawing boards, stretched canvases, and all the paraphernalia of an art student, back and forth from living quarters to classroom. At that time, the botanical garden was a shortcut for me to the Art Department. I trudged through this garden many times, and my one and only intent was to get to class on time. How pleasant to once again return under less-hurried and far different circumstances.

The major portion of the UCLA garden is situated in a small canyon out of the wind in one of the finest climates in North America for growing tropical plants. Everything here is lush and green, with areas near water that are so tropical that the only things missing (thank goodness) are boa constrictors and pythons. Two streams of water flow through the garden terminating in two ponds in which colorful Japanese koi may be seen. Near the south pond I found tree ferns, elephant ears, horsetail, birds of paradise—and just in case this is sounding more zoological than botanical—also bamboos, palms, philodendrons, and banana trees complete with maturing fruit. The plant I chose for this month was there, too.

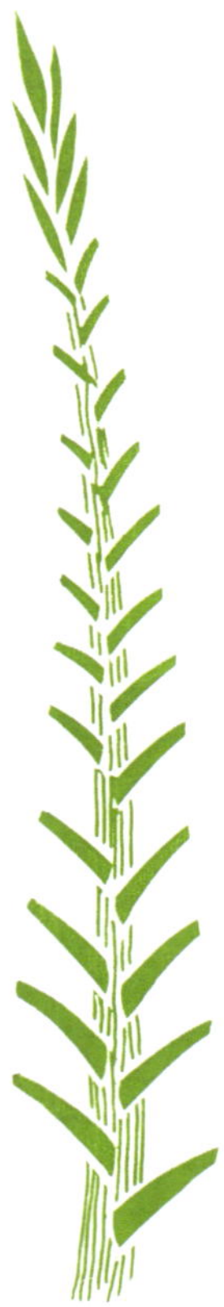

Hedychium gardnerianum

As I walked through I was especially attracted to the handsome clumps of what are commonly known as ginger lilies, growing along the stream banks. The stems of these fleshy-rooted perennials were about five feet tall with luxuriant leaves eight to fifteen inches long and four to six inches wide. These specimens of *Hedychium gardnerianum* were just beginning to bloom in mid-August, but even before blooming the flowering head is attractive. As shown on the envelope art (below) the stem tip is covered with rolled green bracts inside of which is a concealed flower. The flower head is about eight to ten inches long and is composed of fragrant yellow flowers, each with a single red stamen.

The generic name *Hedychium* is a combination of two Greek words meaning “sweet snow,” inspired by the fragrance and color of the flowers of some species. The name honors Edward Gardner, an English colonel serving in Nepal in 1817. Natives of northern India, these plants are found growing as high as 8,000 feet in the Himalayan mountains. *H. gardnerianum* has become naturalized in Hawaii where it is known as kahili ginger, called this because of the flower head’s obvious resemblance to the Hawaiian kahili. A pole with feathers attached to the top, the kahili was carried by early chieftains as a banner to announce their rank.

Revisiting the botanical garden at UCLA was an experience to remember. The day I was there, the overhead circular sprinkler system was in use, the water from it skirting the area where I was drawing. Each time the water came around it was like a sudden tropical squall, and to escape from it I repeatedly sought shelter under a nearby palm frond. This activity was almost diverting enough for me to forget that this lovely tropical setting—the sweet fragrance of moist ginger lilies permeating the air—was in reality located in the midst of a drought-prone, bustling metropolis.

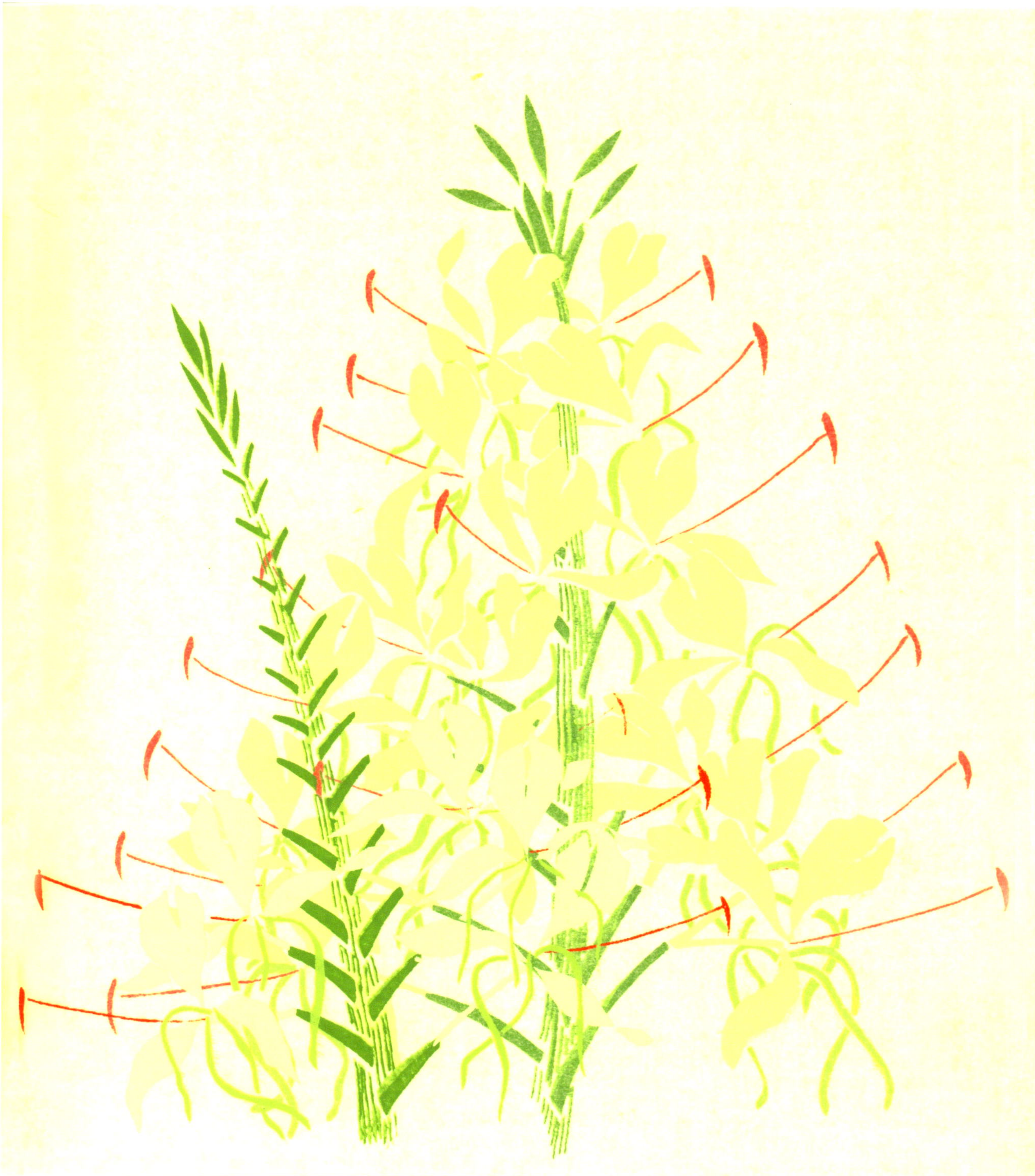

Barberry *(Berberis jamesiana)*

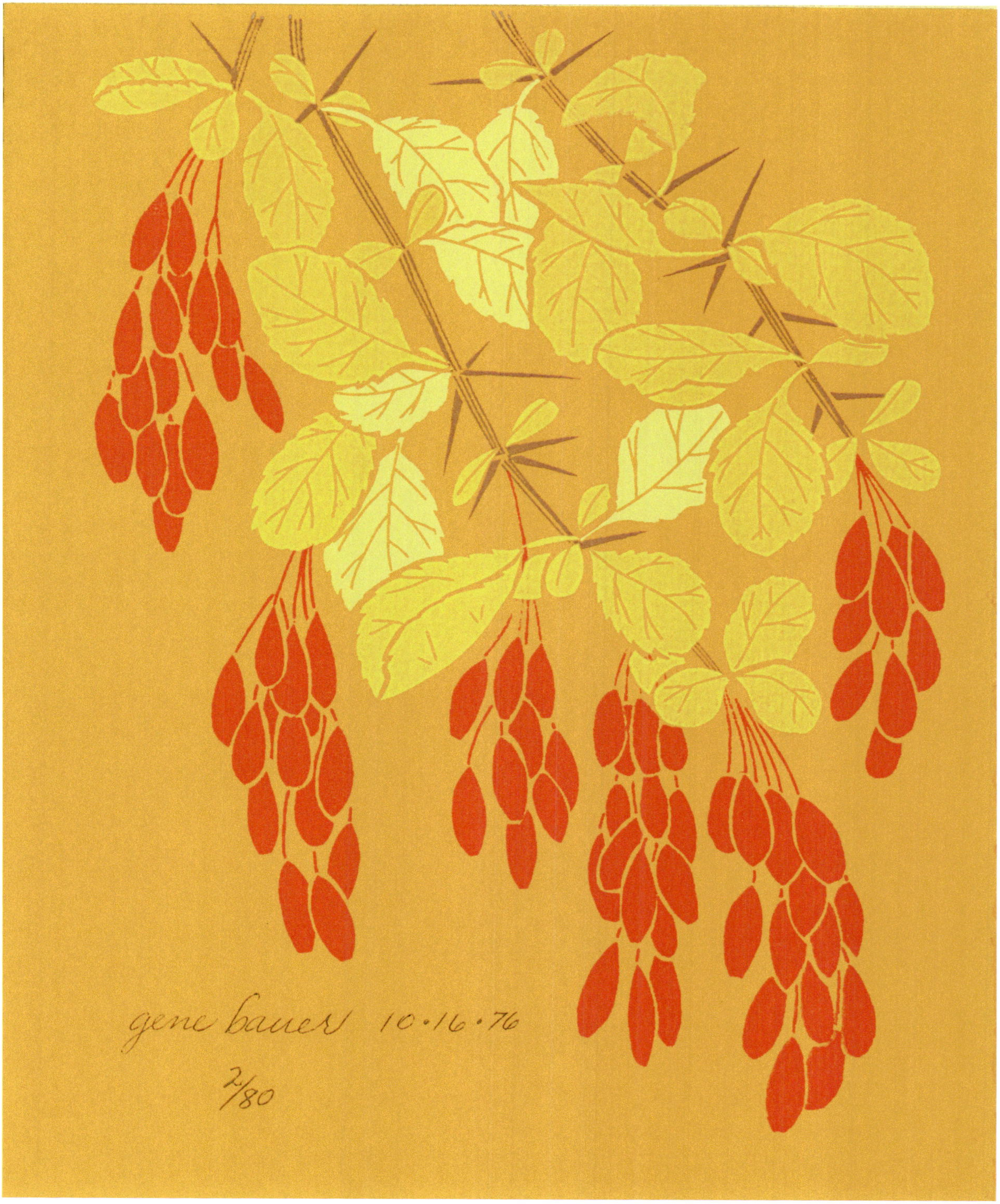
gene bauer 10·16·76
2/80

October 1976

Located in Golden Gate Park in San Francisco, the Strybing Arboretum began in 1937 with funds from the bequest of Helene Strybing, who requested that a collection of trees, shrubs, and plants indigenous to California be a part of it.

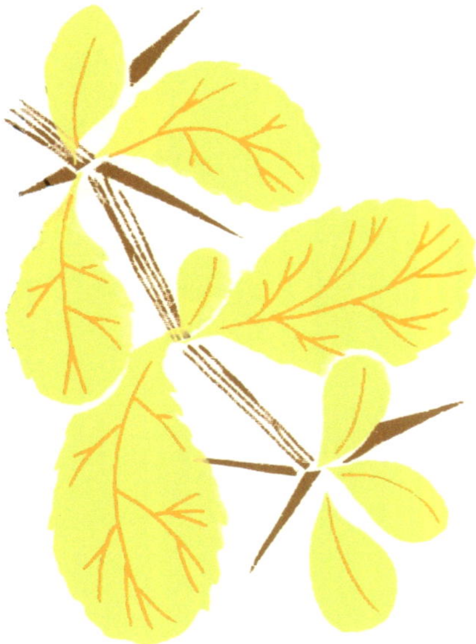

Berberis jamesiana

Quail, squirrels, and even a rich chocolate-colored rabbit? Yes, indeed. When I saw as many as fifty quail and six squirrels in one area, my curiosity was aroused and I simply had to investigate. That area turned out to contain a bountiful harvest for these feathered and furry creatures. They were in the midst of a striking collection of ground covers and shrubs that produce berries at this time of year. I found many different species of *Cotoneaster, Pyracantha,* and *Berberis* growing here and knew immediately that one of them would be my choice for the cover of this issue. I chose *Berberis jamesiana,* a native of Yunnan Province in China.

The generic name *Berberis* is derived from *berberys,* the Arabic name for the fruit of this plant, commonly called barberries. Many different species exist, some deciduous and some evergreen. These plants are known not only for their thorns but also for delightful yellow flowers in spring, followed by a dazzling display of berries in the fall. Many of the deciduous species, such as *jamesiana,* offer the added bonus of exquisite leaf coloration in the fall.

The plant I chose was a handsome, dense shrub about five feet tall. Many of its leaves were still green, but some had turned lovely autumn shades of yellow through yellow orange to apricot. Exquisite three-inch long pendulous racemes of translucent coral-red berries were borne in great profusion, adding much to the attractiveness of this plant.

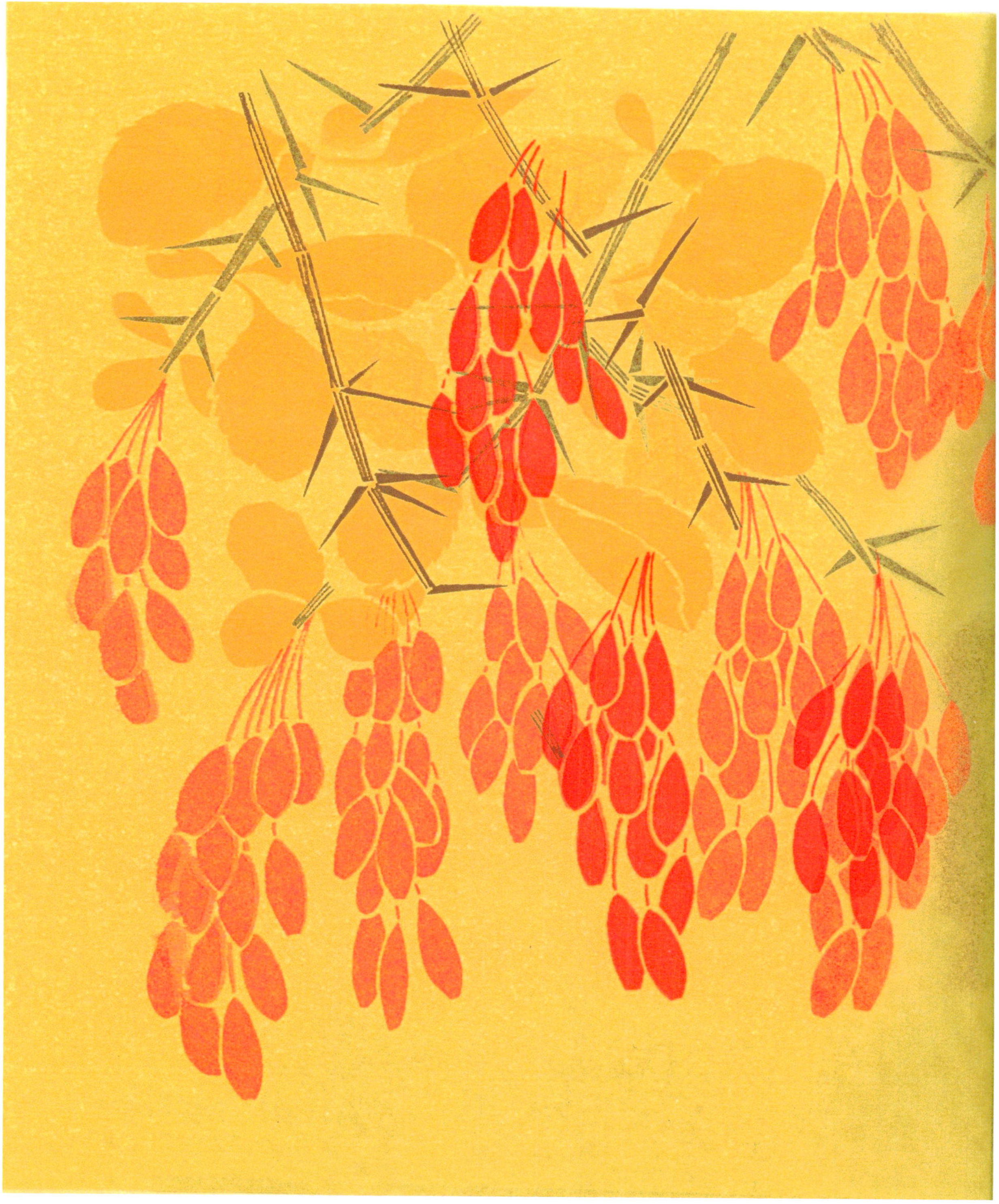

Camellia *(Camellia japonica)*

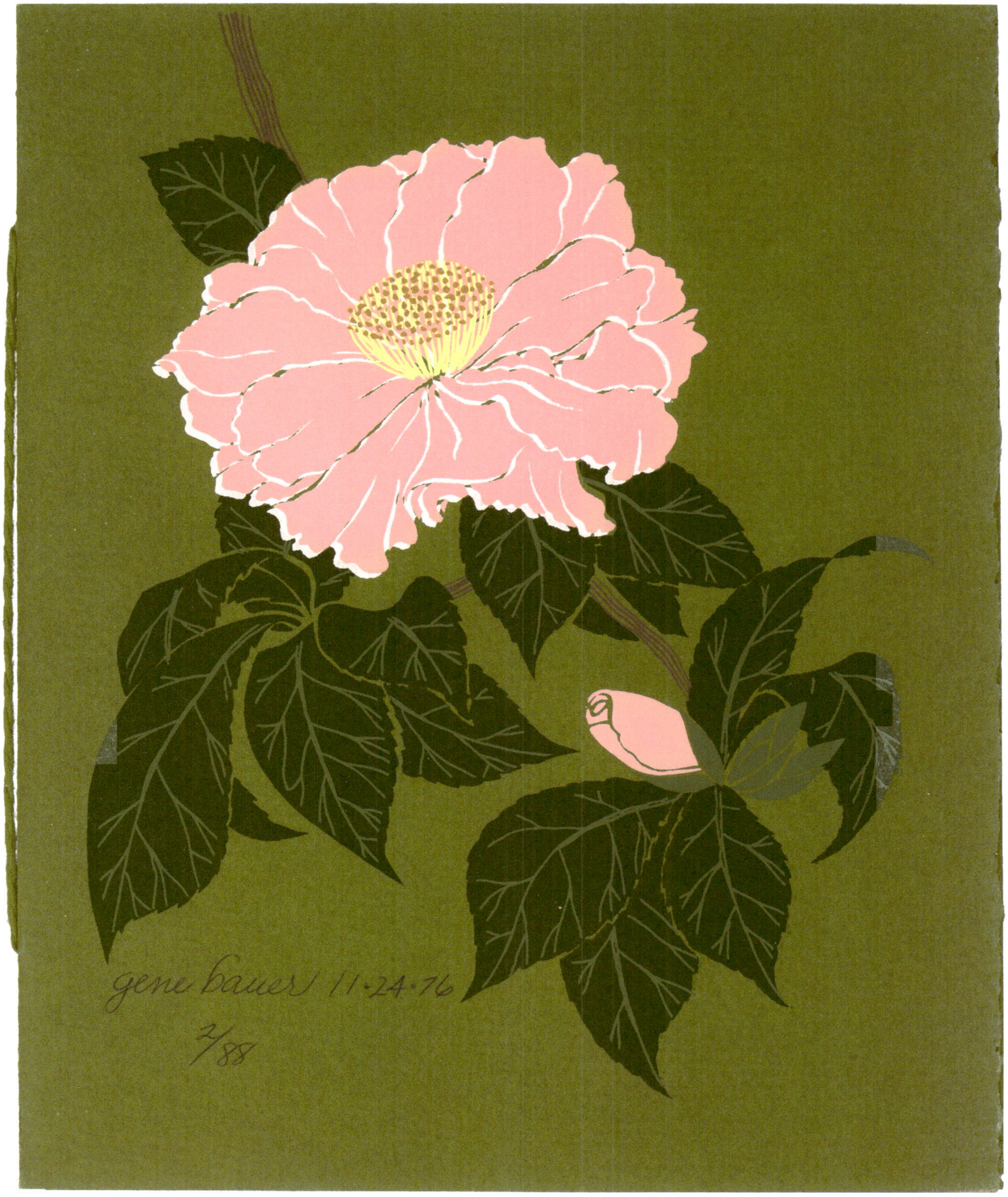
gene bauer 11·24·76
2/88

November 1976

California is a land of great natural bounty. The names of the counties in the Valley Lode region (Alpine, Amador, Calaveras, Mono, San Joaquin, Stanislaus, Tuolumne) indicate what can be found there. The western portion is flat, rich agricultural land full of introduced crops, acres and acres of fruit and nut trees and the vineyards of the great San Joaquin Valley. To the east and a gentle, gradual rise in elevation to the foothills is the picturesque gold country known as the Mother Lode. Continuing eastward and ascending even higher are the peaks of the majestic Sierra Nevada in excess of 14,000 feet. Sometimes we don't know what a gold mine we have until it's gone, but on an early November trip to discover a subject for my drawing, I realized yet again that I am not the only one who thinks so.

Just beyond the town of Sonora is the campus of Columbia Junior College located on two hundred acres of forest land in Tuolumne County. Concern for the natural beauty of the land is apparent at first sight. The large parking area and the buildings themselves obviously were constructed with very little damage to the natural terrain. So often when an area is targeted for development the bulldozers scrape it clean of all the existing vegetation. But here, they have even preserved fifteen acres of the campus in an undisturbed state as an outdoor laboratory for environmental studies and as a sanctuary for indigenous plants and animals. This was a gold mine for my spirits as I walked the Arboretum Nature Trail there—surrounded by trees that were here before many of us appeared on this earth. The experience gave me the perspective I needed (and perhaps the courage) to approach the five hundred different varieties in the Camellia Garden at Micke Grove Park where I arrived later that day. *How did there get to be so many different kinds and how am I going to choose just one?*

Camellia japonica

Camellias are evergreen shrubs that are native to eastern and southern Asia. (The genus *Camellia* was named for George Joseph Kamel, a Jesuit missionary and student of natural history who spent much of his life in the Orient.) Just beginning to bloom in November, these shrubs provide much-needed color throughout our winter months here. Among the several species in the genus, the best known of those growing outdoors in California are probably *Camellia japonica* (to many, this is the *only* camellia), *C. sasanqua, C. reticulata,* and the thousands of cultivars derived from these and other species.

The glossy, dark green leaves provide a complementary background for the flowers, which are produced from November until May. The range of flower size, color, and form is astonishing. Colors range from white through countless shades of pink, rose, and red, and many are variegated with stripings, splotching, and speckles. Some of the various available forms are single, semidouble, formal double, peony-form, anemone-form, rose-form, "rabbit ears," and fimbriata (fringed). Absolute perfection of form is the only way to describe some of the formal double types, with each individual petal exquisitely spaced both radially from the center out to the edge as well as around the flower. From an artist's point of view, some are so perfect they are dangerously close to lacking character.

Camellias have long been a favorite outdoor flowering shrub in this state. Much work has been and continues to be done in the hybridization of this flower. The goals that someday may be achieved are flowers in the color range of yellow, orange, blue, and purple—and the acquisition of that very special but elusive quality, fragrance.

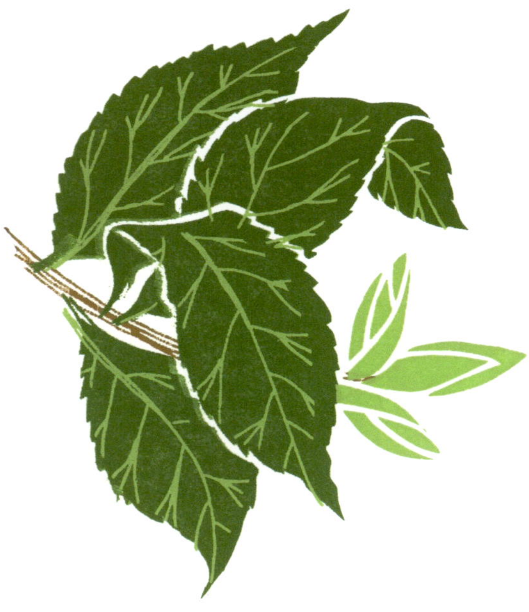

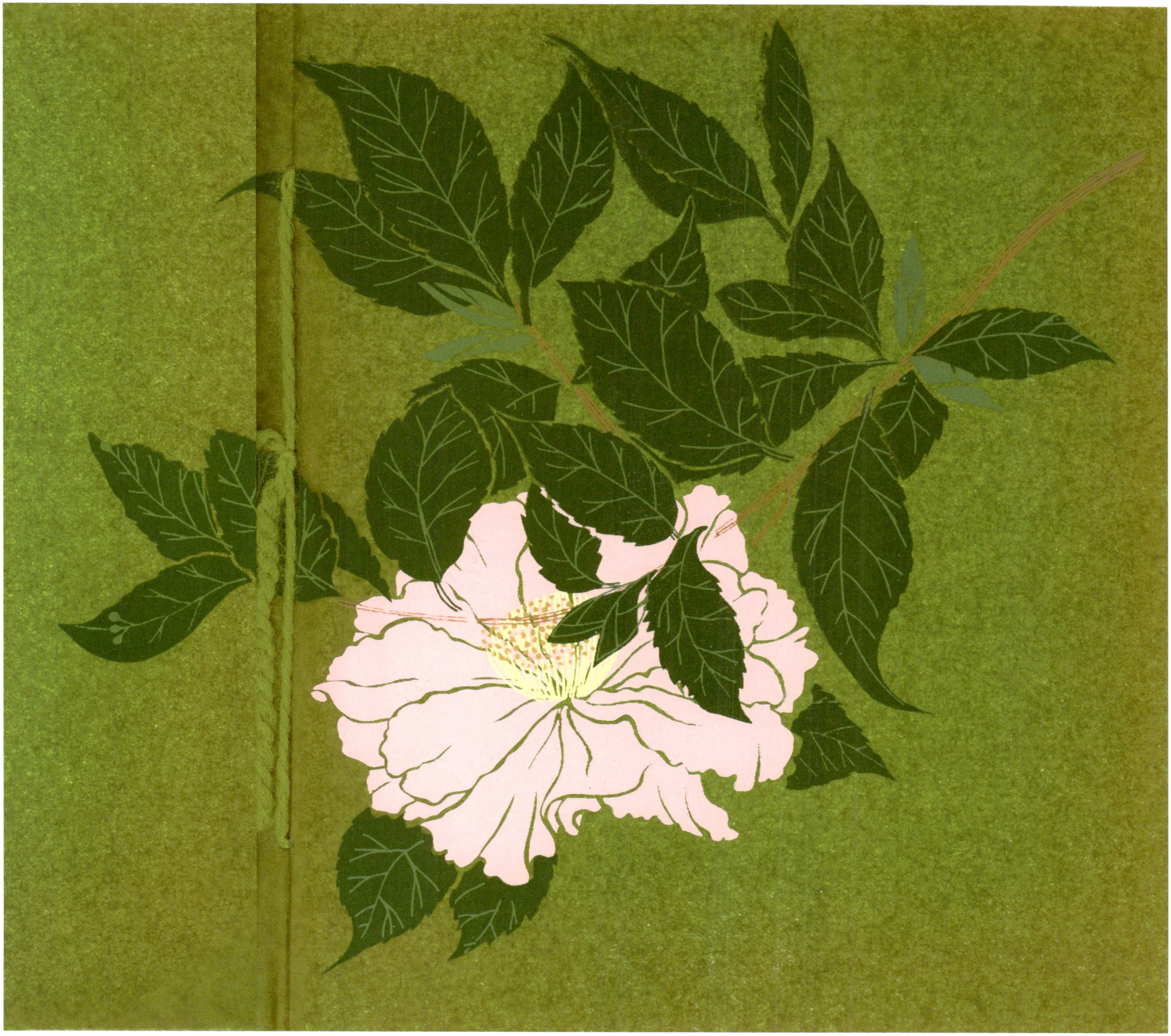

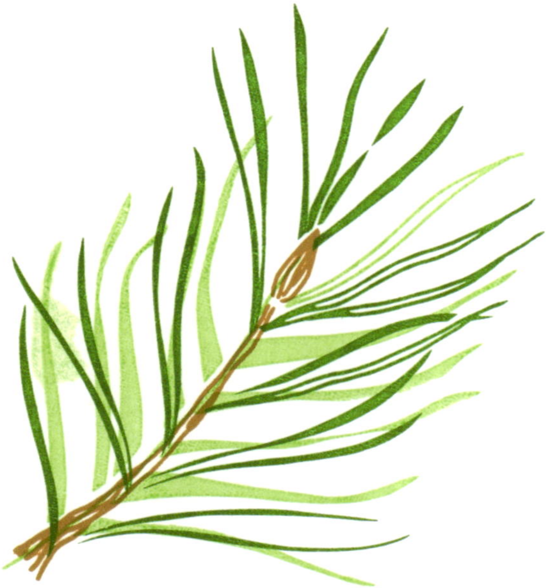

Scotch pine *(Pinus sylvestris)*

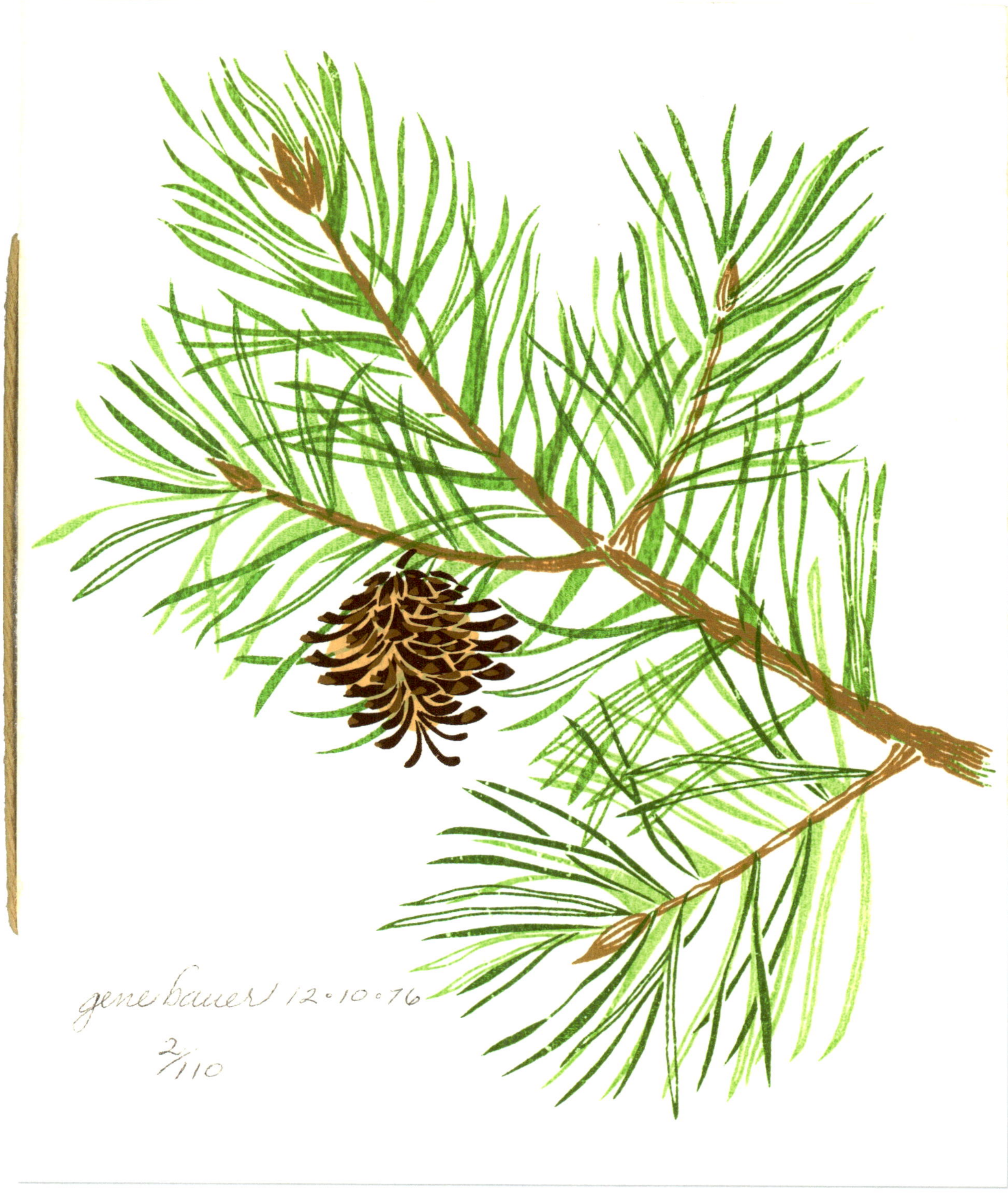
gene bauer 12·10·76
2/110

December 1976

A Seattle lumberman named James G. Eddy had already witnessed firsthand the history of forest depletion in the eastern and southern United States when he saw the same thing happening in the Pacific Northwest early in the last century. Thusly motivated, he wanted to improve forest trees through hybridization, and so established the Eddy Tree Breeding Station in 1925. On 106 acres of land east of the town of Placerville, California, the location was excellent. Eleven different species of pine in natural stands could be found close by, a mixed forest of the Sierra Nevada that became the outdoor laboratory for much of the pine breeding work. Additionally, a wide range of elevation from sea level to 10,000 feet was accessible from the research facility, which changed its name to the Institute of Forest Genetics in 1930.

The raw material for breeding is genetic diversity, so for the first plantings (in the spring of 1926) as many pine species and varieties as possible had to be assembled. Almost all the seeds came from native stands, with their parent-tree origins carefully recorded. Now the Eddy Arboretum contains the first hybrid pines ever produced by controlled pollination, becoming not only one of the best documented pine collections in the world but also a one-of-a-kind genetic resource for tree breeders everywhere.

Almost all of the world's species of pines have been planted here (nearly one hundred), with the surviving seventy plus approaching their fiftieth birthday when I visited. Of particular interest to me were the specimens of the original hybrids planted in 1927. I am quite familiar with some of these crosses—*P. attenuata* × *P. radiata* (knobcone × monterey) and *P. jeffreyi* × *P. coulteri* (Jeffrey × coulter)—they are the same types of hybrids that were planted on government land near my home in 1969. I was pleased to see handsome fifty-year-old versions of my still-young, six-foot California conifer neighbors.

One that I saw, *Pinus sabiniana* or "Digger pine," is worth a note here. It has often been written that this tree is named after the Digger Indians. Actually, there was no such thing as a tribe of Digger Indians. The name Digger was a catch-all name applied to all Indians by the early white settlers. Many native peoples did dig for their sustenance—bulbs, roots, tubers. But each tribe here had its own name and, to me, usually a charming one. The early white men could not be bothered with learning the tribal names and simply lumped all the native tribes together and attached this rather scurrilous designation "Digger" to them. Meanwhile, the original inhabitants of California had only the highest respect, even reverence, for this tree. As one of the "nut pines," no other tree was more valuable and useful to the Indians except possibly the oak. I feel it is most unfortunate that the early white settlers, upon learning that *Pinus sabiniana* had no commercial value as lumber, gave it the name Digger to show their contempt for it. In contrast, it is refreshing to remember how this tree provided food for the original Californians and the many souls who gave thanks in their own way for its existence.

Pinus sylvestris

The particular tree from which I made these drawings grew from seed collected in Scotland in 1928. Commonly known as Scotch pine, it has been widely planted in the United States as an ornamental tree, mainly because of its adaptability to unfavorable climates and soils. The deciding factor in my choice of this pine was what I learned from its label in the Eddy Arboretum at the Institute of Forest Genetics where I found the tree: it is the most widespread pine in the world, with its natural range extending from Scotland to eastern Siberia, and as far south as Spain and Turkey.

The name of this species of the genus *Pinus* means "of the woodland." The needles of *sylvestris* are somewhat stiff and sometimes twisted, and they grow to a length of two or three inches. *Pinus sylvestris* is a two-needle pine, meaning there are two needles in each sheath. Mature cones are about two inches long.

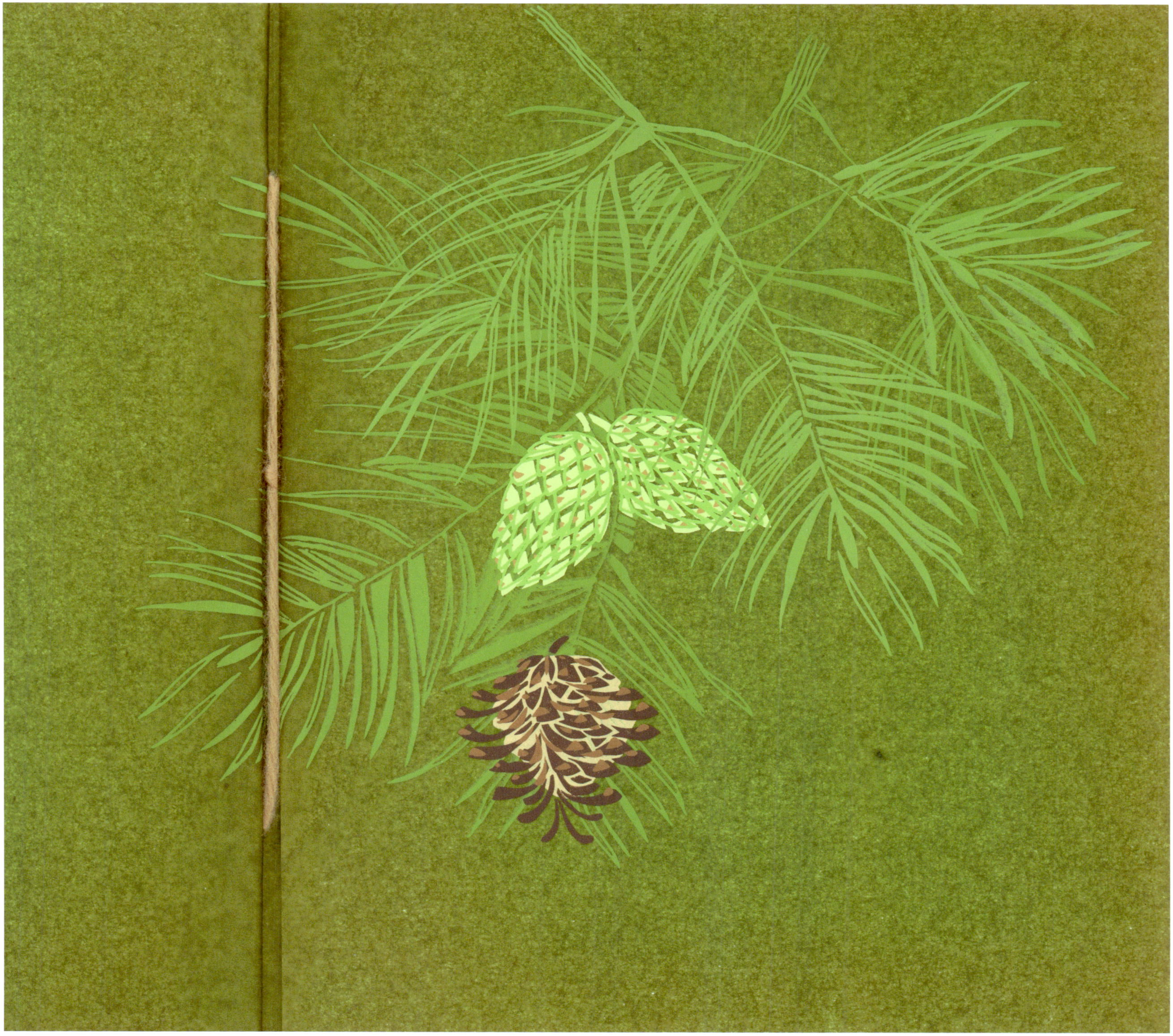

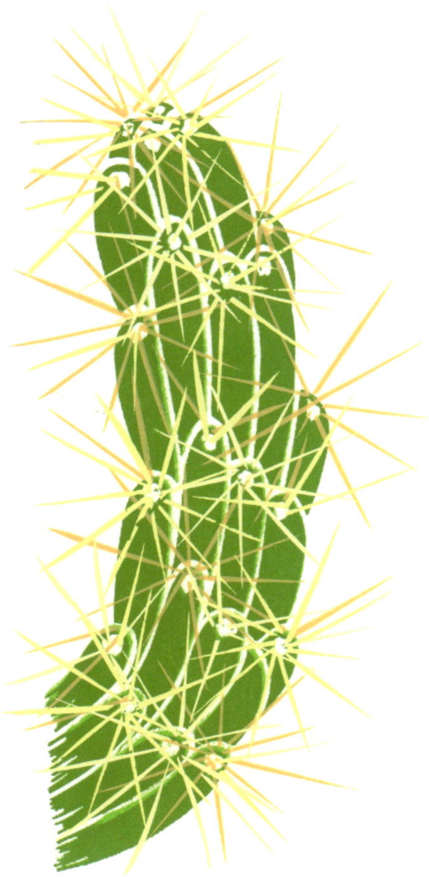

Buckhorn cholla *(Opuntia acanthocarpa)*

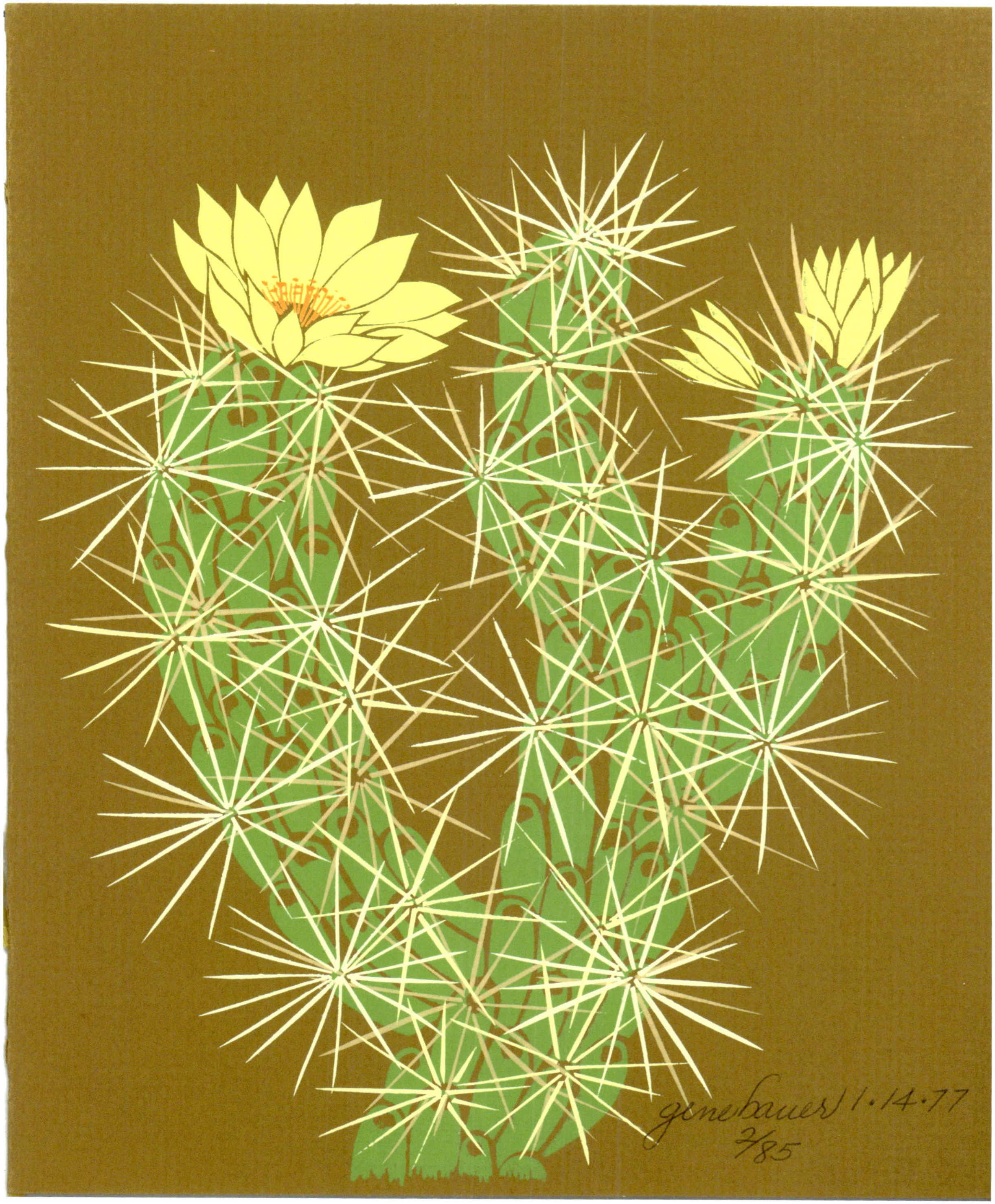
gene bauer 1·14·77
2/85

January 1977

California is indeed a land of great contrasts. With rain forecast for the following day (December 29), I thought it an appropriate time to visit The Living Desert Reserve, one of the most unusual gardens in the state. Leaving the clear, brisk pine-scented air of our mountain home, my husband and I arrived twenty-five minutes later in the town of Redlands in the valley below. He went to his office in town and I embarked on my day's journey, one hour later arriving in that glamorous desert "spa" Palm Springs where people meander about in tennis togs and golfing gear. Fifteen minutes later, I entered The Living Desert Reserve in Palm Desert, spent six hours there, then returned home to a forecast realized. Several evenings later as I am writing this, the full moon shines on our sparkling white world and through an opening in our forest the faraway twinkling lights of Redlands are visible. A land of contrasts indeed. Where else could one leave home in a snow storm and less than half an hour later smell the divine fragrance of orange blossoms and an hour after that be basking in the radiant sunshine of the desert?

More than 14 percent of the surface of the earth is desert—almost eight million square miles of it. There are twelve major deserts in the world, with the Sahara in Africa, the largest, accounting for almost four million. The smallest desert is the Atacama in South America (140,000 square miles). Half a million square miles of the North American continent is desert, including the Yuman, Saguaro, Chihuahuan, Baja California, Mohave, and Colorado Desert. In hosting some form of life from all those North American deserts, The Living Desert is one of a kind. Formed in 1970 as a division of the Palm Springs Desert Museum, it encompasses roughly a thousand acres of the Colorado Desert, all set aside in its natural state for the purposes of preservation, interpretation, and education. Like so many public gardens and arboretums, it is totally dependent on funds from donations and membership dues, but unlike them it lends equal attention to fauna as well as flora. Here, not far from the peninsula bighorn sheep *(Ovis canadensis cremnobates),* the bobcats, and sundry slitheries, wigglies, and squirmies (my names for reptiles, rodents, and spiders), grows the object of my attention for much of my six-hour visit.

Opuntia acanthocarpa

Located in an area of delicate and wondrous beauty, this plant adds oddly to it. Under clear blue skies and an occasional fleeting cloud, its desert home is ringed on the far distant northwest by the San Bernardino Mountains, on the west by the nearer San Jacinto Mountains, and on the south, snuggling close, by the Santa Rosa Mountains (the most northern limit of the previously mentioned and rare species of bighorn). The varying distances and ruggedness of these three distinct mountain ranges are responsible for their exquisite coloration—from the softest of blues and mauves through every imaginable red/blue hue to burgundy and the deepest purple. But don't spend too much time with your eyes on the horizon—you might trip over the species of *Opuntia* cactus whose specific name *(acanthocarpa)* means "spiny fruit" (in Greek).

Of all the cacti found in the southwestern United States, the *Opuntia* may be the most prevalent. A large genus, *Opuntia* encompasses around three hundred species, many of which are nearly impossible to identify because of the variability fostered by different environments and by a great amount of natural hybridization. Basically, divided into two large groups, those with flat joints are called prickly pears and those with cylindrical joints are chollas. Many of the cacti belonging to the prickly pear group have edible fruit.

A member of the cholla group, *Opuntia acanthocarpa* is known as buckhorn cholla or staghorn cholla, the names deriving from the branching pattern. The species grows from two to five feet tall and is openly branched. The numerous and sharp spines are straw-colored and about 1½ inches long. The flowers are 1 to 1½ inches in length and are quite variable. Some plants produce yellow flowers, while others bloom in yellowish green; still others produce flowers of a reddish shade.

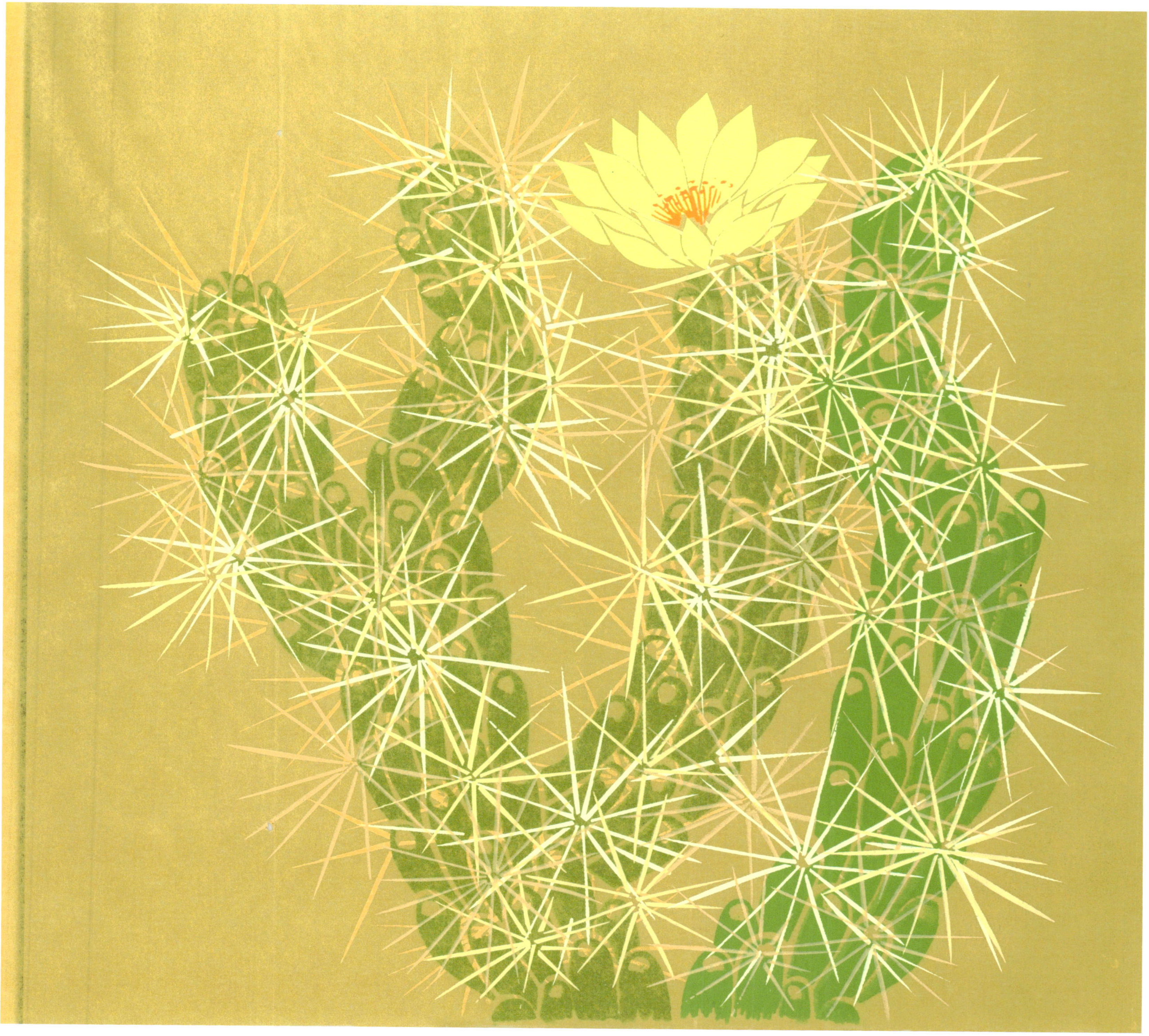

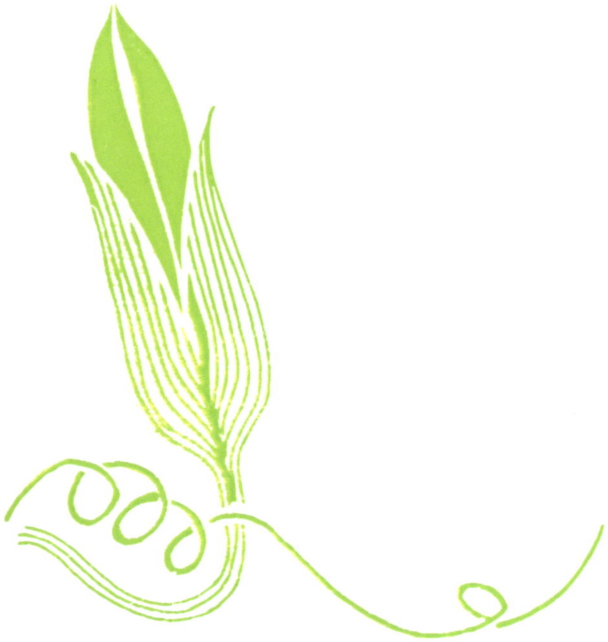

Passion flower vine (*Passiflora jamesonii*)

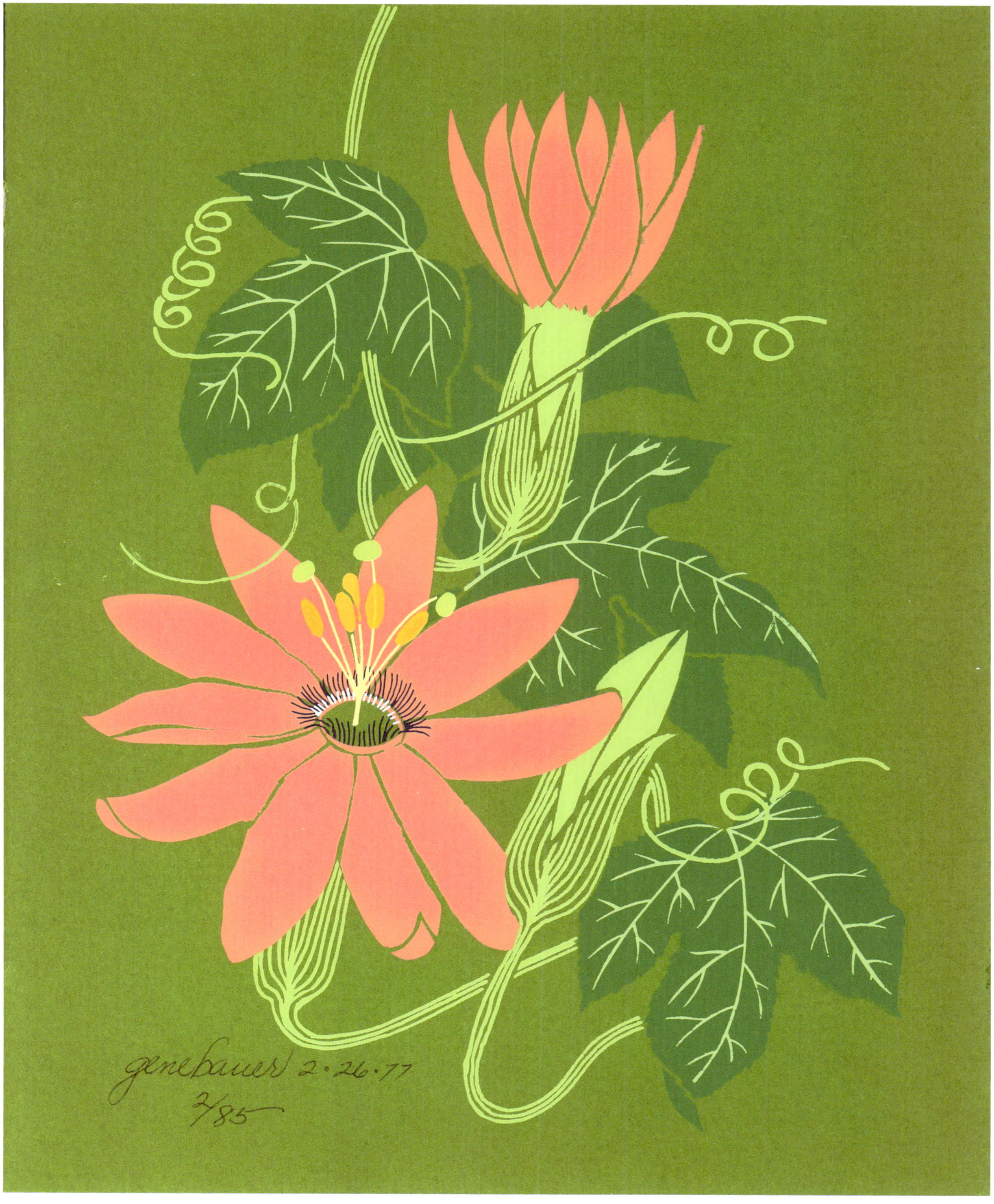
genebauer 2·26·77
2/85

February 1977

The University of California figures very strongly in any study of the arboreta and botanical gardens in this state. To be considered truly an arboretum or botanical garden, the entity must be education-oriented. This does not mean that it must be a part of an educational institution, yet often they are. Nearly all of the University of California campuses are situated in particularly stunning settings, and on several campuses, exotic plant material has been incorporated in the basic landscaping and labeled so that the grounds themselves almost constitute an arboretum. UC Berkeley's botanical gardens began in 1891 on the university's main campus, before being relocated to its present site in Strawberry Canyon, securely nestled in the Berkeley Hills. Thousands of species from all parts of the world, arranged according to their geographical origin, have been grown there from seed collected in the wild. Although much newer, holding its own is the Santa Cruz campus of the UC system. It is one of the most magnificent settings of any university, and it includes both a botanical garden and an arboretum.

The two-thousand acre site, formerly the S. H. Cowell Ranch, is located in the meadow-to-forest transition zone of the Santa Cruz Mountains. After leaving the entrance kiosk, the gradual-incline road approaching the various colleges is one of sublime tranquility. Reminiscent of the Austrian Tyrol, the grass-covered, rolling hills are dotted with contented cattle, grazing under sparkling sunshine.

The serenity of this pastoral setting is only the first treat to be discovered here. The road ascends, winding gracefully through the hills, and coming into view from this lofty height are Monterey Bay and the mighty Pacific Ocean beyond. Upon arriving at the first college buildings, one is struck by the grandeur and splendor of a natural redwood forest. Bordered on the west by coastline, and inland by mountains of the South Coast Ranges, the region itself contains the most southerly natural range of *Sequoia sempervirens* (coast redwood). Imagine, an institution of higher learning set amongst a forest of towering redwood trees! One can feel in a state of grace after witnessing such spectacular vistas; as if aware of that, the sign greeting visitors at the entrance to the garden is unusually gentle, a refreshing change from the negative signs cluttering up public gardens ("No Trespassing," "Please don't pick the flowers," "No food, please," "Don't walk on the Grass"):

> *Welcome to The Garden*
> The Garden is always open. Come to browse or work anytime.
> Flowers should be cut in the early morning.
> Please cut flowers only where you see the symbol 
> We ask that all dogs be leashed, please.
> If you are unsure of anything, kindly inquire.

A student project covering approximately four acres of land, the garden is terraced with clearly marked walks and always abundantly alive with color. I observed students casually strolling through picking not armsful but one or two flowers, treating them with great respect as if each was a priceless treasure. Continuing through the central part of the campus, eventually three geodesic domes appear, marking the site of the arboretum.

Located in the rolling hills–meadow area, the arboretum includes South African and Australian sections. There is a fine collection of *Grevilleas* from Australia, many in bloom that day. Although their flowers are striking, these plants are equally valued for their handsome foliage, some species green while others blue gray. The South African section features a superb collection of *Proteas, Leucospermums,* and *Leucadendrons.* When in bloom, usually in February, these are extremely attractive plants, presenting a dazzling display of fascinating forms and colors. Finally, a large area of the arboretum is devoted to a fine collection of the conifers of the world.

Passiflora jamesonii

Visiting the campus before, in November, I had been attracted to a vine covering a ten-foot-high chain-link fence adjacent to the garden. The fence itself, about two-hundred feet long, was protecting a rather unsightly bit of electrical apparatus, and as if rewarding it for that, the vine's lush green foliage blanketed the fence entirely. Sensational salmon-coral flowers from the vine adorned this embrace. The lustrous vine, still covered with flowers when I visited again four months later, was *Passiflora jamesonii.*

The genus *Passiflora* contains more than four-hundred species, most of which are indigenous to tropical America, particularly Brazil, Colombia, Ecuador, Peru, and Venezuela. The genus was named by Jesuits in the sixteenth century who found a trailing, scrambling plant by the seashore in Brazil (possibly *P. caerulea*) that represented to them a symbol of the crucifixion. Many of the flowers of this genus incorporate the colors blue, purple, white, and green in an intricate pattern, and much lore is connected with them. The species *jamesonii* lacks the showy filaments associated with most passion flowers. This species is particularly spectacular in the Carmel, Monterey, and Santa Cruz areas but is rarely seen south of this region.

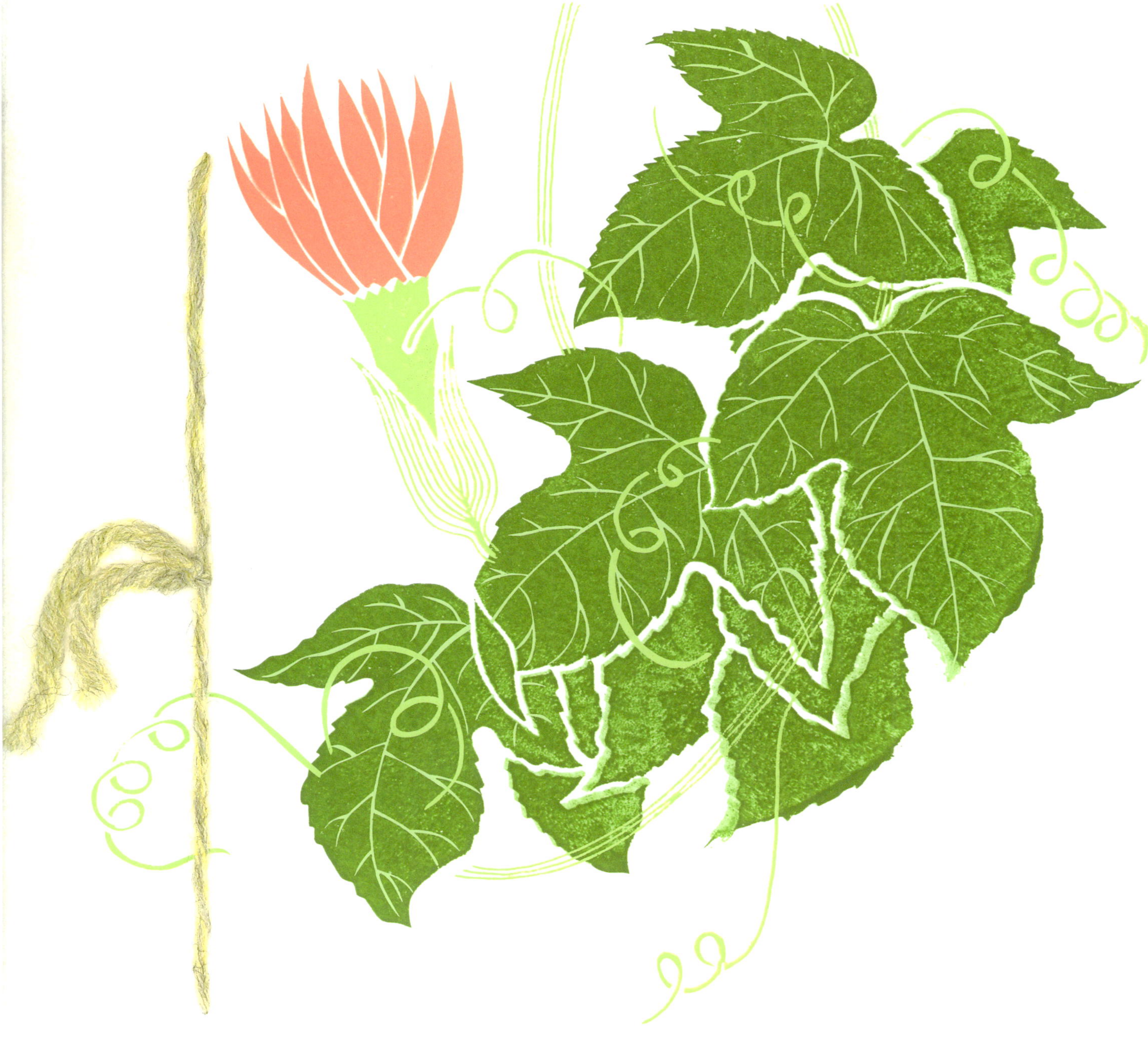

Purpleleaf plum (*Prunus cerasifera* 'Thundercloud')

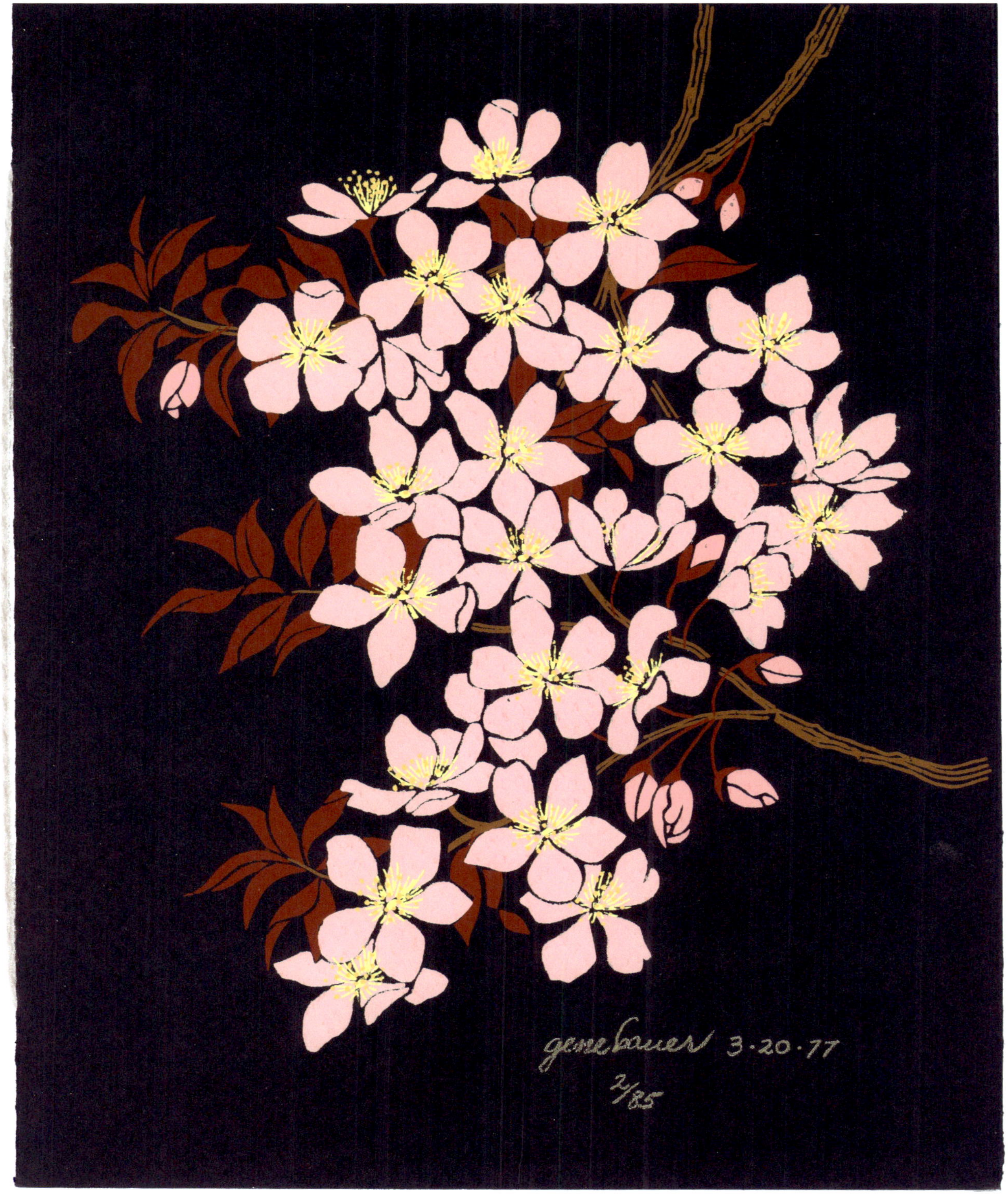
gene bauer 3·20·77
2/85

March 1977

> "I firmly believe, from all I have seen, that this is the chosen spot of all the earth, as far as nature is concerned. The climate is perfect… everything is like a beautiful spring day all the time."

Luther Burbank wrote this in a letter home describing the Santa Rosa Valley, fifty miles north of San Francisco, where he settled and spent his life in research and development toward producing more and better varieties of cultivated plants. Born in Massachusetts in 1849, he came to California in 1875 with the one-hundred-fifty dollars he received from developing the Burbank potato. The Luther Burbank Garden, where I found the specimen depicted here, is located in the heart of the town of Santa Rosa, where Burbank worked for more than fifty years developing original hybrids and improved cultivars. He intended for life of plant breeding to enrich the welfare of mankind. The fact that the official California Arbor Day, March 7, commemorates his birthday may be some indication that he was successful in that.

Among the more than one hundred thousand experiments he conducted, many were focused on dethorning plants with thorns. Here in the garden, large specimens of his spineless cactus, thornless berries, and thornless rose quietly testify to his success in that, too. His efforts produced some very unusual trees. I saw his original cross between a plum and an apricot (resulting in a plumcot) here, as well as one of his best-known and tastiest introductions, the Santa Rosa plum. Many of his experiments involved plums, prunes, cherries, and apricots—both fruiting and flowering varieties—one of which I chose for this month.

Prunus cerasifera 'Thundercloud'

"On and on we go; one step backward, sometimes; then two steps forward; marking time awhile; then onward with a spurt." To me, Burbank may just as well have been talking about springtime, nature's repeatable seasonal experiment, as the reawakening time it is for us all thanks to the flowering fruit trees. What a joyous sight when the buds burst open on flowering cherries, nectarines, peaches, apricots, and plums. And their variety reminds us that nothing stays the same: colors range from white through all shades of pink to salmon coral to warm reds, while flower form ranges from single to semidouble to full double.

Burbank hybridized the ornamental purpleleaf plum (represented here) in 1919. Its flower form is single, its color a very pale pink, and its fragrance is mild. The deep, coppery-red leaves unfurl at the same time the flowers open, contributing immensely to the overall aesthetic value of this tree. The name fits it well. Like silence broken with a wild clap of thunder from the sky, this tree's rich, dark leaves contrast strikingly with its pale pink cloud of soft, billowy flowers.

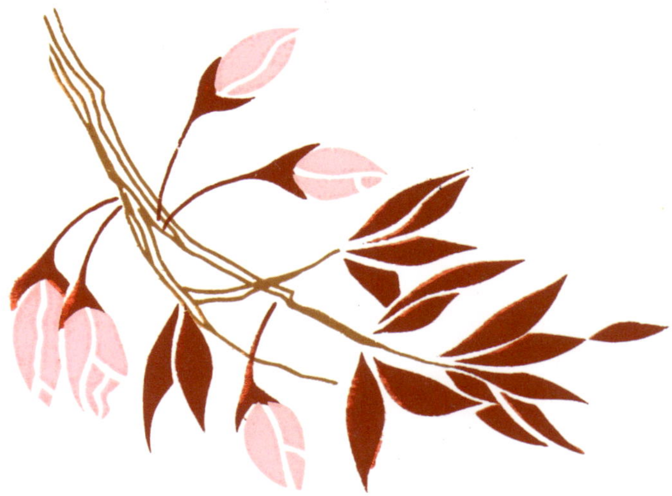

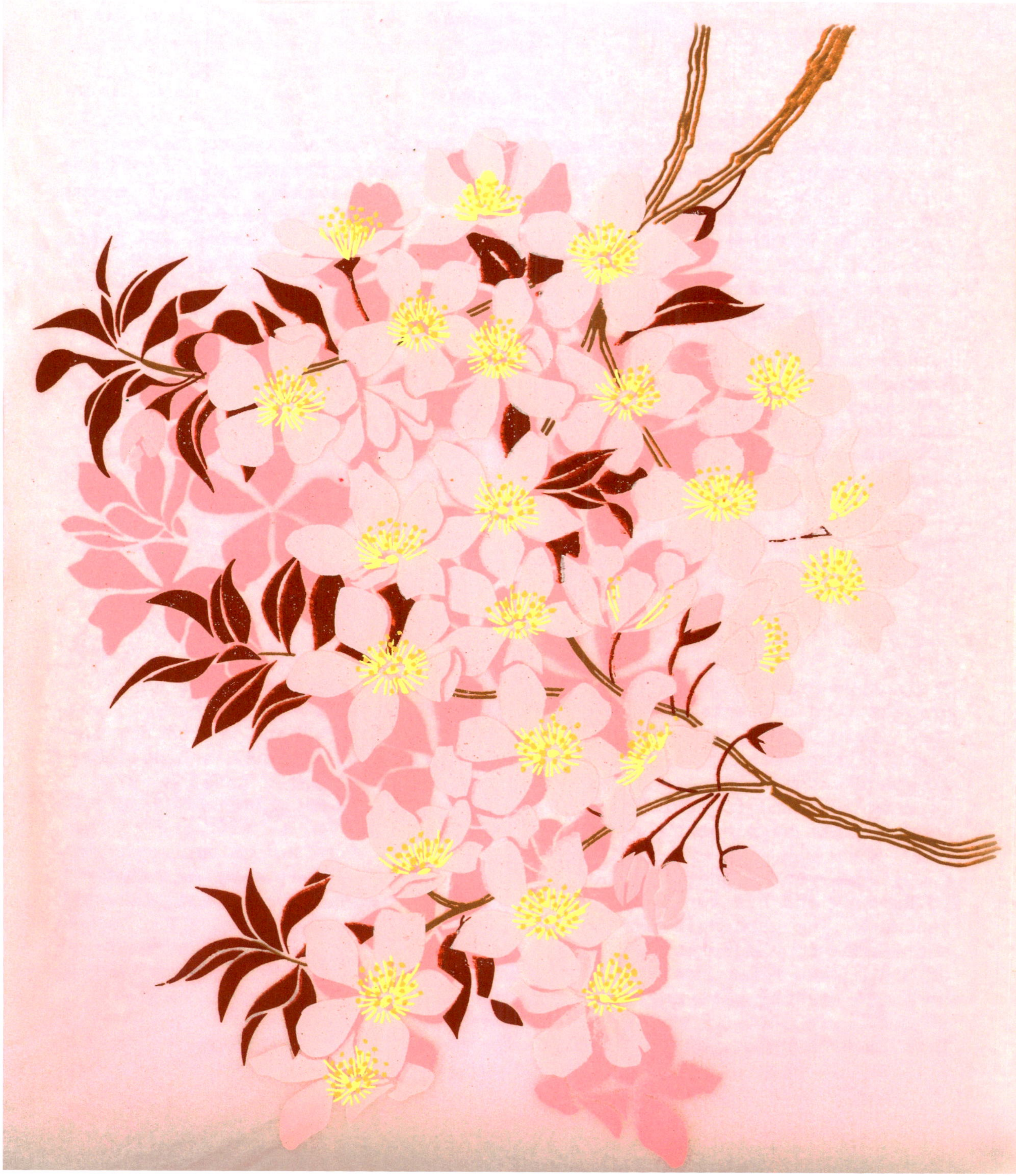

Chinese wisteria *(Wisteria sinensis)*

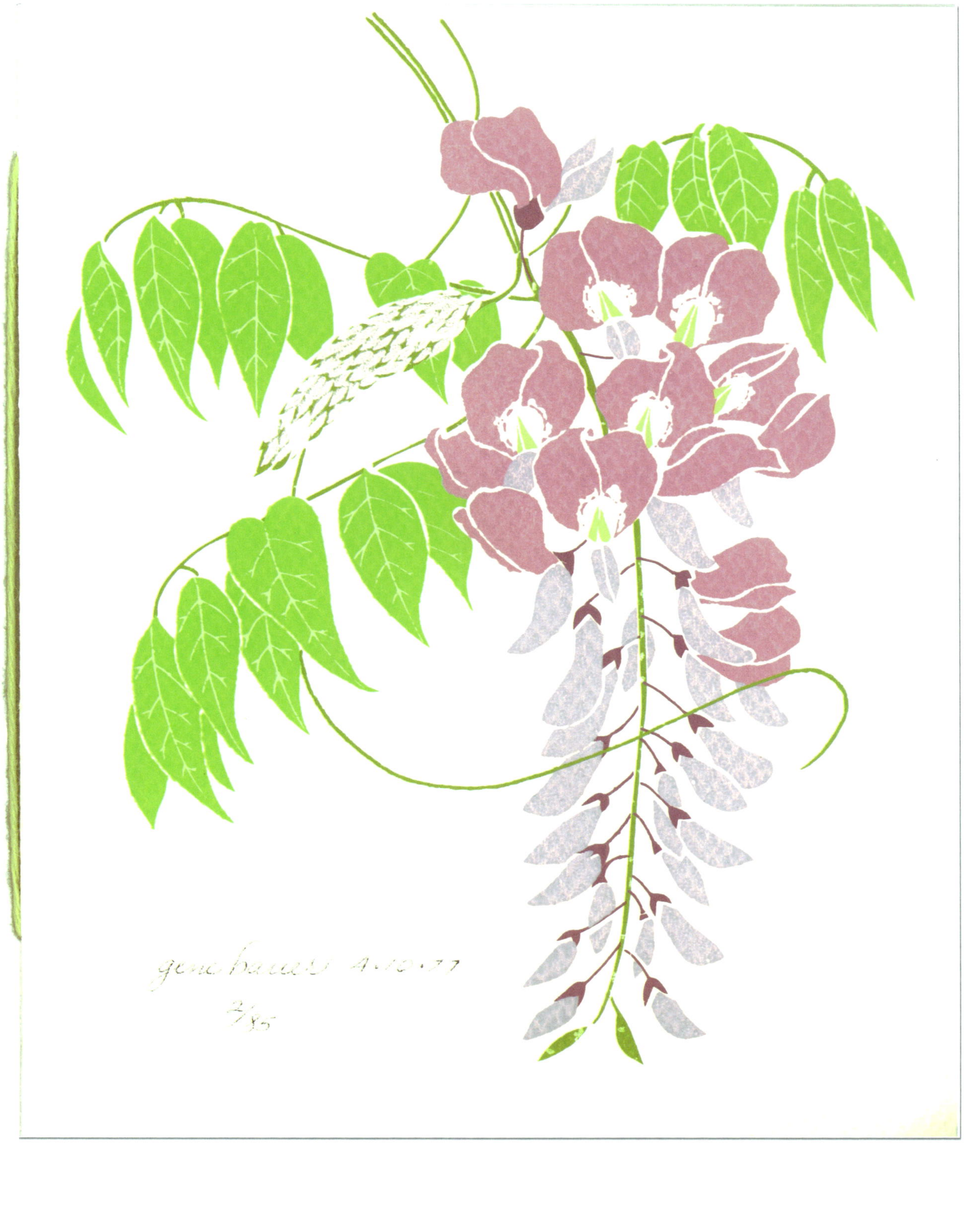

April 1977

Fight, love, and *life*—three strong words of the English language—were combined to form the name Filoli, the gardens growing the plant you see here. Filoli is located in a setting of unsurpassed Northern California beauty in Woodside, in the Crystal Springs Watershed area of San Mateo County. With spectacular views of the South Coast Range Mountains, Filoli was built here in 1915 for William B. Bourn because the site reminded him of the Irish countryside. With seventeen acres of ornamental gardens, Filoli is a masterful blending of the finest of two worlds, the natural or indigenous and the exotic or introduced. It is this way because the original designer Bruce Porter was a painter. He greatly admired the California landscape and was intent on planning the more formal gardens to be in harmony with the natural beauty that already existed here. This is exactly the way I feel gardens should be, and so it is not at all surprising that I fell in love with Filoli the moment I saw it.

Great care has been taken to retain the wild beauty here, in the way its natural quality flows into the cultivated areas, so gently, so gradually. Yet there is unexpected excitement beyond every bend in the path. These surprises come because the gardens are divided or segmented by hedges, so that one is repeatedly delighted by the newly discovered treasures just around the corner. Even among such magnificent flowering trees, Irish yews, holly trees, Italian stone pines, rhododendrons, azaleas, and camellias, in the month of April the most lavish display of color is provided by the plant I chose to draw.

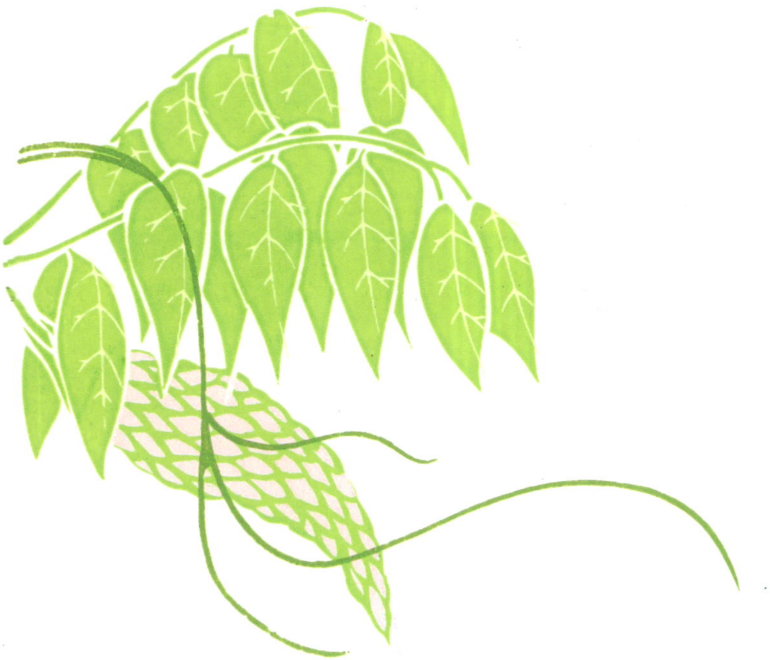

Wisteria sinensis

Few plants are as versatile as the wisterias. Vigorous, woody twiners, they may be trained as a vine, shrub, or small weeping-type tree. Their most popular use is as a vine to cover porches, trellises, and buildings. The nine species in the genus *Wisteria* are native to the eastern United States and eastern Asia. The species most often planted in California—and so spectacular this day at Filoli—was *Wisteria sinensis.* The generic name honors Caspar Wistar, a professor of anatomy at the University of Pennsylvania. Its species name means "Chinese" or "of China." It is commonly known as Chinese wisteria.

The many hybrids and named varieties within this genus encompass the color range from white through lavender and pink to red-violet and blue violet. The pea-like flowers, hanging in long, pendulous racemes, are borne in great profusion. The length of the flowering raceme ranges from six to twelve inches on *W. sinensis* to an astounding sixty inches on the *W. floribunda* variety *macrobotrys* growing under ideal conditions in Japan. The vines grow to great size and are long-lived, and when in flower this wisteria's exceptional beauty is matched only by its fragrance, sweet and spicy.

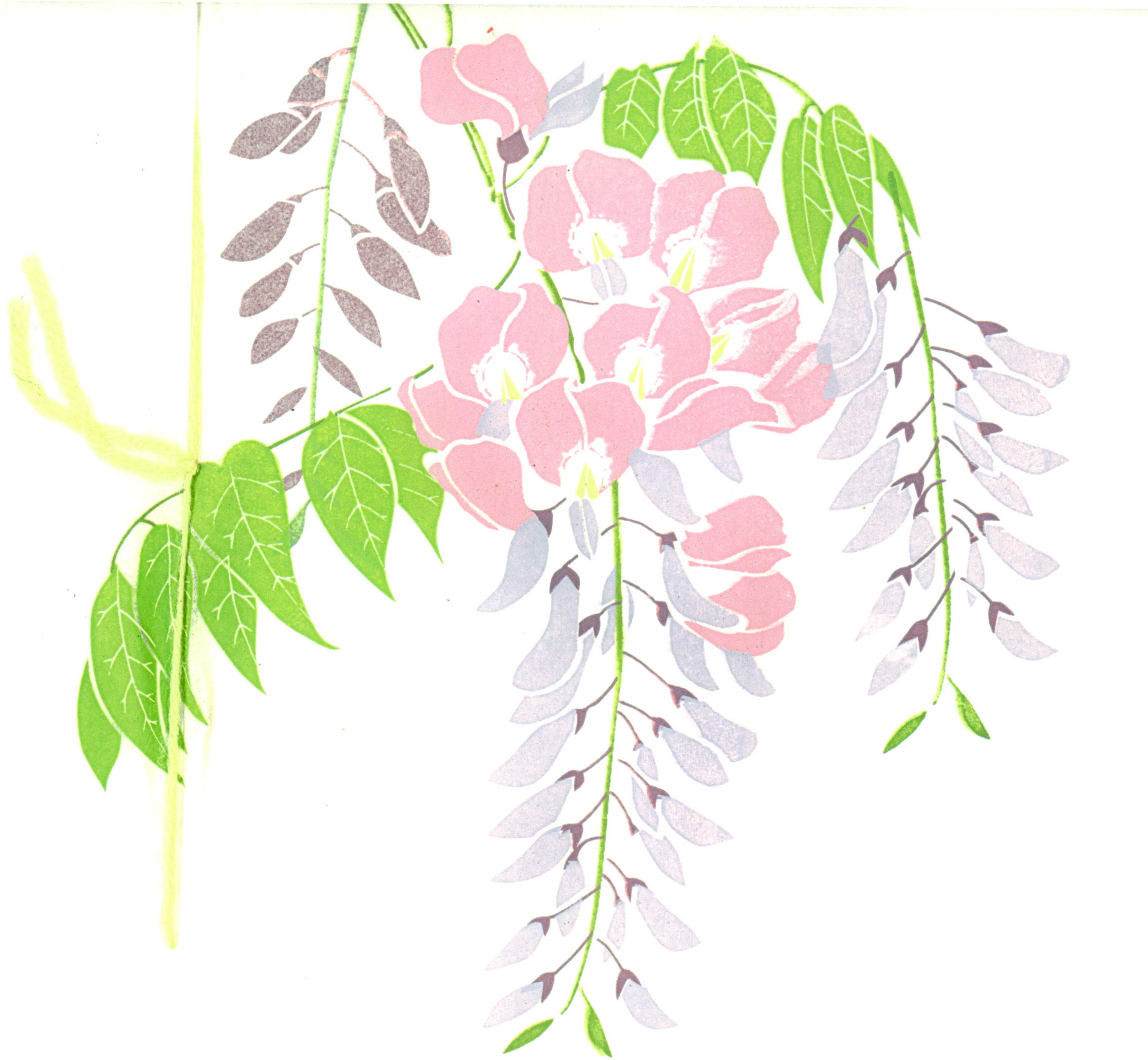

Flannel bush (*Fremontodendron californica* 'California Glory')

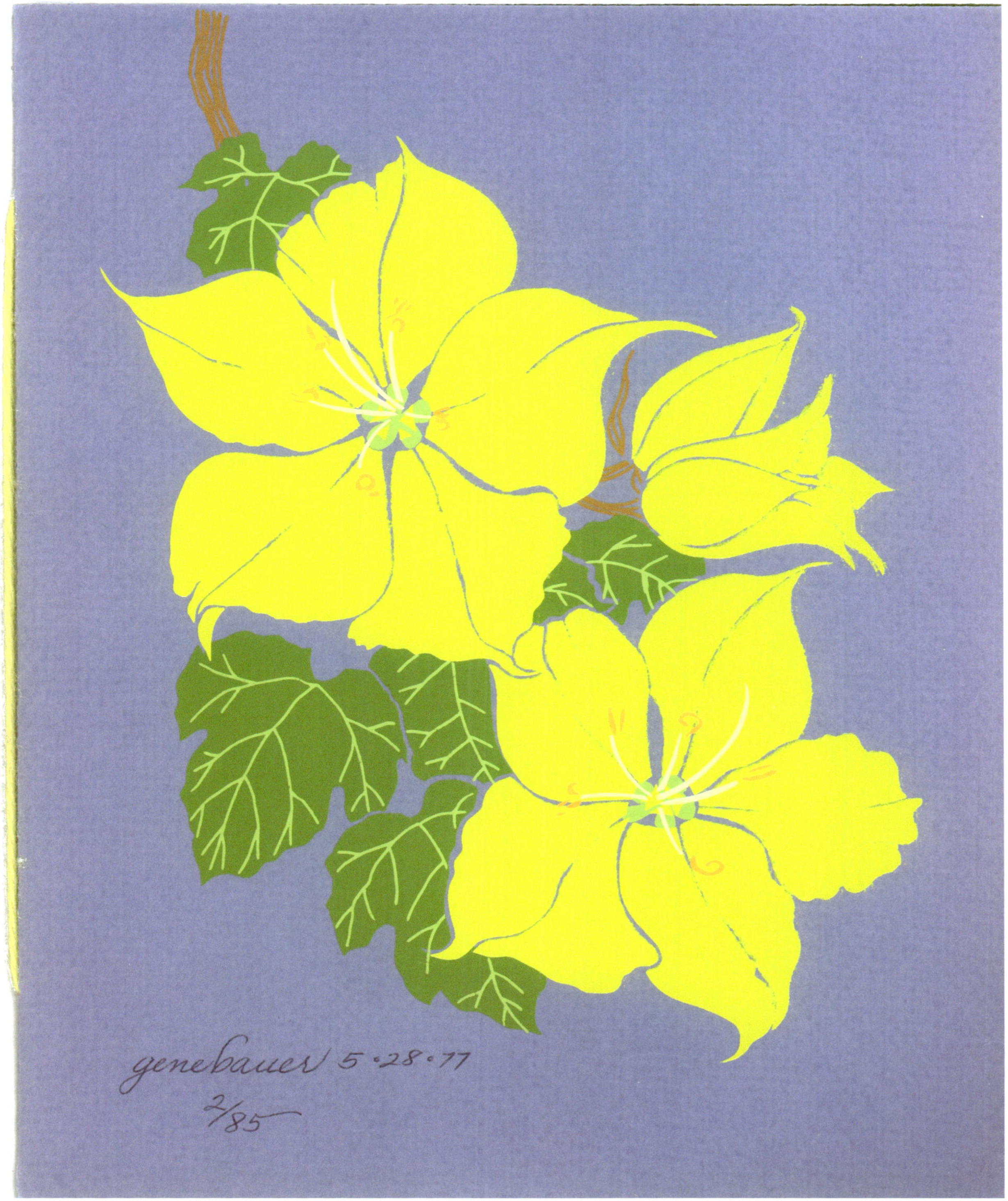
genebauer 5·28·77
2/85

May 1977

How full can the human heart become? If size limits capacity, then continually falling in love with botanic gardens in this state is perhaps not such a good idea for me. The Santa Barbara Botanic Garden, for example, has already taken a piece of my heart, and I never saw it coming. This garden was established in 1926 for the study and preservation of flora indigenous to the state of California. Because I was already familiar with the garden, I thought just a short visit to refresh my memory would suffice. But I was wrong. Till now, none of my previous visits were during the month of April. To see the native wonders of this large state all together in springtime turned out to be an incredible opportunity. Arriving in the early afternoon I remained until the sun set in the west, then made arrangements to stay overnight in town so I could return early the next morning.

Occupying approximately sixty-five acres, the garden is ideally situated in a natural canyon that includes natural woodland and chaparral vegetation at various elevations and exposures. The garden is divided into ten sections, each representing very distinct environments found in California. These sections have been given the following names: Arroyo, Canyon, *Ceanothus,* Chaparral, Desert, Forest, Island, Meadow, Redwood, and Woodland. Yet the entire garden is medium-size relative to other gardens—in other words, I had the chance to experience an exclusively California-native-flora spring. Irresistible.

So I was back in the morning to witness the sparkling, new-day freshness of the most enchanting plants of this state. Scattered through the garden and in full bloom that day were many handsome specimens of the plant I chose for this month. But two especially superb ones were espaliered on either side of the entrance to the library, and I decided to make my original drawings from them.

Fremontodendron californica 'California Glory'

Fremontodendron are almost always included in lists of the ten most outstanding shrubs native to California—and in the top-ten lists for North America just as often. In the genus *Fremontodendron,* formerly known as *Fremontia,* are but three species and all are indigenous to California. The generic name honors General John C. Fremont, an early western explorer and army officer whose name is deeply woven into the early history of this state.

Fremontodendron 'California Glory' is a hybrid between the two species, *F. californicum* and *F. napensis.* A good-looking evergreen shrub or tree, it grows to twenty feet tall. It is particularly beautiful in the spring—usually March until June—when it is covered with three-inch flowers of blazing gold so profuse that they almost hide the foliage.

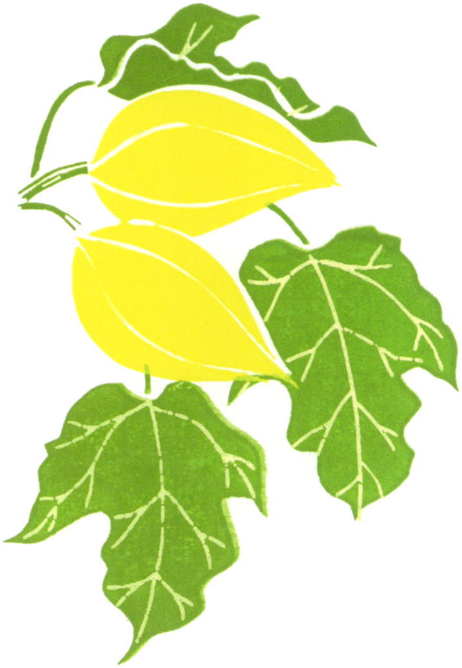

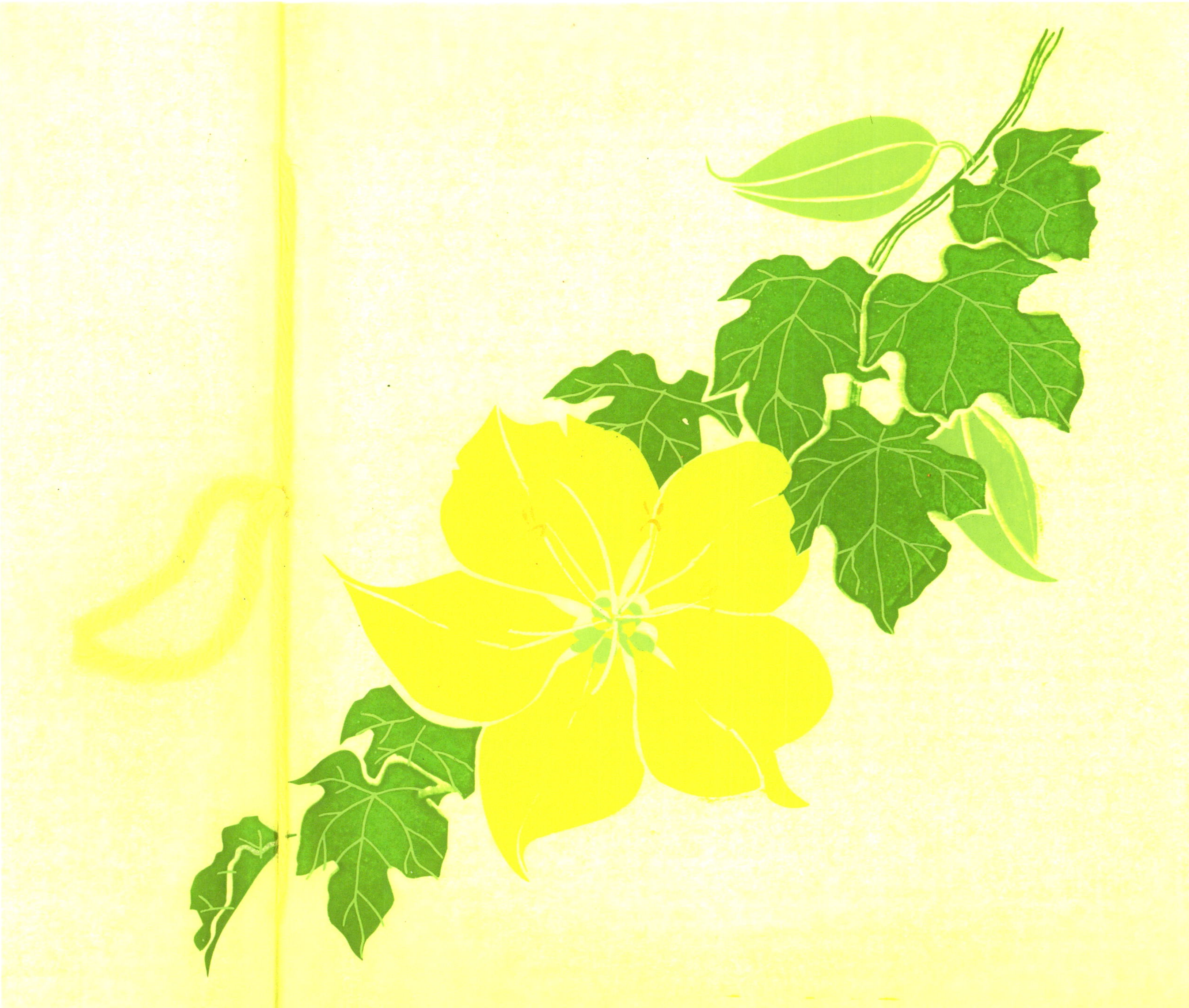

Sweet orange tree *(Citrus sinensis)*

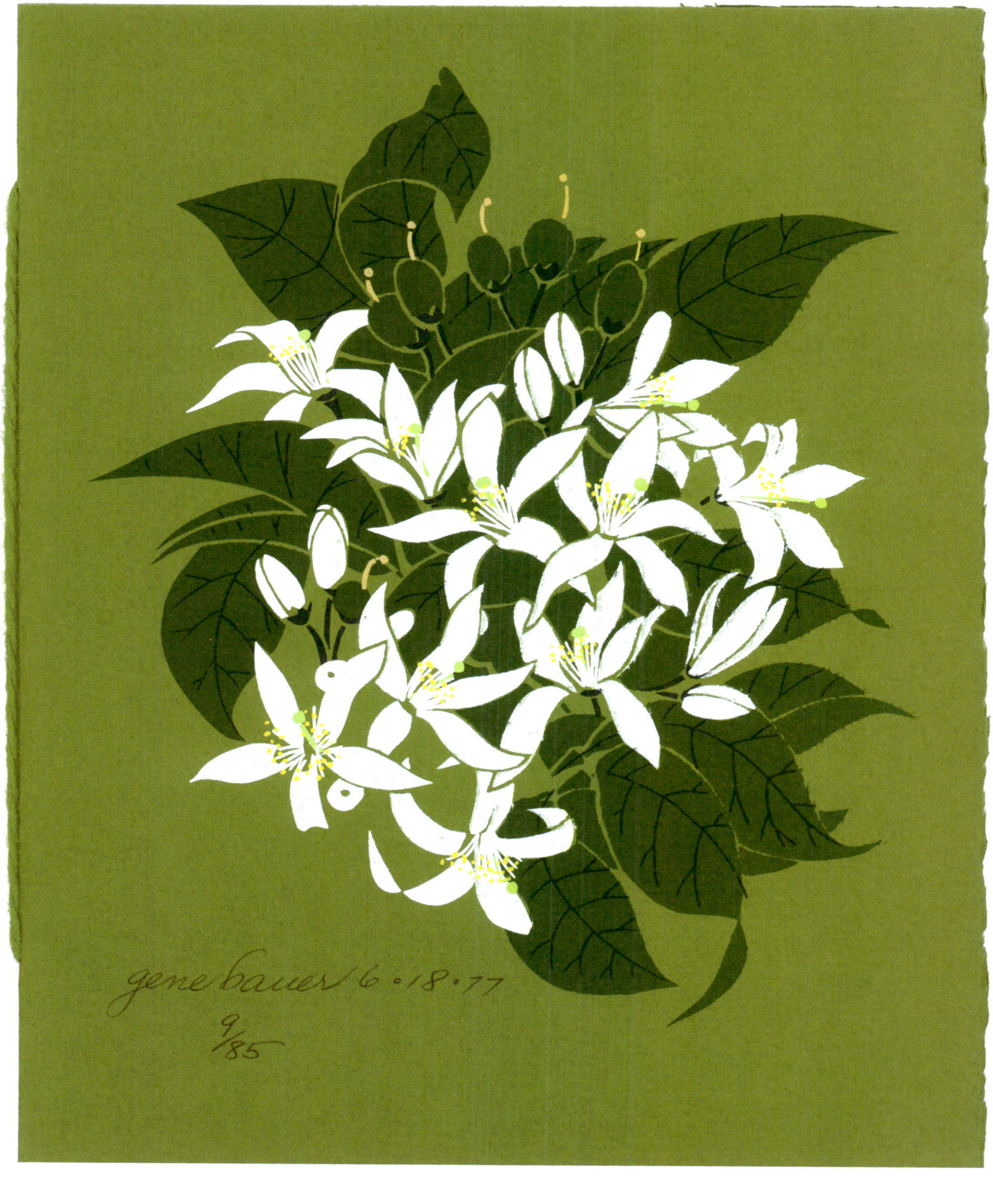
gene bauer 6・18・77
9/85

June 1977

Citrus sinensis—the sweet orange tree—originally came from China. Many people who grew up with the fragrance of orange blossoms—virtually everywhere in Southern California during much of the twentieth century—are surprised to hear that it is not a native plant. However, it is among the earliest to be introduced, and in a way we have the Franciscan missionary, Father Junipero Serra, to thank for our familiarity with these sun-kissed beauties. The oldest gardens in California—those originally established at the Franciscan missions and subsequently restored—cannot be considered true botanic gardens, but because of their tremendous influence on the agriculture of the state and therefore its history, they are noteworthy. Many plants were introduced to California to provide food—avocados, grapes, figs, olives, walnuts, pomegranates, and the most important among them, the genus *Citrus*.

Wending its way from the Mexico/California border in the south northward just past San Francisco to Sonoma is El Camino Real, along which Father Serra and the mission padres who came after him established twenty-one Franciscan missions, each within a day's journey from one to the next. When first named in honor of the king of Spain, El Camino Real (The King's Highway) was a trail covering about 650 miles. Now known as Highway 101, it still skirts the Pacific Ocean in many areas and now and then turns inland through some of the most picturesque parts of the state. The first mission was founded in San Diego on July 16, 1769, marking the beginning of European settlement in California. As have the buildings themselves, some of the original mission gardens have been restored with the intention that they might appear today as they did two hundred years ago. In a garden of fragrances at Mission San Luis Obispo de Tolosa, atop a little hill in the midst of the busy present-day town, I found many plants associated with early California, including the one here.

Citrus sinensis

The genus *Citrus* includes about twelve species and many, many hybrids. Among the species are *C. limon* (lemon), *C. paradisi* (grapefruit), *C. reticulata* (tangerine), *C. medica* (citron), *C. aurantifolia* (lime), and *C. sinensis* (meaning "of China"). Groves of the latter, the sweet orange, were established in southern Europe, and from there Christopher Columbus brought seeds to the New World on his second voyage. These were grown in what is now Florida, Central America, and Mexico. Father Serra obtained seeds, no doubt from Mexico, planted them, and from this humble beginning in a mission garden eventually emerged a highly profitable crop industry in California. Until 1942 California was the leading producer of oranges. That honor now goes to Florida, which has chosen the orange blossom as its state flower.

With glossy, evergreen leaves, exquisitely fragrant flowers that are waxy and white, and sweet, juicy fruit (technically, a berry), the trees are a joy to behold. Many varieties, especially the Valencia, are resplendent in brilliant orange, bright white, and rich green in the same moment: this is because the tree is producing a crop of ripening fruit, blooms for the next season's crop, and small green fruits—all at one time! A tree exhibiting all these various phases of fruit development at once is truly a wondrous sight, but unfortunately, few take the time to notice then look and really see.

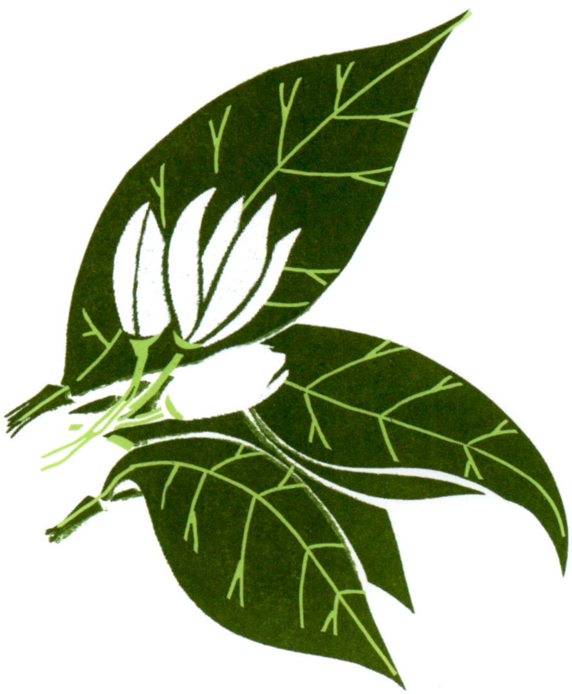

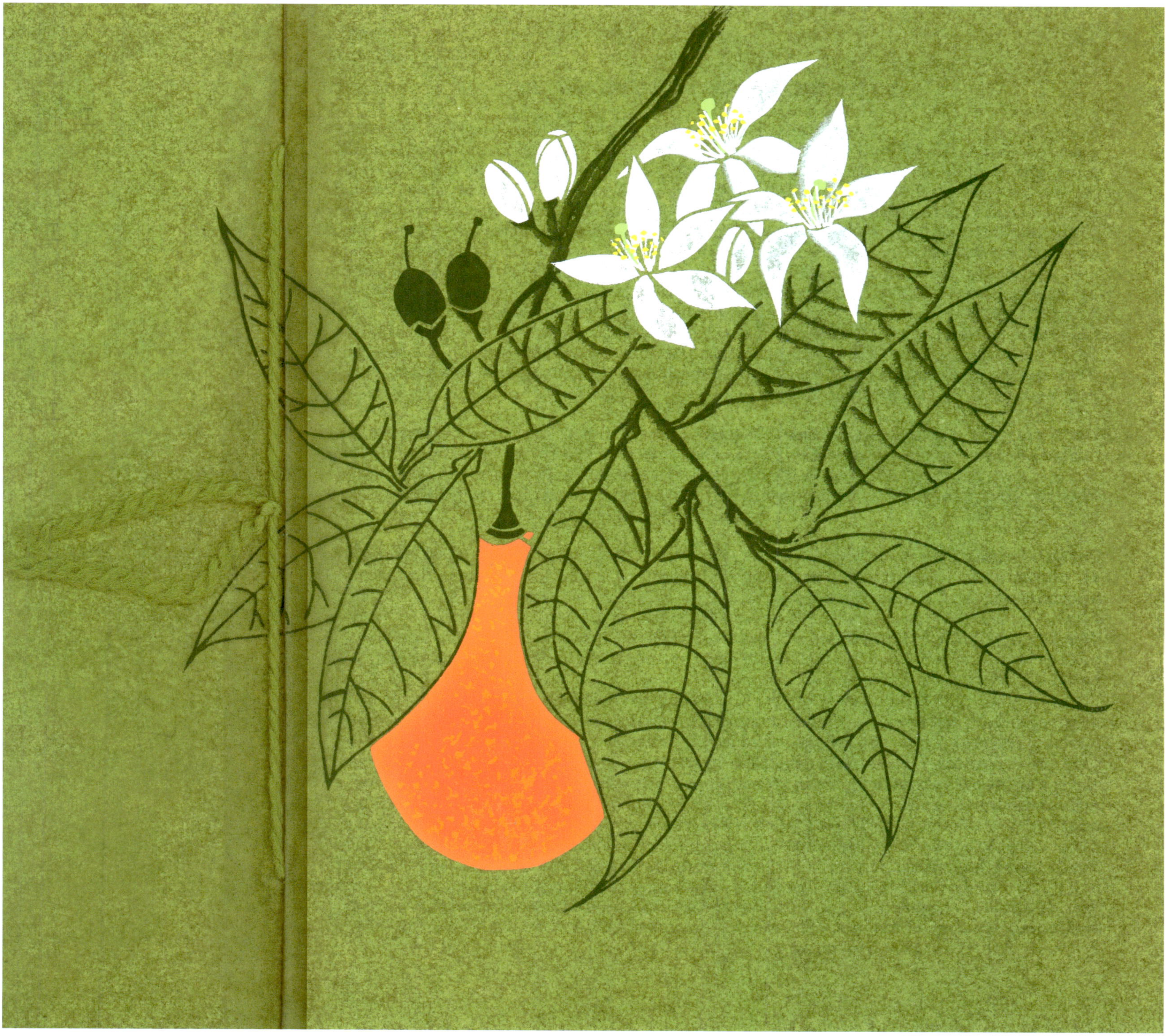

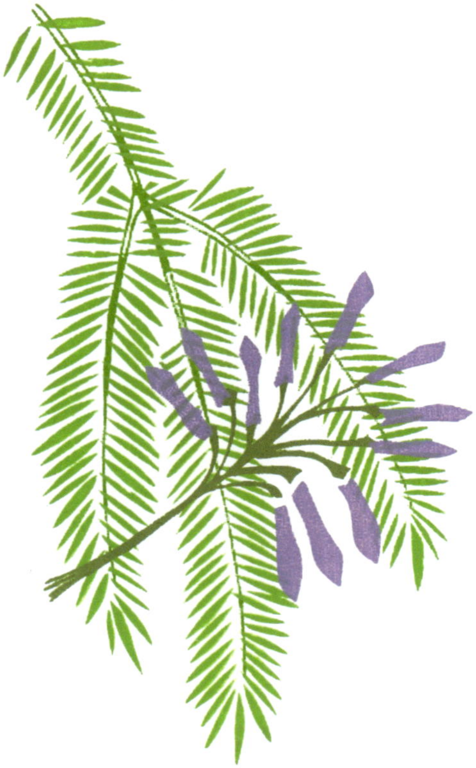

Jacaranda tree *(Jacaranda acutifolia)*

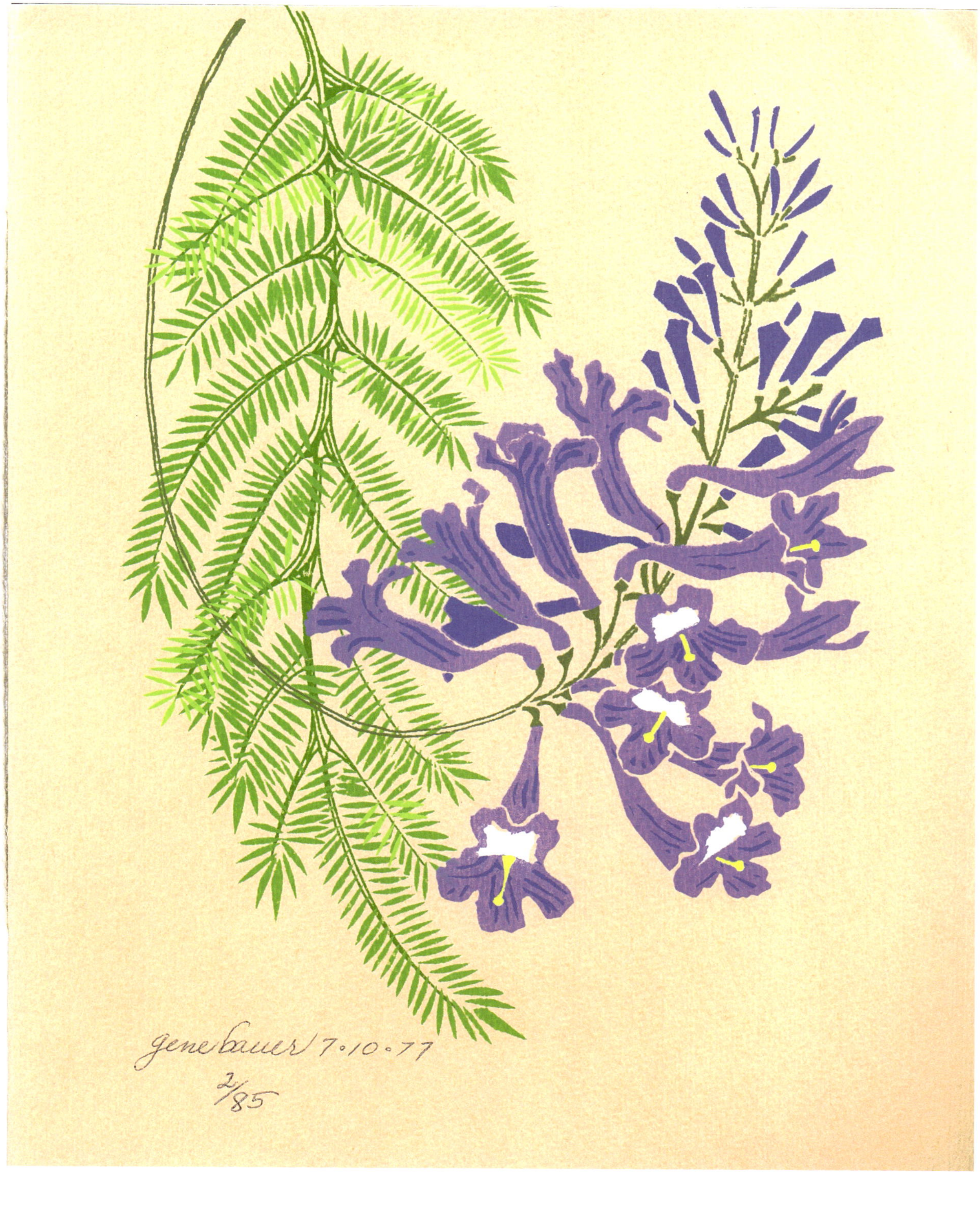
gene bauer 7.10.77
2/85

Early July 1977

I visited Mission San Fernando Rey de España on an extremely warm day in late June, and how pleasant it was to step inside the thick-walled adobe buildings where it was many degrees cooler, and incidentally, without benefit of electrical air conditioning (adobe is good that way—cool in summer, warm in winter). The inner courtyard was particularly lovely and serene with a stream of water flowing through it under the cool shade provided by large trees.

Founded on September 8, 1797, this mission was seventeenth in the chain established by the Franciscan padres, and dedicated to a Spanish king and saint, Ferdinand III, who reigned during the early thirteenth century and was canonized in 1671. In close proximity to the little pueblo of Los Angeles, the mission became a favorite stopping place for travelers along El Camino Real, and consequently, one of the largest of all the missions. It was also among the wealthiest, with cattle raising as its dominant industry. Located in a fertile and beautiful valley with an abundant water supply, the padres were able to irrigate orchards and gardens, growing some of the finest crops in Alta (high) California (at the time, still part of Mexico; south of the border, Baja California remains part of Mexico).

Grapes became one of the most successful crops. Vines of wine grapes *(Vitis vinifera)* from Spain via Baja California were planted here the year it was founded. By 1832, thirty-five years later, there were 32,000 vines bearing large, bluish-black grapes that produced a red wine similar to claret. The wine was used with meals, as medicine, for mass, and in exchange for provisions.

Now the area is known as the San Fernando Valley, located in the northwestern portion of Los Angeles county, extending from the north face of the Santa Monica Mountains. Neither a desert nor a coastal region, it is an inland valley, very warm and still very densely populated. On the day I visited, many blooming flowers populated the area in front of the mission and the courtyard. In addition to oleanders, palms, pines, olives, figs, pomegranates, bananas, and oranges, there were six huge and handsome Moreton Bay fig *(Ficus macrophylla)* trees in the Mission San Fernando garden. Most refreshing on such a hot day, however, was the tree I chose for the cover of this issue.

Jacaranda acutifolia

Jacaranda trees compose a most familiar sight in the southern part of this state, dotting the skyline of almost any town, but especially those more inland from the immediate coast. They appear as ethereal, wispy, billowy, pale lavender-blue clouds, and to encounter them is an especially soothing experience. Perhaps the best known and most admired of all the flowering trees in California, it inspires much curiosity from visitors wanting to take one home. Unfortunately, they must be told that the jacaranda is for frost-free areas only: it can be very seriously damaged at temperatures of 27 degrees and below.

The genus *Jacaranda* includes about fifty species, all native to Central and South America and the West Indies. *Jacaranda acutifolia* is native to Brazil, and the word "Jacaranda" is a Brazilian name. *Acutifolia* means "acutely-leafed" or "sharp-leafed," but the leaves are not really sharp or stiff as the specific name might indicate. They are finely cut, fern-like, and exceedingly graceful. Each leaf may be from ten inches to two feet long and is made up of many, tiny (quarter-inch) leaflets. With the leaves dropping in midwinter, the trees are sometimes referred to as semievergreen.

Jacaranda trees may grow to be fifty feet in height. Open and rather lacy in appearance, when in full bloom they are a breathtaking sight. The tubular-shaped flowers, each about two inches long, are borne in large, loose clusters and often fall to the ground in masses forming a lavender-blue carpet beneath the tree. Flowering is a bit erratic with some blooms appearing as early as April, but the heaviest bloom is during the summer months, and some even appear as late as Thanksgiving (late November). There is also a form with white flowers that is sometimes seen in special plantings. Unusual but very decorative are the woody disk-shaped seedpods that follow the blooms.

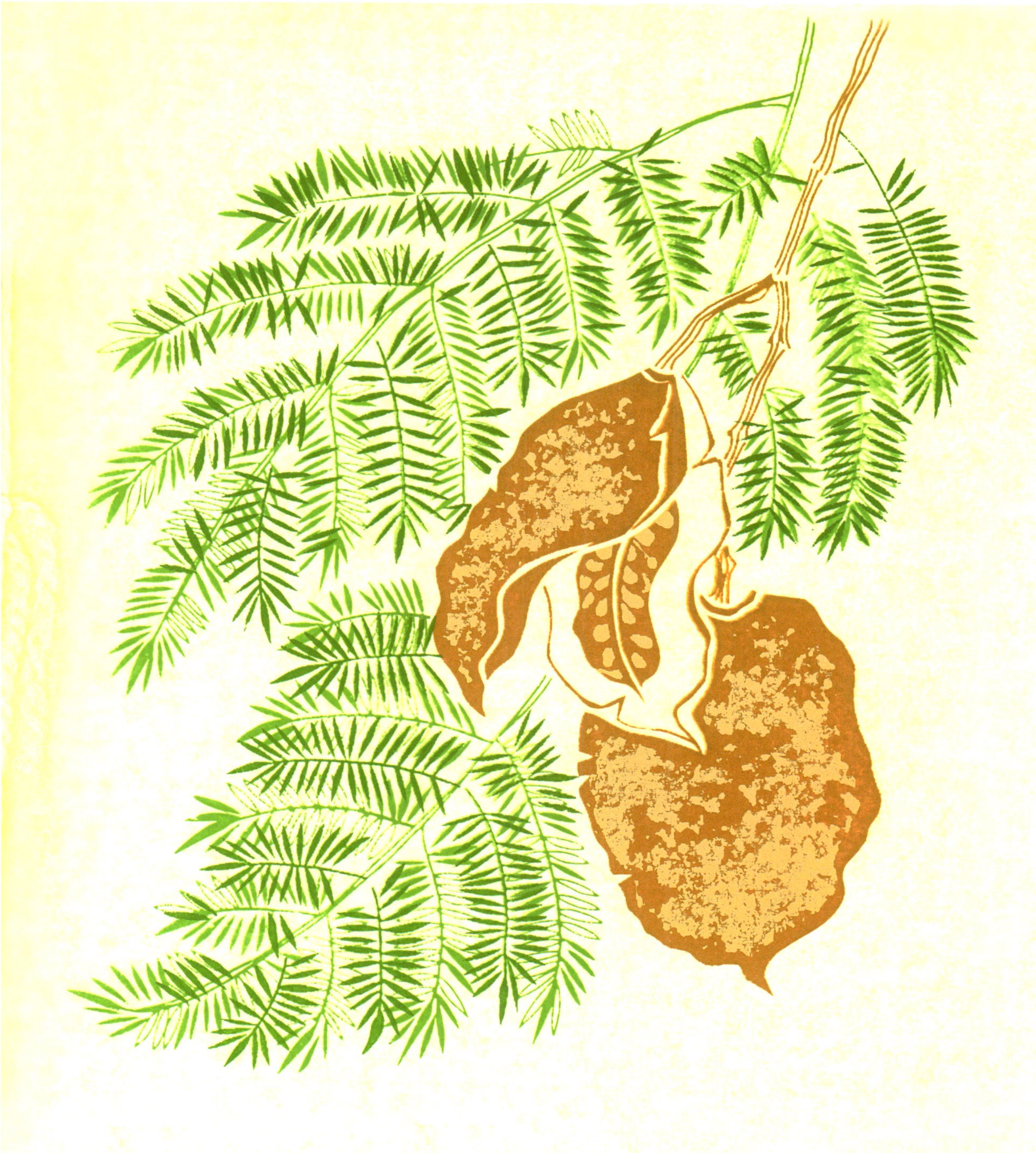

Peace rose (*Rosa* [Hybrid Tea] 'Peace')

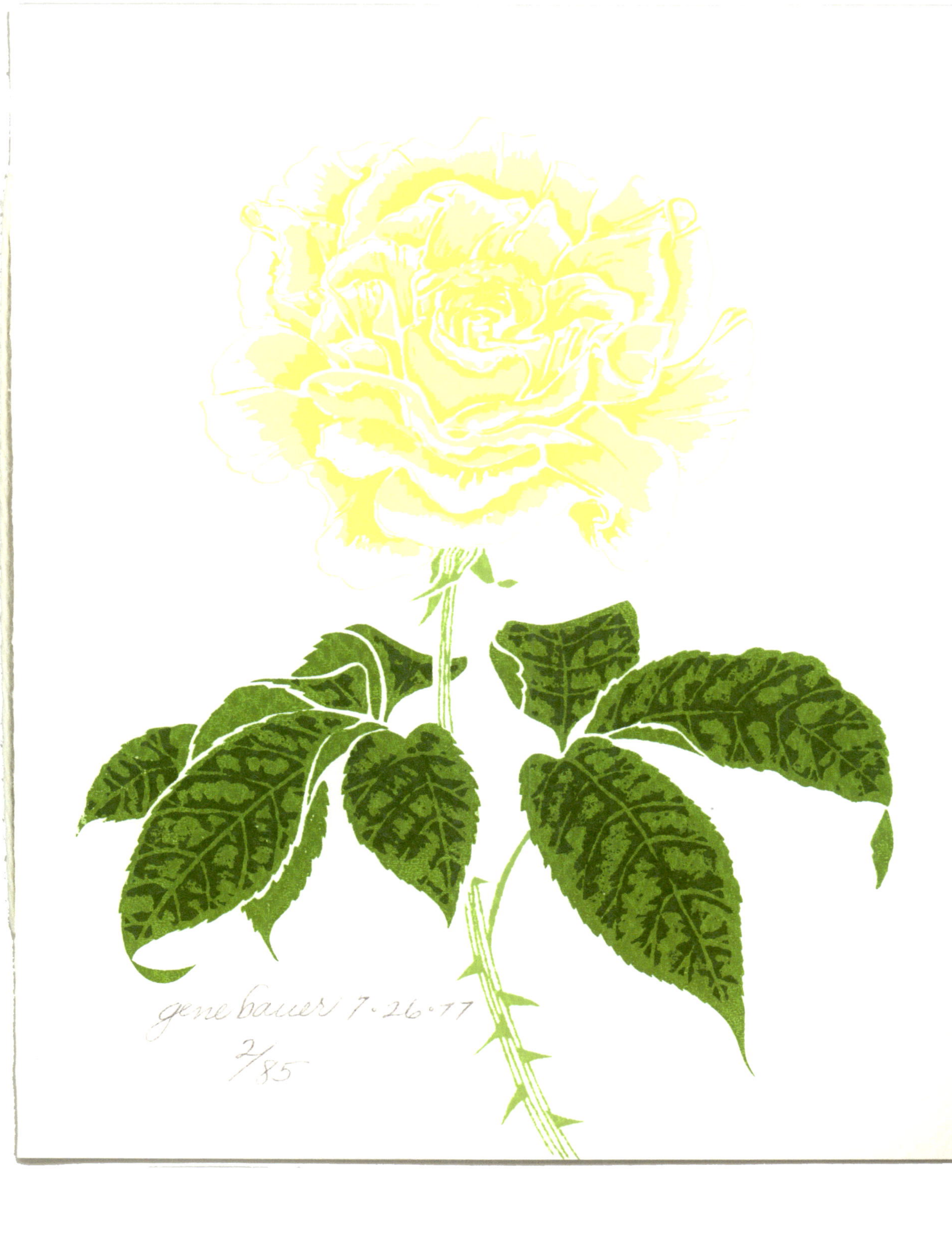
gene bauer 7·26·77
2/85

Late July 1977

Among the most favored flowers of course is the rose. One of the flowers in cultivation the longest, the rose has inspired poets, artists, and musicians and has long been symbolic of youth, beauty, and love. What might be called characteristics in other flowers become facets of personality in the rose: the satiny texture of its petals; its delicate then sometimes dazzling color; its simple then sometimes complex (but always elegant) form; its strong, sweet, then sometimes subtle fragrance—all these things endear it to all who behold the rose.

From the Latin, *Rosa* is the generic name. This genus is a large one, including more than one hundred species found in the temperate parts of the northern hemisphere, especially Asia, southern Europe, and North America. Many California cities take pride in offering a public rose garden. Plants of this genus are usually referred to as prickly shrubs, although some roses are thornless. The shrubs are often of upright, sturdy growth, but sometimes they are climbing or trailing. The distinctive leaves appear in clusters of one, three, five, seven, or nine leaflets, but the most attractive part of the plant of course is the flower, which is usually fragrant in varying degrees. Most of the true species are single and show little resemblance to what is known as "modern roses." They ordinarily bear five petals in a single layer with stamens and pollen quite exposed, no doubt to increase their chances of reproduction in the wild.

No flower has received more attention over the years in terms of hybridization. The rose is now available in an almost endless variety of colors, sizes, shapes, and fragrances. Roses are classed according to their form and growth habits into basic types: Hybrid Tea, floribunda, grandiflora, polyantha, miniature, shrub, and rambler.

Rosa (Hybrid Tea) 'Peace'

With hundreds of roses in an almost limitless color range from which to choose, I selected 'Peace'. Many experts consider this rose to be the best ever developed. On a scale of one to ten it is ranked 9.6 by the American Rose Society. Not only does it have flawless form, exquisite coloring, mild fragrance, and strong, vigorous growth, but its name and the circumstances concerning it are especially interesting.

In 1939 in Lyon, France, Francis Meilland found a particularly sturdy plant with a lovely pale yellow flower among the seedlings he had grown, and realizing it was quite special, he named it after his mother, Mme. Antoine Meilland. He sent cuttings of this rose to Italy, Germany, and the United States. Meilland knew nothing of what happened to his rose during the years of World War II that followed. Without his knowledge, the rose was named 'Gioia' in Italy and 'Gloria Dei' in Germany. Cuttings of it were flown out of France aboard the last American plane before the Nazi invasion. Delivered to a Pennsylvania rose grower, the flower was intensively propagated here during the war years, and at the Peace Conference in San Francisco in 1945 these roses were placed on the table of every delegate. To commemorate the war's end, this rose was named 'Peace' and has been known by this name since that day.

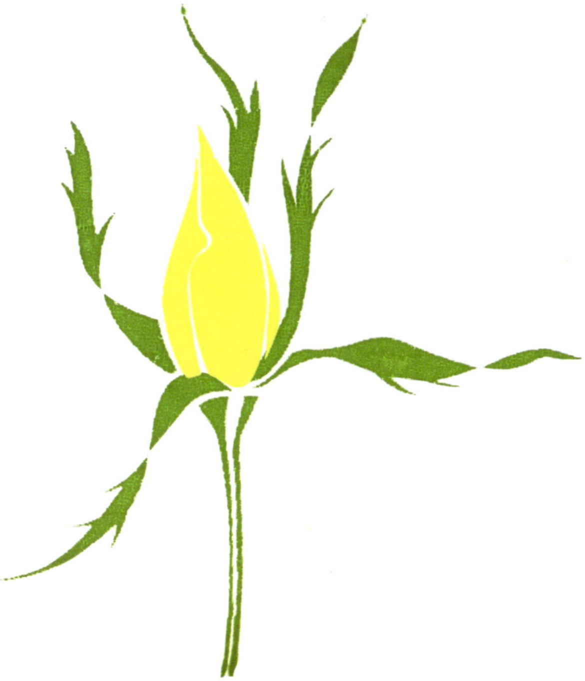

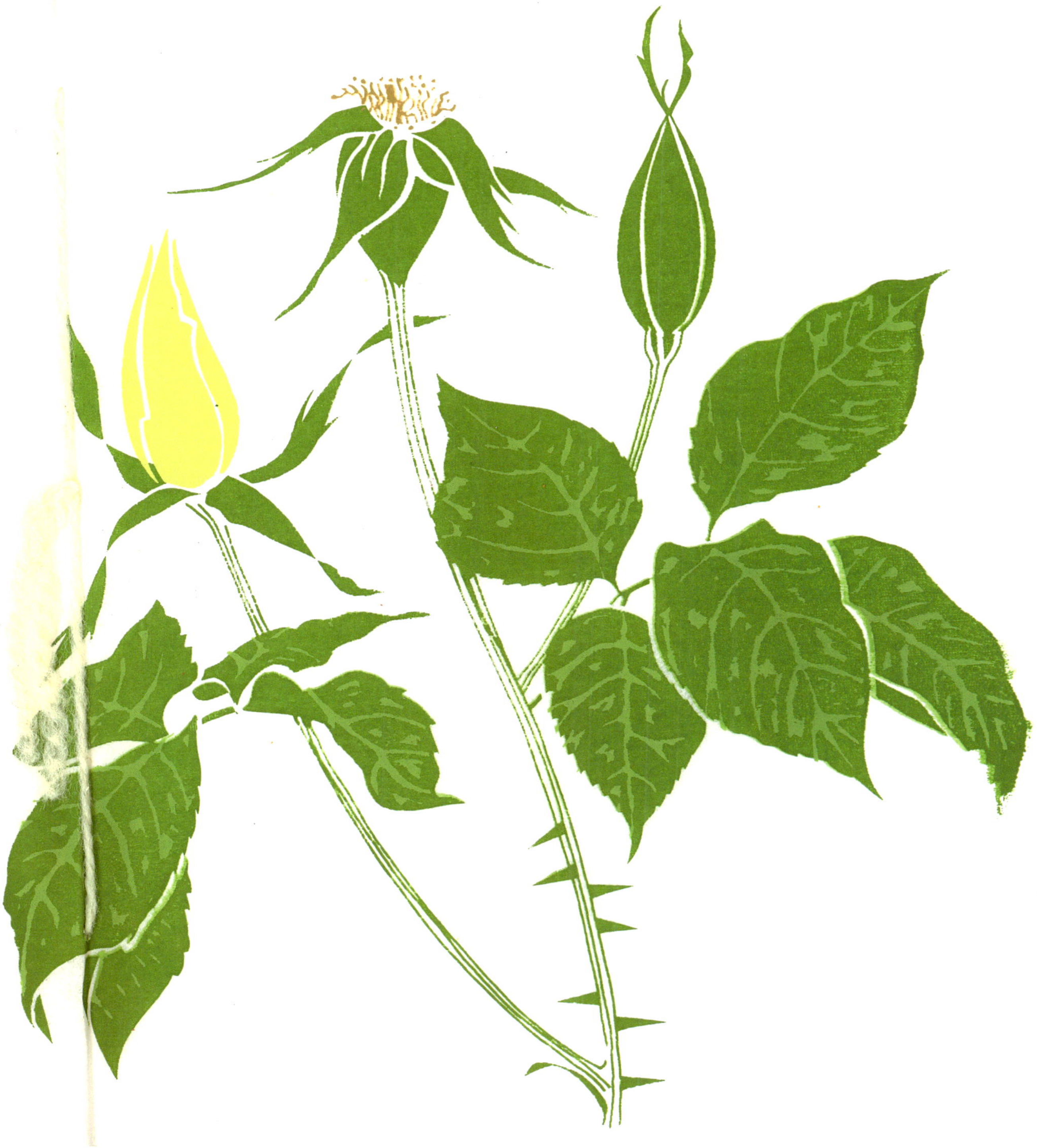

Oleander *(Nerium oleander)*

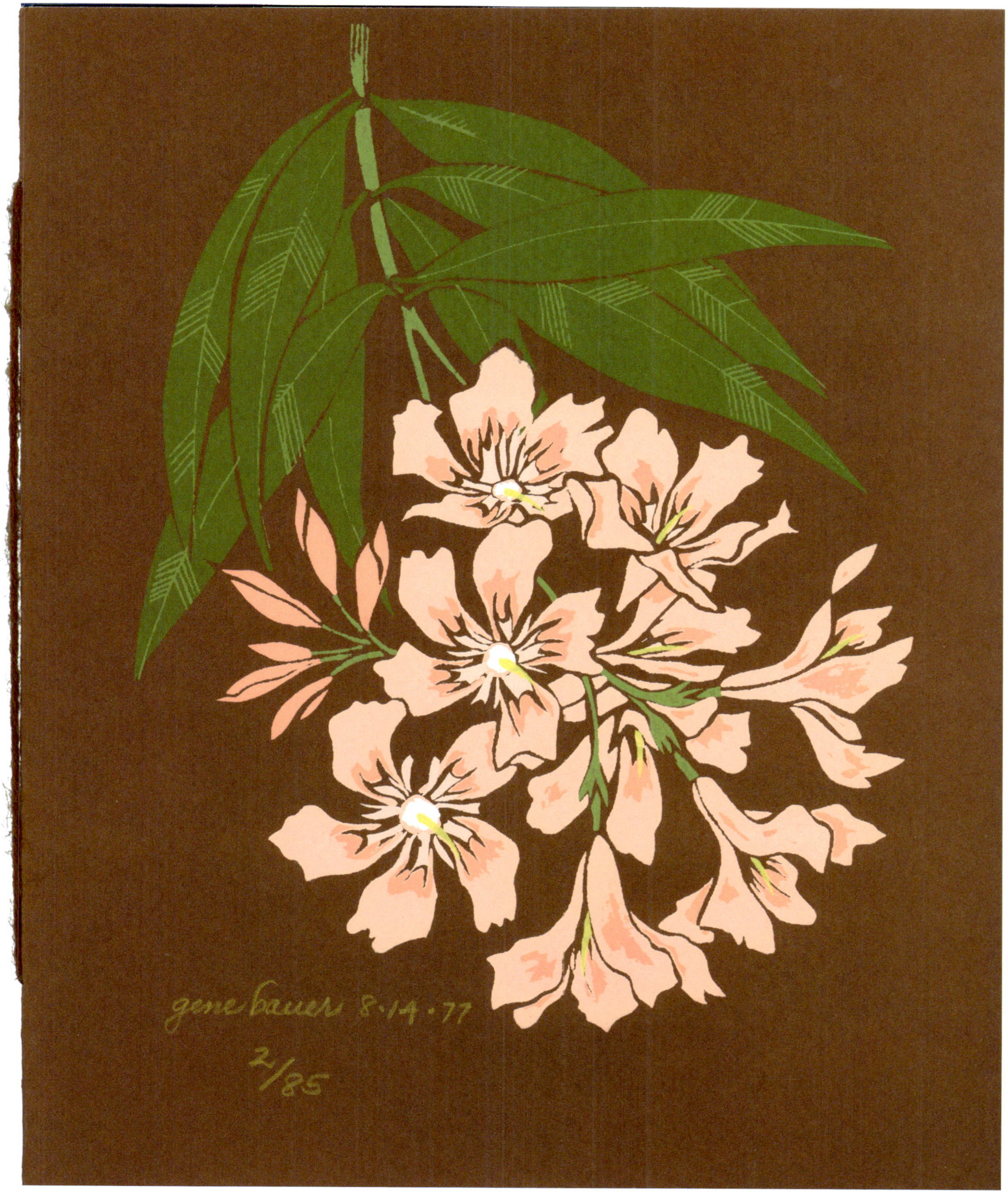
gene bauer 8·14·77
2/85

Early August 1977

I came upon my choice for this month's drawing within a delightful oasis for residents of the warm inland city of Merced called Applegate Park. Two lovely specimens of oleander, one a multistemmed small tree with pure white single flowers, the other a smaller shrub with pink, semidouble flowers, were in full bloom there. They sparkled in a small clearing, perhaps enjoying how much the rich, deep green foliage of neighboring plants set off their own beauty. But I had another reason for choosing the oleander.

The southern portion of Highway 5, constructed straight—and I do mean straight—north through the center of the state, may be considered an achievement insofar as it is the shortest distance between two points, but wow, it is a dull, drab, dreary freeway. Mile after identical mile of monotonous nothing (those who drive it know what I mean). By comparison, the older and somewhat meandering Highway 99 is a joy. Beginning north of Los Angeles, every inch of its median strip for the two hundred miles north to the city of Merced (even beyond to Sacramento) is a solid mass of oleanders. Pink, salmon, rose, red, and white. White, pink, rose, salmon, and red. As the highway skirts each town along the way, the maze of overpasses is tastefully planted with large solid groupings of oleanders. A cluster of ten small trees about fifteen feet tall smothered with white flowers joining a smaller group of pure pink, then a mass of red and another group of whites. All in full bloom during the warmest days of summer in one of the driest years in the history of this state. Conditions along this highway include extreme heat, drought, dust, wind, smoke, and smog. What other plant can withstand such treatment? For every handful of grit it gets, the oleander gives beauty back. Highway 99 is only one example of a well-traveled road brightened with oleanders. The plant graces many, especially throughout the warmer sections of the state.

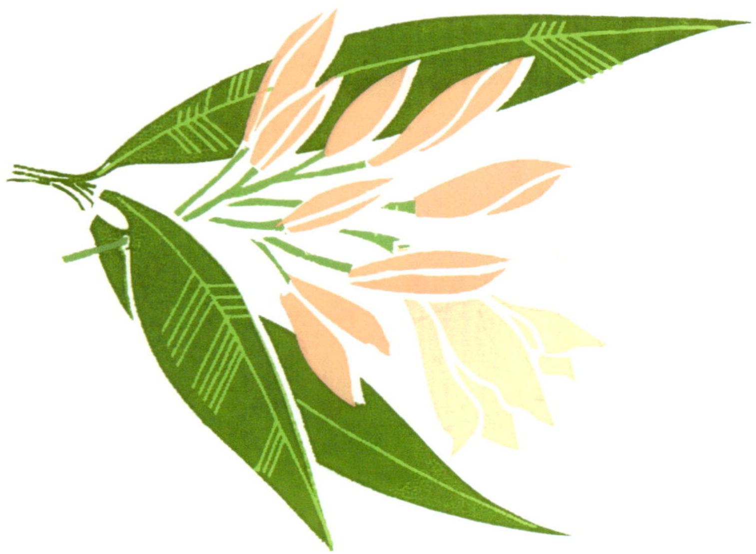

Nerium oleander

In the genus *Nerium* are three species native from the Mediterranean region to Japan. The word *Nerium* is the Greek name of the oleander; *oleander* refers to the resemblance of the leaves to the olive.

Oleanders are evergreen shrubs or small trees growing to a height of ten to twenty feet. Their thick, leathery, narrow leaves look good all year, and the plant even has an extensive blooming period, bearing flowers usually from late April all the way to October. The individual flowers are one to two inches across and borne in terminal clusters. Some oleander varieties produce single flowers, while others bear semidouble; still others grow double flowers. Flower color ranges from pure white, creamy white, and pale yellow to all shades of pink and red, and some are even variegated. Frequently they are available only by color, but many varieties have been given names.

Oleanders have only one very unfortunate and objectionable attribute: all parts of the plant are very poisonous if eaten. Except for this one trait, the oleander is a top choice for companionship whether on the highway or in a private garden. Few plants provide such lively summer color under such adverse conditions. Summer would hardly be summer in California without these magnificent splashes of cheer playfully provided by the hearty oleander.

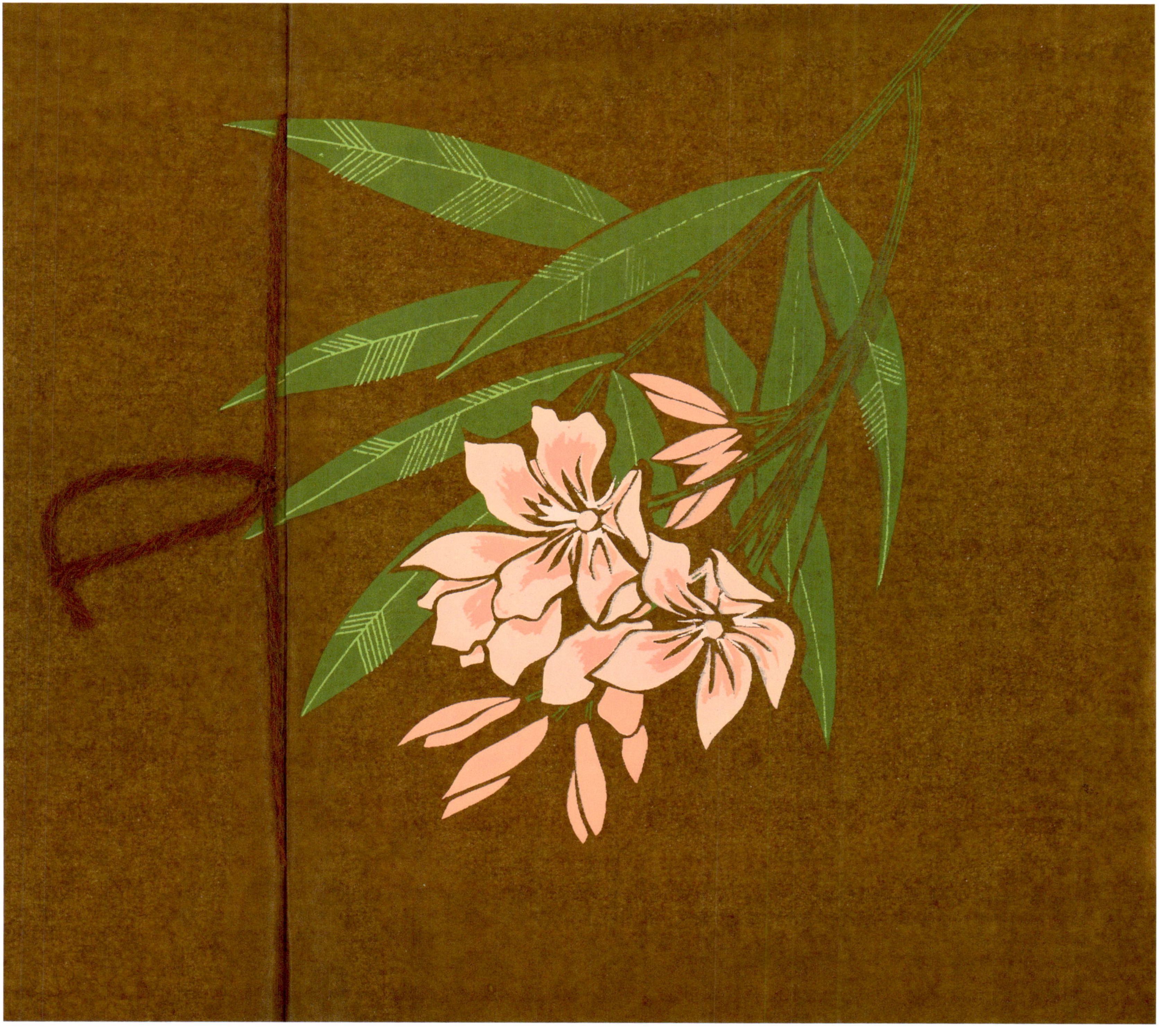

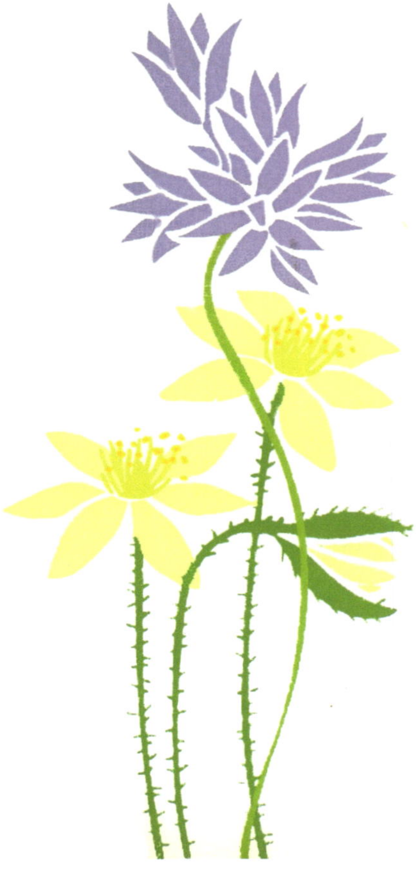

California poppy, blue dicks, cream cups
(Eschscholzia californica, Brodiaea capitata, Platystemon californicus)

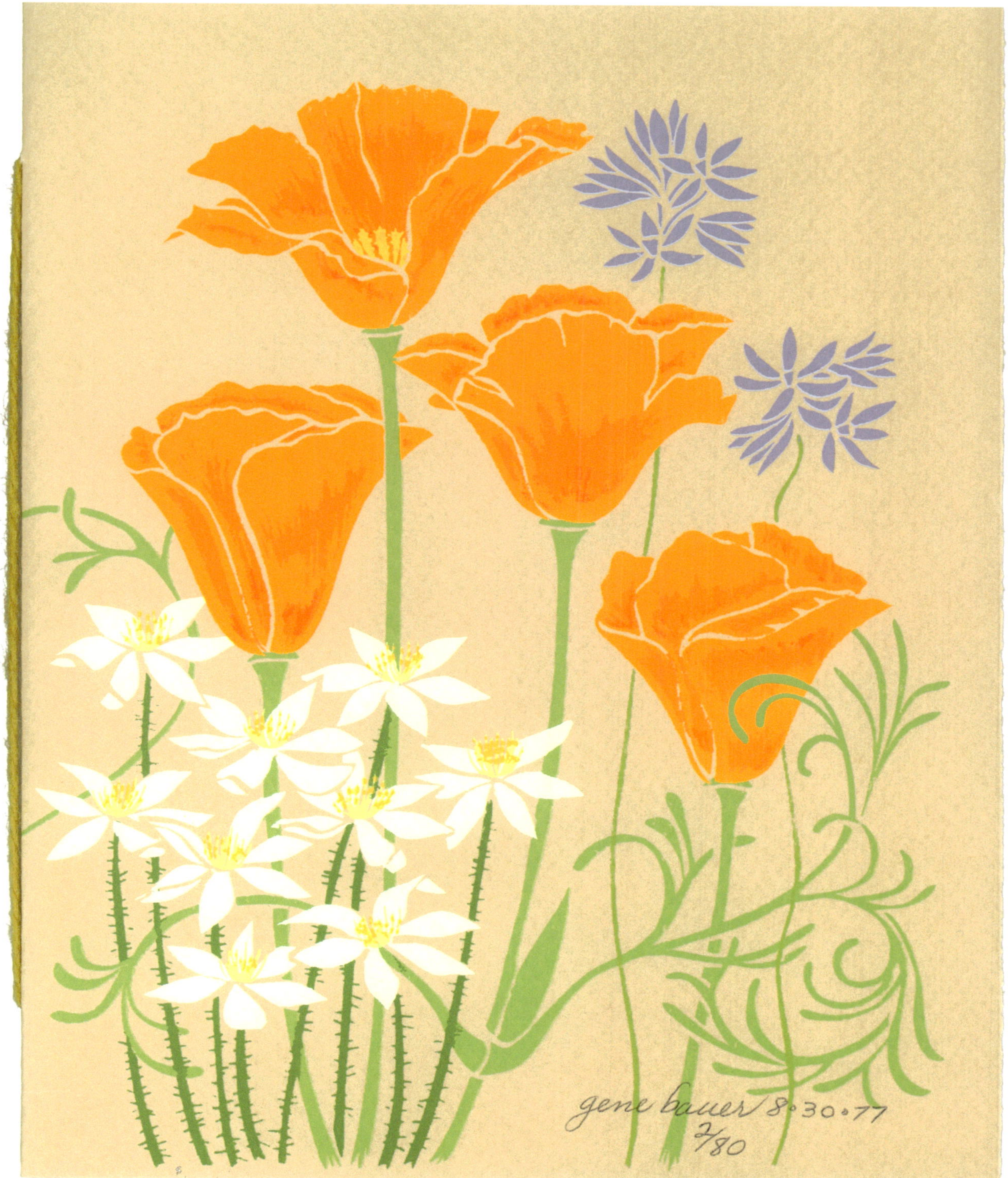
gene bauer 8·30·77
2/80

Late August 1977

The fifty-mile width of the Great Central Valley of California, including the Sacramento Valley in the north and extending five hundred miles down to include the San Joaquin Valley in the south, was at one time a vast, natural wildflower garden. Acres and acres of wildflowers in solid seas of orange, yellow, blue, lavender, white, pink, and red were abundant through this inland valley. Agriculture, industry, and housing have steadily encroached on this wildflower paradise until today the Antelope Valley is one of the few remaining areas where the state flower, *Eschscholzia californica* flourishes in great masses along with other indigenous plants.

In 1971, the Lancaster Woman's Club formed the Wildflower Preservation Committee, and with the help of community and statewide organizations (including California Garden Clubs, Inc.), two thousand acres of "prime poppy land" were purchased for the Antelope Valley California Poppy Reserve. As part of the Bicentennial Year, a dedication took place on the site on April 24, 1976. With more contributions, the reserve could possibly expand to accommodate the return of the wild antelope, so abundant in earlier years and for whom the valley is named, and they could roam there again.

I had never visited the reserve, but I knew the ideal time would be in the spring. July is hardly the month to look for wildflowers on the high desert—the temperature exceeds 100 degrees, but it was a convenient time for me and I was determined to see for myself the exact location. The road there flows smoothly and gradually up until it is capped by the two Antelope Buttes and Fairmont Butte at elevations around 3,000 feet. The distant rugged peaks of the Tehachapi Mountains, at elevations more than twice as high, form a dramatic background for this serene butte country. From here in the gentle tranquility of spring may be seen the magnificent panoramic vistas of California wildflowers.

The term wildflower may beg for some clarification here. Except for one or two European escapees, all the flowers found here are not only indigenous to California but also native to this area. Not introduced from some other part of the state, they exist under natural conditions only—no one waters them, no one feeds them. Sometimes they are browsed by animals and some years, due to the weather, they are not as spectacular, but this is true of many things. Because this is the one remaining area in the state where one might still see what a glorious garden the whole state used to be, I have chosen as my subjects three wildflowers that are in great abundance at the poppy reserve.

Brodiaea capitata

All species in the genus *Brodiaea* are native to this state. The generic name honors James J. Brodie, a Scottish cryptogamic botanist. *Capitata* means "collected into a head" and in this case refers to the flowers, known as blue dicks. Half-inch flowers of a soft, translucent, lavender blue appear in a tight cluster of seven or eight atop a slender, sometimes wiry stem about one foot high. Rarely are all the flowers in the cluster open at the same time. The few leaves are grasslike and soon wither leaving the stem appearing quite naked. The flowers are produced from a small, fibrous, coated corm. Compared to larger and more flamboyant flowers these may appear insignificant, yet they are among the best known and dearly loved flowers in California. The flowers, stem, and leaves appear very delicate, almost too fragile, but their abundance shows how capable they are of withstanding the rigors of the open field.

Platystemon californicus

Derived from the Greek, the generic name *Platystemon* refers to the many broad *(platus)* stamens *(stemon)* in the center of these flowers. There is but one variable species and it is named for the place it is found (California), but commonly known as cream cups. These wildflower plants are annuals growing about ten inches high with many stems from the base, each bearing a small flower about one inch in diameter. The leaves appear on the lower part of the slender, hairy stems. A lovely, creamy straw color, the flowers are produced in great numbers. When the plants are found in large masses their color serves as a soft, mellow contrast to their more brilliantly colored neighbors.

Eschscholzia californica

There are about eight species in this genus, all native to the state (see page 107). Annual or perennial depending on its location, *E. californica,* commonly known as California poppy, grows twelve to eighteen inches high with delicately cut, blue-green, branching foliage. The slender, graceful stems support cup-shaped flowers two to three inches across whose four golden petals emanate a satiny sheen. The flowers open in the warmth of morning sunshine, but by late afternoon the petals have gently folded together. Don't worry. They will open again the next morning.

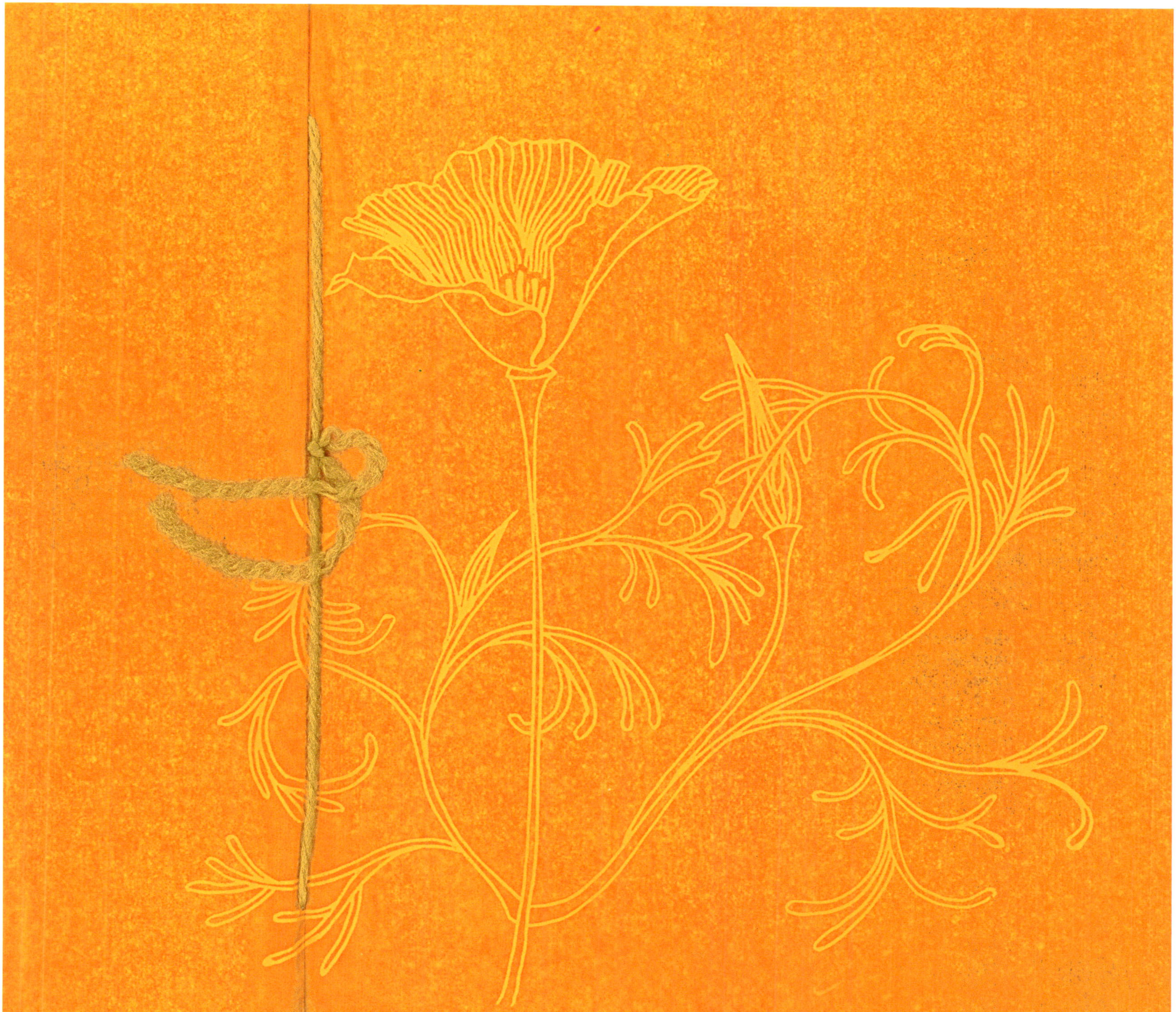

Western azalea *(Rhododendron occidentale)*

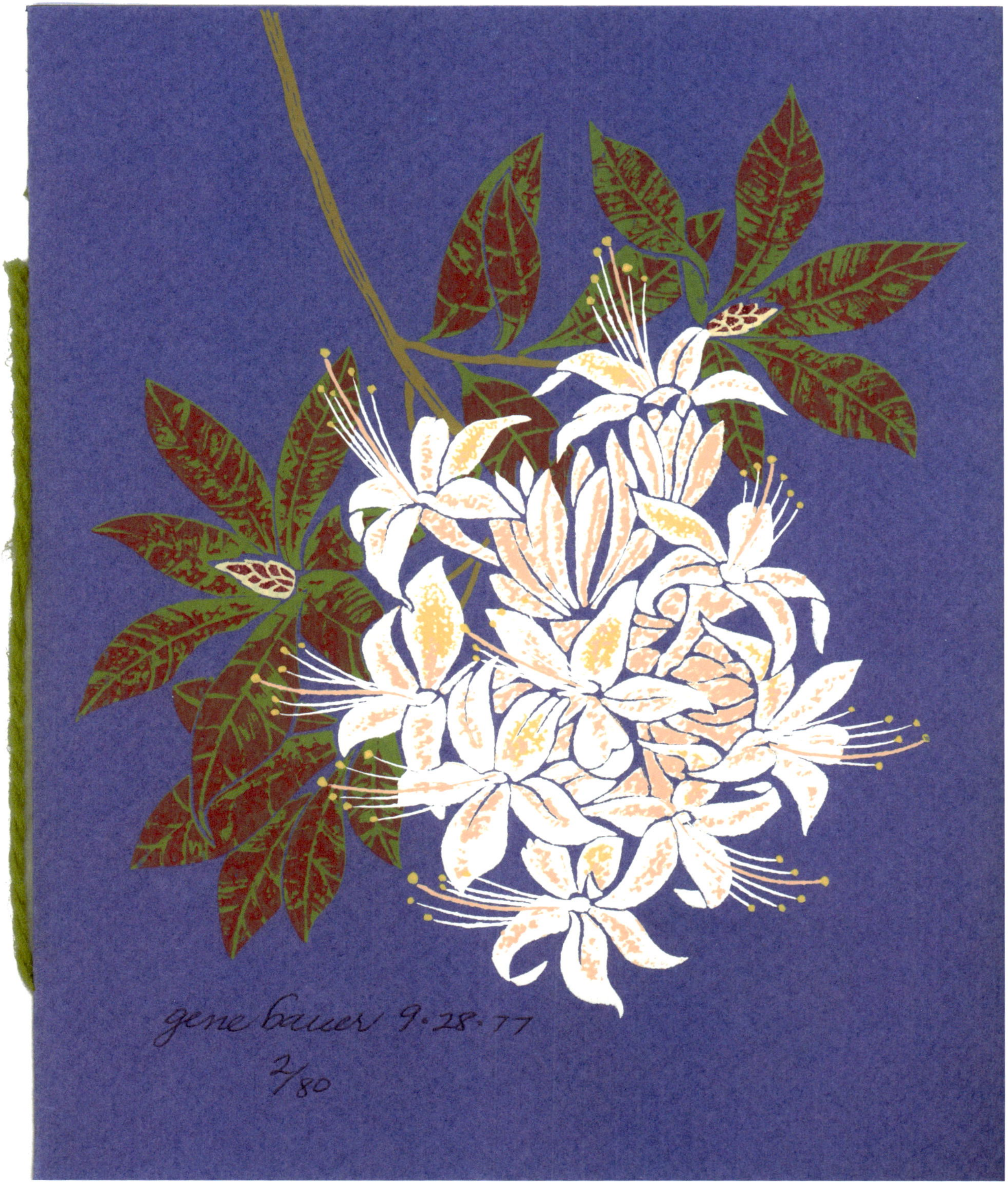
gene bauer 9·28·77
2/80

September 1977

I had not intended to feature the azaleas until next spring; it seemed more appropriate to focus on them during their normal flowering time. Also, I prefer drawing from living specimens and I believed no flowers would be available for my purposes until that time. However, as I wandered through the Azalea State Reserve, just northeast of the town of Arcata in Humboldt County, I was both surprised and delighted to find one large shrub in full bloom. An added bonus was the foliage that was just beginning to assume fall colors. What a striking combination—white flowers against wine red leaves. Although blooming in September is considered a bit unusual, I couldn't resist this rare color combination. Do not expect to find this particular mix of contrast unless you too are lucky enough to find one as late as I did. Remember that the peak of their bloom is late May and June.

The genus *Rhododendron* is an extremely large one including more than nine hundred species, with plants ranging in size from small ground covers to some that might be called trees. In fact, the generic name is from the two Greek words *rhodos,* meaning "rose," and *dendron,* meaning "tree," hence, a rosetree, though commonly known as the azalea. These are predominately evergreen or deciduous shrubs, found in many parts of the world but mostly in the northern hemisphere. Much confusion has surrounded this genus, but finally some order was achieved when botanists arranged these many species into series and subseries. There are now forty-three series, each containing closely related species, and those flowers commonly called azaleas have been assigned their own series, the Azalea Series.

Rhododendron occidentale

The specific name *occidentale* means "western," so this is the western azalea. Native to moist areas in Oregon and California, *R. occidentale* is an upright-growing deciduous shrub six to fifteen feet tall. The funnel-shaped flowers are borne in terminal trusses of six to twenty-five and vary considerably in color. On many of the plants the flowers are pure white with just a bare touch of pink. Others bear pink flowers in a variety of shades; some plants even have red flowers. The upper lobe of the flower usually shows a golden blotch or flare. On some western azaleas the flower petals are smooth-edged, while the flowers of others have frilly or scalloped edges. All the azaleas in this extremely variable species have at least one thing in common, though: they are delightfully fragrant.

Abundant along the Northern California coast where it may be seen in all its variations of color, form, size, and growth habit (especially in the Stagecoach Hill area), the western azalea also grows in the Coast Ranges as far south as Santa Cruz County. Inland, *R. occidentale* may be found in moist places below 7,500 feet in the Sierra Nevada and, in the more southerly part of the state, in the Cuyamaca and San Jacinto mountains. On that same day in mid-September, I saw many spectacular specimens of this plant growing among salal, bramble, *Ceanothus,* ferns, and an almost endless number of wildflowers along a two-mile strip of land called Stagecoach Hill. A group of azalea fanciers gave Stagecoach Hill its name because of a road along its ridge that was used by stagecoaches during the early part of the twentieth century. Standing in this place overlooking the blue Pacific, with gentle mists flowing in from the sea, I saw thousands of unusual specimens of *Rhododendron occidentale.* An official of the American Rhododendron Society said about them, "No two are alike."

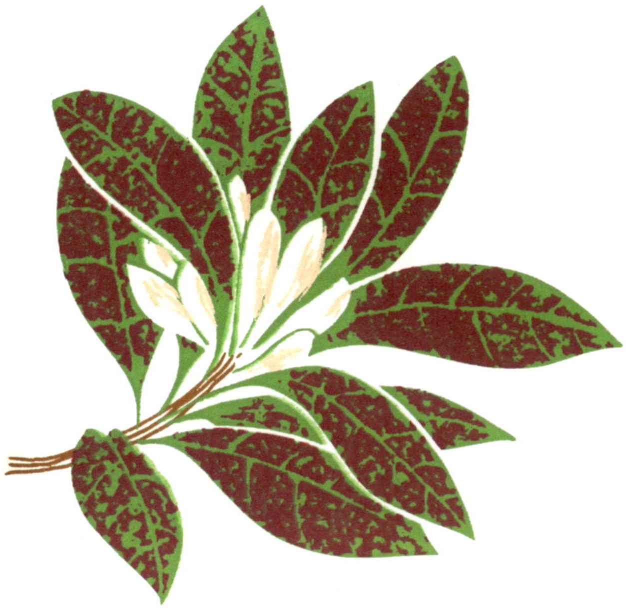

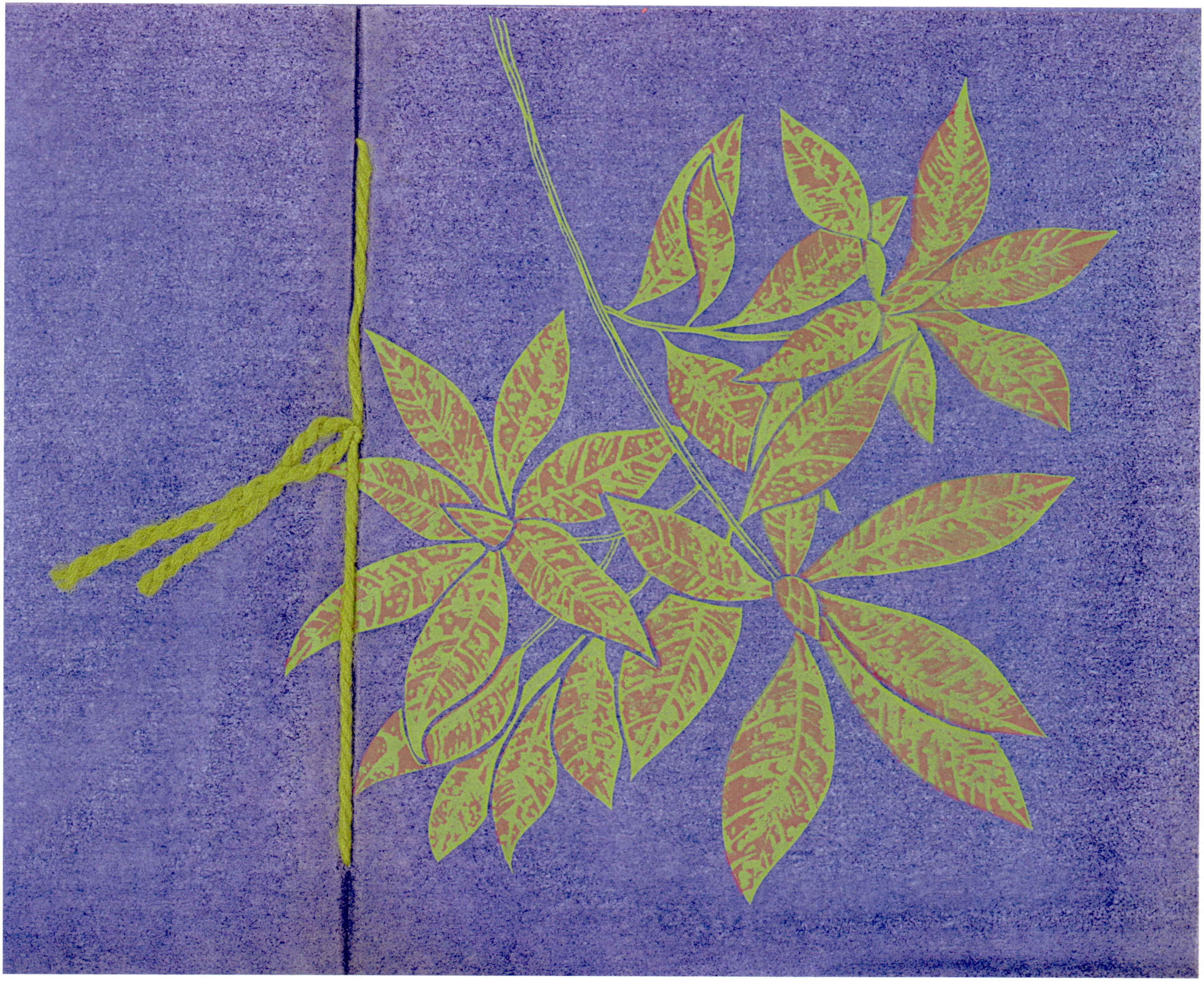

Colchicum *(Colchicum autumnale* 'Waterlily')

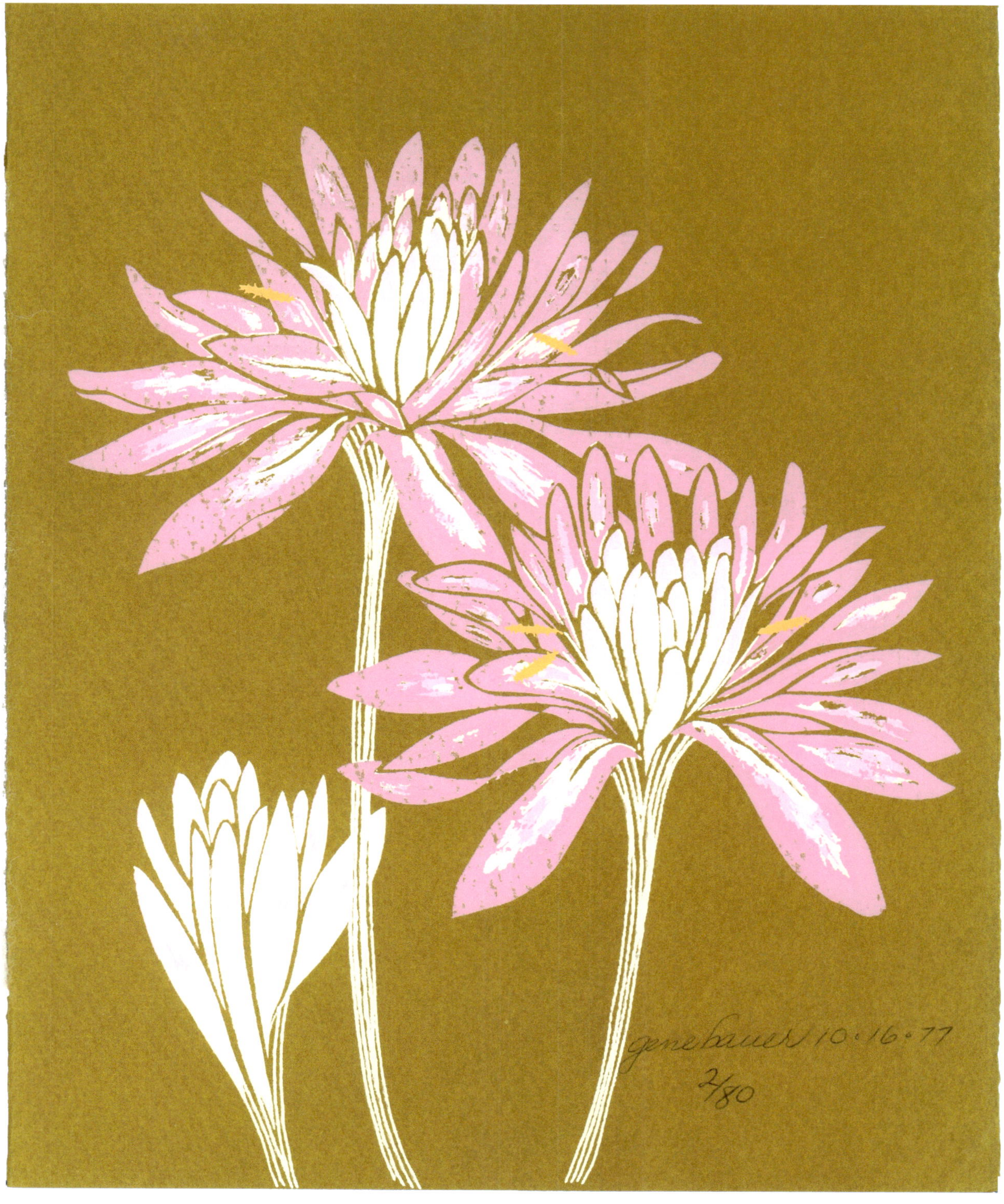
genebauer 10·16·77
2/80

October 1977

Color, form, texture, and an element of surprise—in all these attributes members of the genus *Colchicum* qualify as outstanding. In the state of California they are not seen very often, and so they are neither well known nor understood. This is unfortunate because colchicums are among the easiest and most carefree of all the bulbous plants.

The name *Colchicum* is derived from Colchis, an ancient country along the Black Sea where these plants grew in great profusion. There are about sixty species in the genus, native to Europe, Asia, and North Africa. Most of the *Colchicum* flowers are single with six petals arranged in a smooth, very elegant goblet shape. The four- to six-inch petals open out wide during the warmth of the day but close at night. Technically they are stemless, the flowers produced at the end of an elongated tube thrust up quite naked from the ground. About six to eight inches tall, they bloom in September and October unaccompanied by foliage. The large, rather spinach-like leaves appear in early spring unaccompanied by flowers. Flower color ranges from white to pinky lavender to purple. Many are tessellated or checkered with a darker shade of the same color.

Colchicum autumnale 'Waterlily'

Colchicums happen to be one of my favorites, and passersby our home are always astounded to see these striking flowers in their magnificent glory in October. Many people erroneously think they are autumn crocus and some even mistake them for tulips! I am particularly fond of these flowers for the obvious reasons, but also because the odd-shaped corm seems to be unattractive to rodents and takes advantage of this good fortune to become persistent and long-lived. Some have been in the ground for fifteen years with no care from me and they still produce exquisite flowers, sometimes as many as ten per corm.

Few of the species of the genus *Colchicum* are available for planting and consequently are seldom seen except in collections of *Colchicum* fanciers. The two species *C. autumnale* and *C. speciosum,* and their many varieties, appear to be the best known. It is from crosses of these two that some of the most beautiful and charming hybrids have been developed. Among the better known and more readily available hybrids are 'The Giant', 'Lilac Wonder', 'Violet Queen', 'Autumn Queen', and 'Waterlily' (which you see here).

'Waterlily' is, of course, a double, and it almost outshines real water lilies. I am not drawn to the double forms of many flowers, but 'Waterlily' is an exception. I have these planted under a tree that provides filtered light. When the sun shines through the leaves onto these flowers, their petals opening out to six inches across, it is as breathtaking a sight as I have ever witnessed.

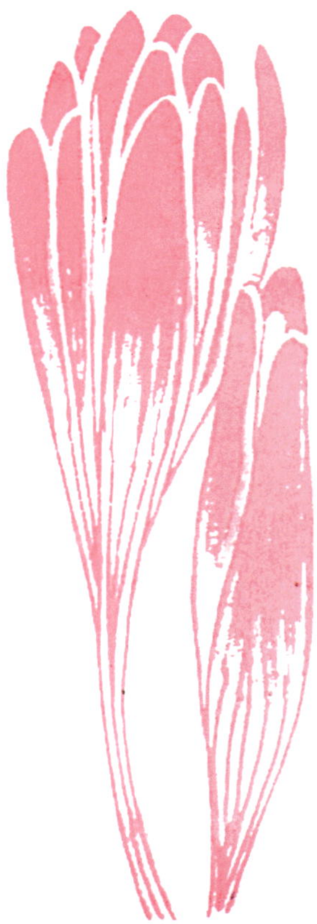

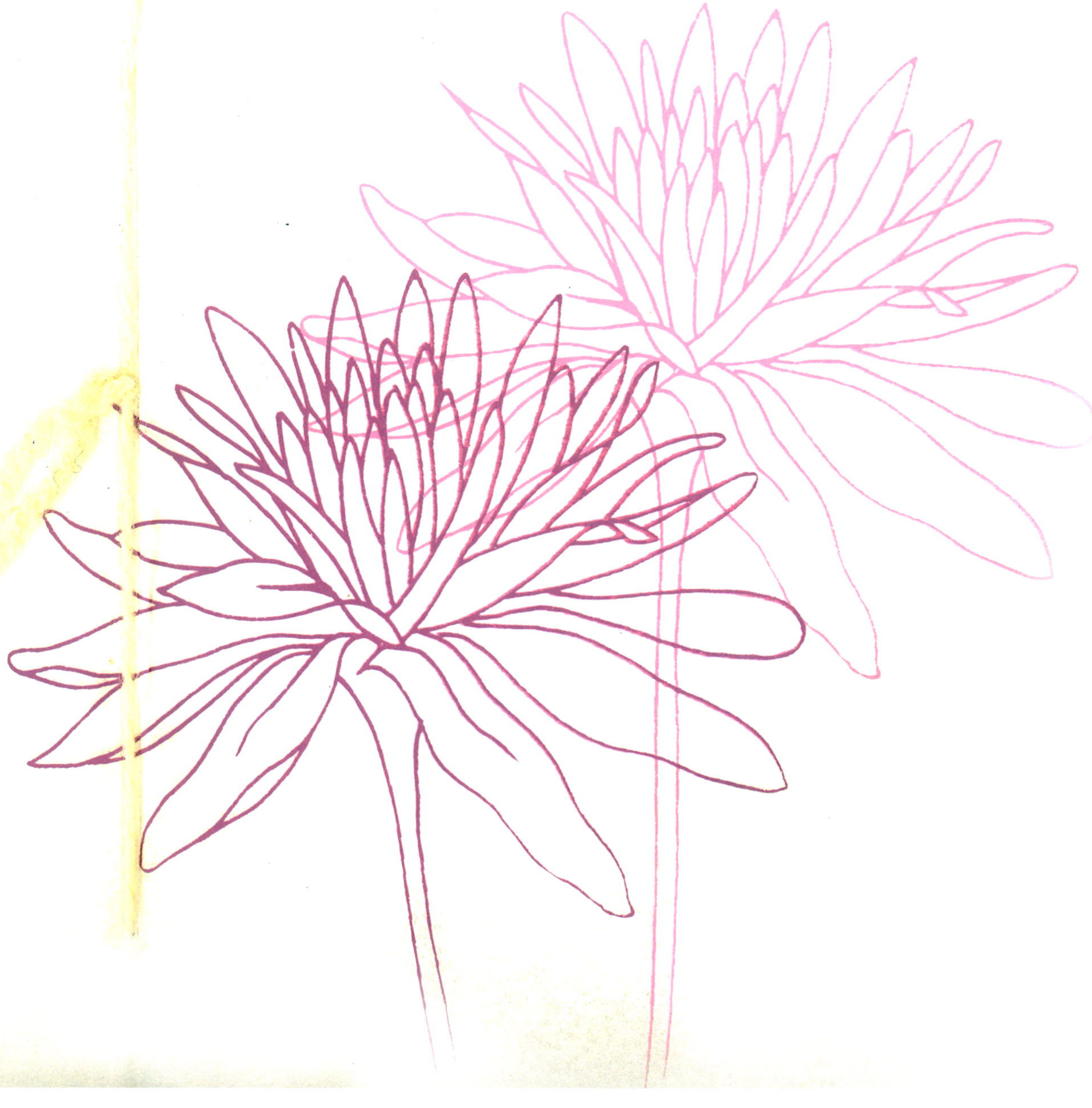

Bitter cherry *(Prunus emarginata)*

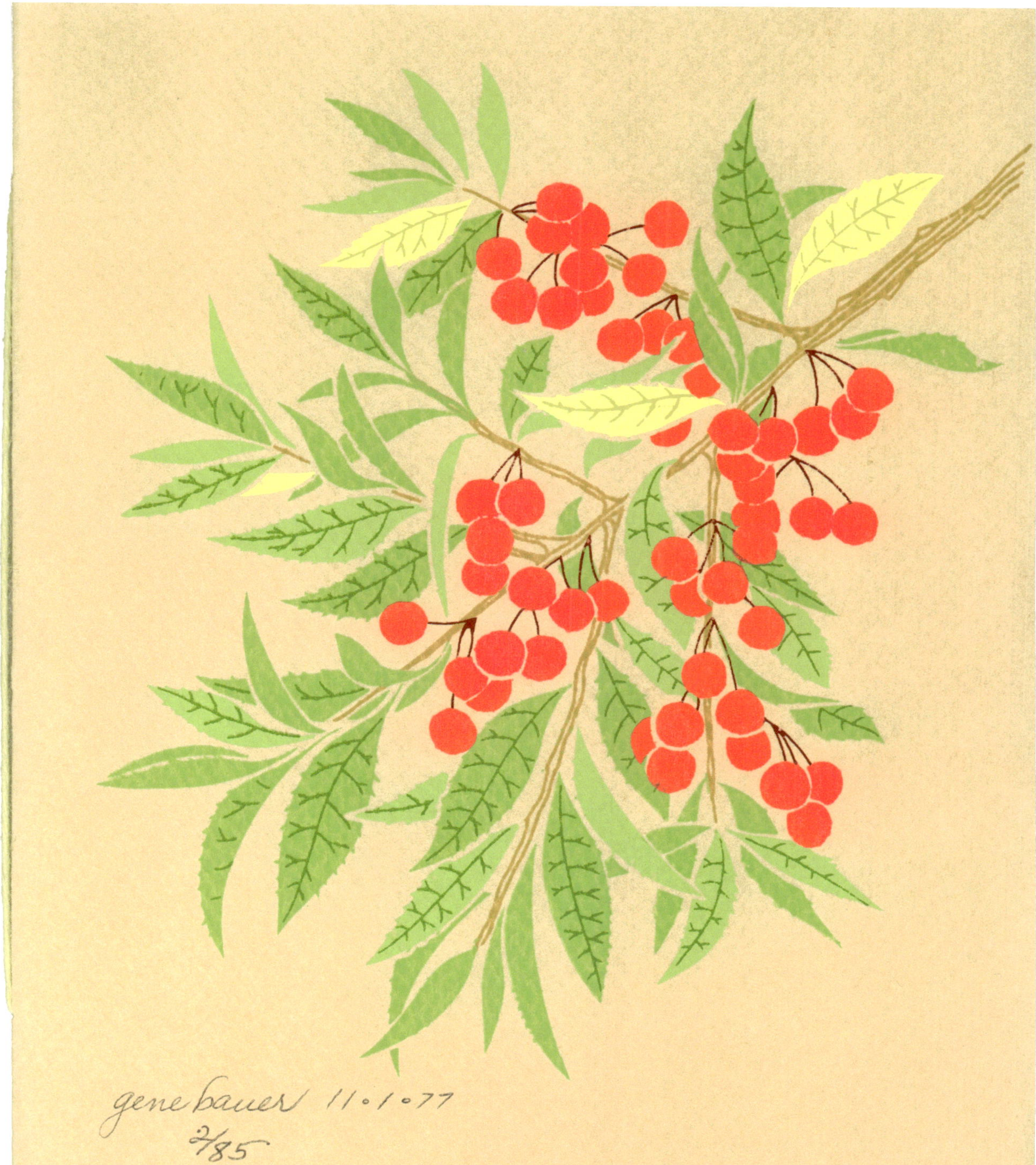
gene bauer 11·1·77
2/85

Early November 1977

If the freshly blacktopped spur road to it was any indication, the John Everitt Vista Trail was so new on the day we came that I believe we were its first visitors. "We" on this day included the horticulture chairman for California Garden Clubs, Inc., Corrine Bromberger, who accompanied me to see some public gardens in the Top O' the State District. It was great having Corrine along on these excursions because the plants I couldn't identify, which were many, she could, and the ones she couldn't identify, which were few and far between (mostly high-elevation natives), I identified. But there was one plant neither of us could figure out. We saw it nearly everywhere we went, but even in long-established places we were not able to discover its name. I took to calling it the "coral-berried goodie." So how ironic that this still-wild, brand-new nature trail was the place that taught us its real name.

A plaque designates the John Everitt Vista Trail, established in the Shasta National Forest as a memorial to John S. Everitt, a forest supervisor of the early Shasta National Forest who died at age forty while fighting a 1934 forest fire on Mt. Shasta's lower slopes. Just beyond the plaque are some well-chosen snags that, although scorched and charred, possess an exquisite patina. Much restraint was exercised in arranging these snags so skillfully and so subtly that I suppose people could walk right by without noticing them. But having experienced three forest fires in the area where we live, I understand their significance as sculpture; simple and honest, nothing could be more appropriate.

We were 5,700 feet high, in the transition zone between the yellow pine forest and the red fir forest, heading toward the Vista Plateau. Corrine reached the base of it first and cried out, "You're not going to believe this. It's growing everywhere here, and I do think I see a label!" She didn't have to tell me what she meant by "it"; names are so much nicer than calling things "it," but not having the option of its name, "it" had been on my mind all day long. What a treasure to find at the base of this magnificent vista—the spreading thicket was like a crown set with jewels—and the jewels were named *Prunus emarginata.*

Prunus emarginata

This large genus contains about two hundred species, most of them native to Asia, Europe, and North America. Many species of *Prunus* are prized for their edible fruit and some for their edible seed; others are highly valued for their ornamental beauty, especially that of their spring blossoms. We had found growing wild a member of the same genus to which the plums, cherries, apricots, peaches, and almonds belong. Its species name *emarginata* refers to the finely serrated margins of the leaves. Known as bitter cherry (and known at last by me), it is native to several of the western states.

The fruit comes from the clusters of dainty white blossoms that cover this fifteen- to twenty-foot tree in springtime. The blossoms develop into quarter-inch cherries of a luminous and clear coral red. The cherries are extremely bitter to the human palate, Corrine and I can attest, but birds and other animals love them, devouring them in great quantities. In addition to providing spring flowers and colorful autumn fruits, these shrubs—forming dense thickets on rocky, damp slopes—are valuable in controlling rapid runoff from heavy snowpacks.

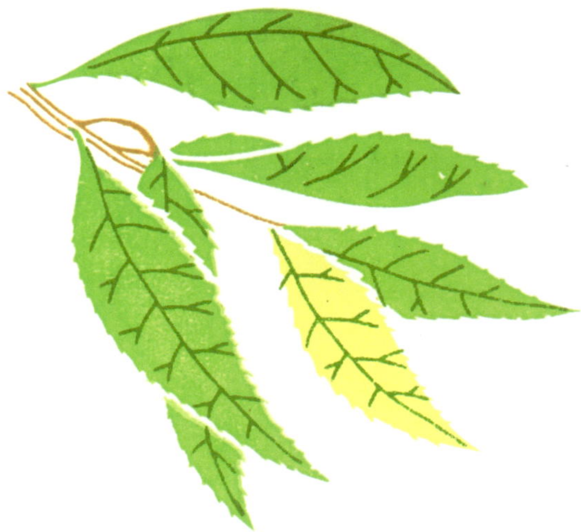

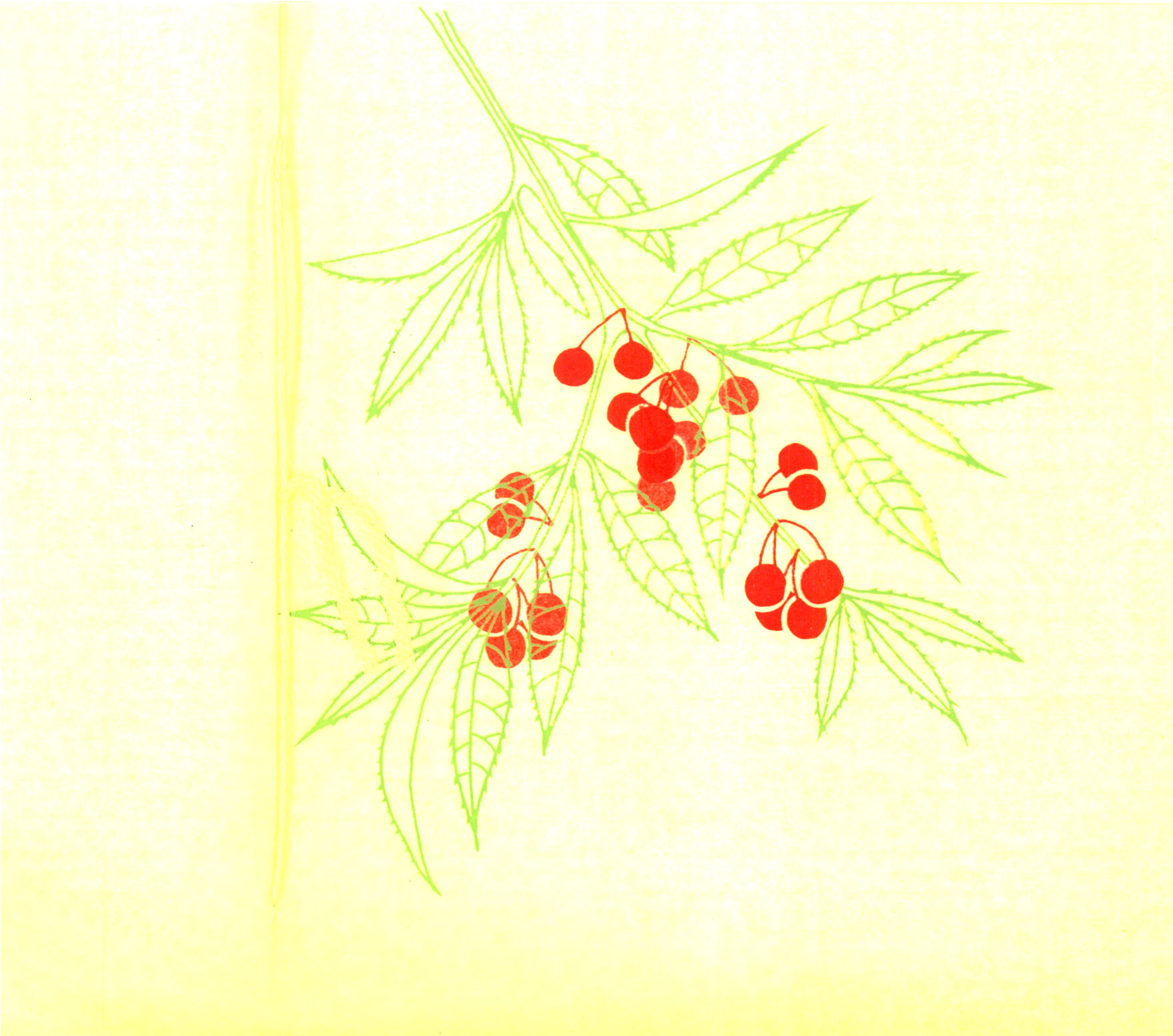

Chinese pistache *(Pistacia chinensis)*

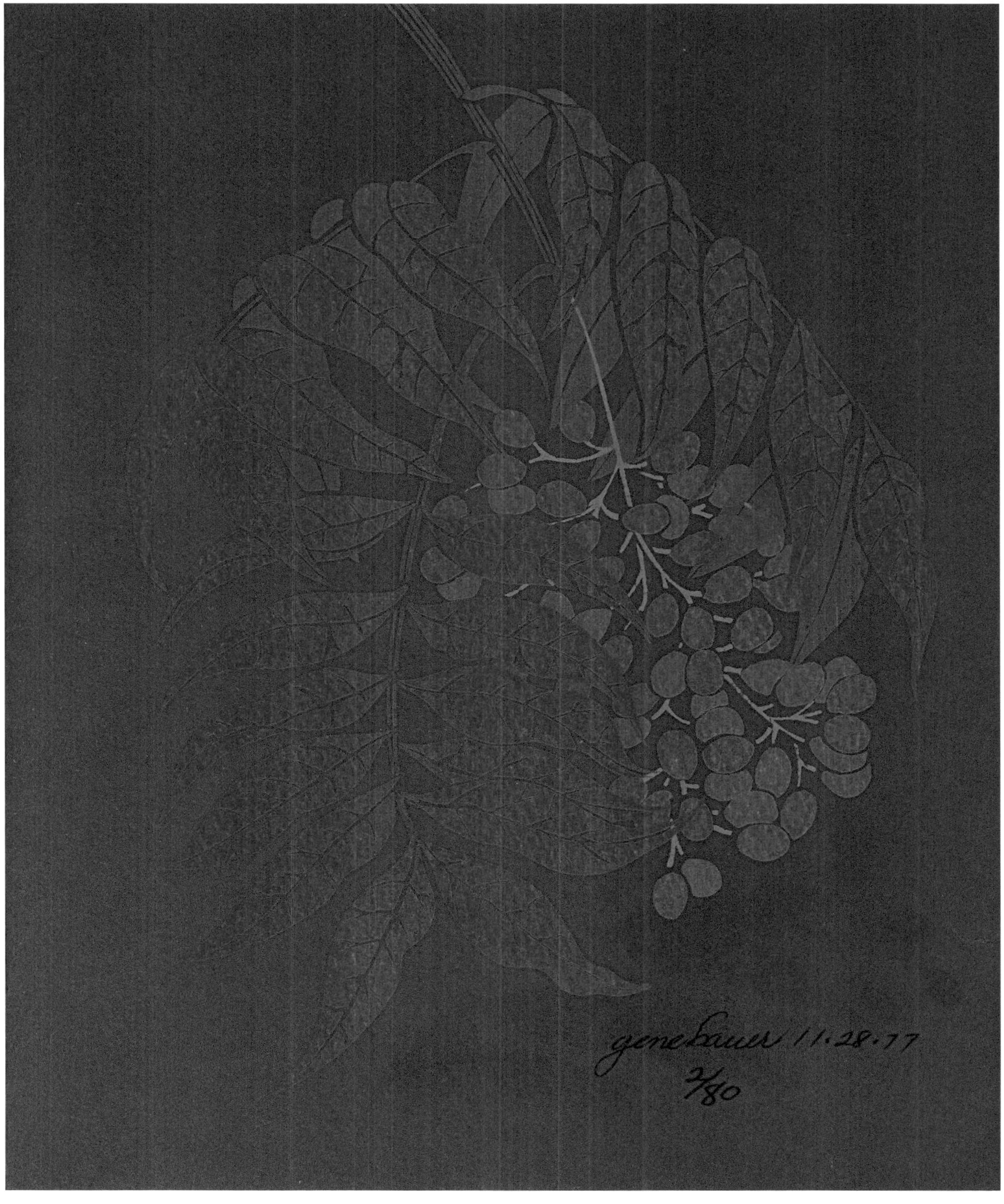
gene bauer 11·28·77
2/80

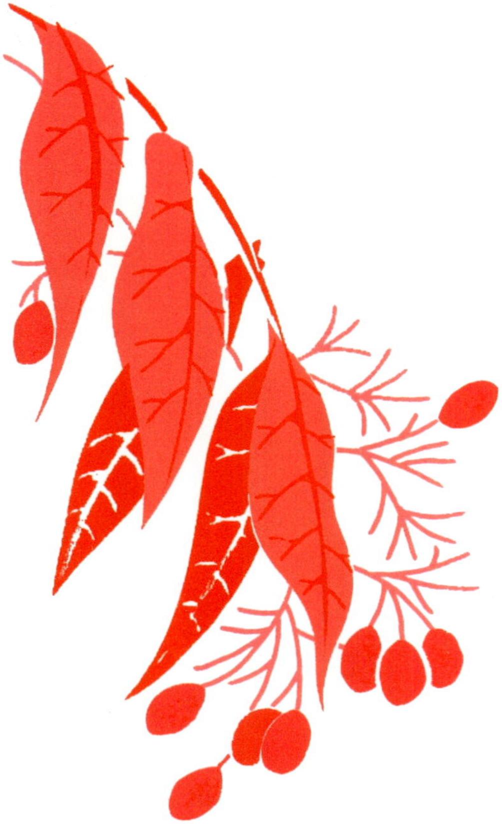

Late November 1977

Local Chinese laborers built the Oroville Chinese Temple and Garden in 1863 with funds provided by the Emperor and Empress of China. Originally attracted to the area by the gold mines of nearby Bidwell Bar, Chinese immigrants had established a large colony here by 1850. The Chinese brought with them very few material possessions but were well-endowed with courage, strength, industry, and the rich heritage of the civilization they left behind. The temple they erected served as a place of worship for a community of ten thousand of various faiths, including Taoism, Confucianism, and Buddhism. [A California Landmark, the temple is now listed on the National Register of Historic Places.] The Garden Courtyard at the temple, intended to be an authentically Chinese garden both in structure and plantings, was designed as a place for meditation. The plants in the courtyard were chosen not only because of their Chinese origin but also because of their strong symbolism.

Planted in attractive groupings on the exterior grounds of the temple are a number of trees, all of Chinese origin. Three of the trees seen here have been introduced to and thrive in nearly all parts of California. Widely used as street trees, specimens, and even container subjects, they have become familiar to people but remain little known. Two are truly living fossils: *Metasequoia glyptostroboides,* which grew millions of years ago but was only discovered in 1941; and *Ginkgo biloba,* so distinctive it cannot possibly be mistaken for any other tree. The sole survivor of a once-large genus, it is often referred to as a fossil tree; paleobotanists have traced this tree back 175 to 200 million years to when dinosaurs roamed the earth. No known wild stands of this tree exist today, so the world is indebted to the Chinese for preserving it; they used it as a tree to plant near temples and tombs. Descendants of those trees have been planted throughout the world, first introduced to Europe around 1730 and to the United States in 1784.

The genus *Ginkgo* contains but one species, *biloba,* which refers to the shape of the leaf formed in two lobes. It is commonly known as the maidenhair tree because its beautiful fan-shaped leaves, turning a lovely clear yellow before falling, resemble those of the maidenhair fern. Like some palms, and the Joshua and boojum trees, *Ginkgo biloba* appears so primitive it almost seems as if it might be a prototype, “the result of a tentative and not quite successful design for a new kind of plant,” as Andreas Feininger described it in *Trees.* Oddly enough, it is astoundingly successful. Widely grown as an ornamental or shade tree, this is one of the finest of all to plant for tolerating city conditions. Seemingly immune to polluted air as well as pests and diseases, it has been used as a successful street tree in many of the world’s large cities.

Pistacia chinensis

Pistacia chinensis is perhaps the least known of these three Chinese trees. Of the twenty or so species in the genus *Pistacia* (native from southern Europe to Asia, Mexico, and the state of Texas) only two seem to be known at all here, *P. vera* for the edible nuts (pistachio) it produces and *P. chinensis* (meaning “of China”) for its ornamental value. People who do know and grow the tree are usually quite enamored of it, and I chose it for the cover of this issue because of its dependable and stunning fall foliage. It is one of the few trees whose leaves turn scarlet in the desert areas of California. Widely planted throughout the state, it grows very well in the warm interior valleys where it is heat and drought tolerant. It does not produce spectacular flowers, nor is its basic structure particularly outstanding, yet by late fall few trees in California command more attention.

This tree may grow to be sixty feet tall and fifty feet wide. Leaves are composed of five or six pairs of leaflets arranged feather-fashion. The leaves are attractive from spring when they unfurl with a rosy hue, through summer when they are green, until fall when they are stunning in brilliant shades of orange, apricot, scarlet, and wine. Often the entire tree is aflame with only a bare hint of green deep inside. The female trees produce dense clusters of red berries that are as appealing as the brilliant-colored foliage. This tree thrives with little care, and yearly, from late October to mid-November, displays a magical color extravaganza envied by many far more demanding trees.

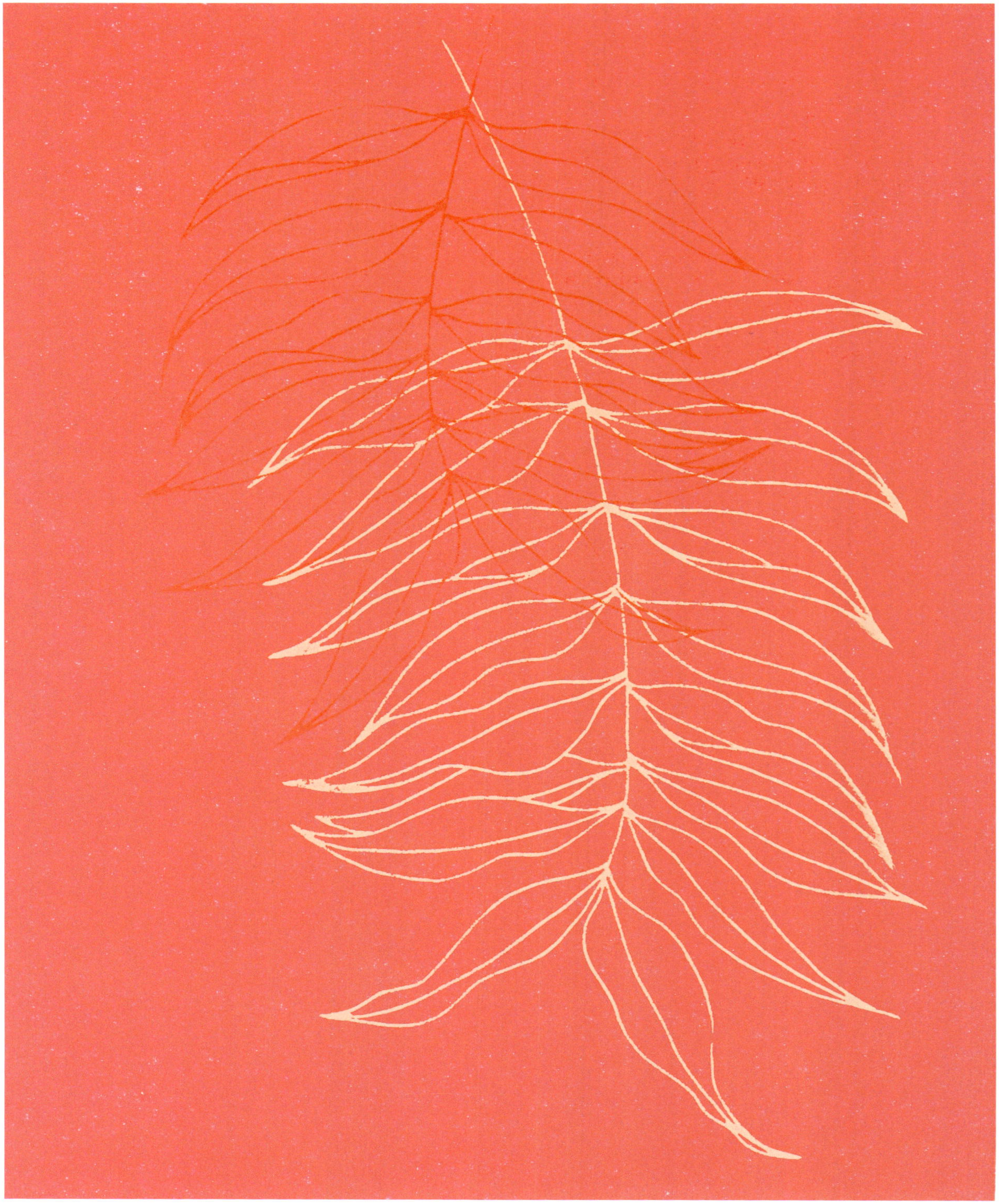

Poinsettia *(Euphorbia pulcherrima)*

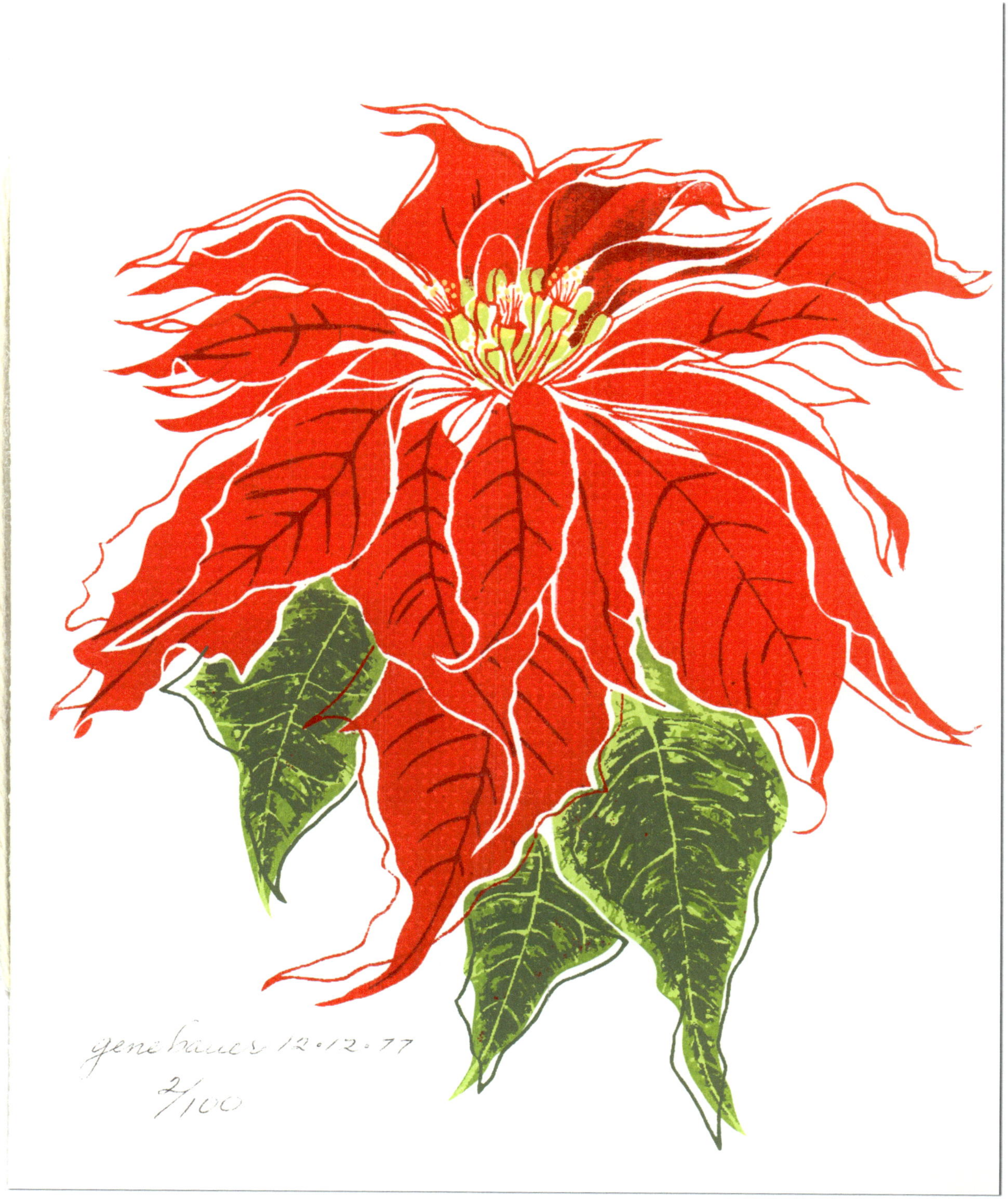
gene bauer 12·12·77
7/100

December 1977

It was Christmas 1825 and the first U.S. Ambassador to Mexico Joel R. Poinsett, MD, was walking through the markets of Mexico City when he saw the large, exotic flowers people were buying to decorate their homes and churches. He was fascinated by these unusual flowers, and being an avid botanist, believed if this wild beauty could be tamed it would become a universal symbol of Christmas. Dr. Poinsett returned home to Greenville, South Carolina, and in his greenhouses began propagating this plant that he called "painted leaves." Now this jungle beauty from tropical Mexico is indeed the traditional Christmas plant, with its colors symbolic of the season: red for warmth, friendliness, and life; green for the continuance and renewal of life. The name poinsettia honors this man, but the flower is known in its native Mexico by more descriptive names: *flor de la Nochebuena* (flower of the blessed night), *flor de Pascua* (Christmas flower), *flor de fuego* (flower of fire), and *hojas de fuego* (leaves of fire).

In its native habitat as well as in gardens in frost-free areas of Southern California, poinsettias are bushy shrubs ten to fifteen feet tall, the leaves a deep, rich green and the showy bracts a brilliant red. After much work with this flower, some of it accomplished at the Ecke Poinsettia Ranch in Encinitas, California, small, multiflowered plants perfect for pot culture were developed. These flowers are very durable, remaining in excellent condition for many weeks and with a reasonable amount of care even for many months. Poinsettias are now available in other colors in addition to the brilliant red, including pink, white, and bicolor forms, but these colors may never achieve the popularity of the red for this season of the year.

Euphorbia pulcherrima

The genus *Euphorbia* is a large one containing more than fifteen hundred species widely distributed around the world. *Euphorbia* is the old classical name and *pulcherrima* means pretty or beautiful. *E. pulcherrima* is native to Mexico and Central America, and while all euphorbias are similar insofar as they contain a milky, acrid juice, the similarity ends there. Members of the genus are strikingly dissimilar in their characteristics. Many are cactus-like succulents whose flowers are not at all showy but are surrounded by highly colored bracts. *E. pulcherrima* is not at all cactus-like, however. Its large, brilliantly colored parts are actually leaf-like bracts surrounding the tiny, inconspicuous true flowers in the center, which are greenish yellow.

The beauty pictured here is the result of a trip taken to Balboa Park, an enormous fourteen-hundred-acre wonderland of rubber trees *(Ficus)*, eucalyptus trees *(Eucalyptus)*, and joggers *(Homo sapiens)*. Here these introduced trees, planted for the 1915–1916 Panama–Pacific Exposition celebrating the opening of the Panama Canal, are almost as awe-inspiring as the indigenous redwoods of the northern part of the state. Balboa Park is situated right in the heart of the city of San Diego, about as far south as you can go in California before reaching the border we share with Mexico. The poinsettias were stealing the show in the park's Botanical Building. Within this high-arched structure of Spanish Colonial architecture built in 1913, poinsettias were growing in elaborate hanging baskets, their brilliant scarlet repeated in the masses of potted poinsettias placed at ground level.

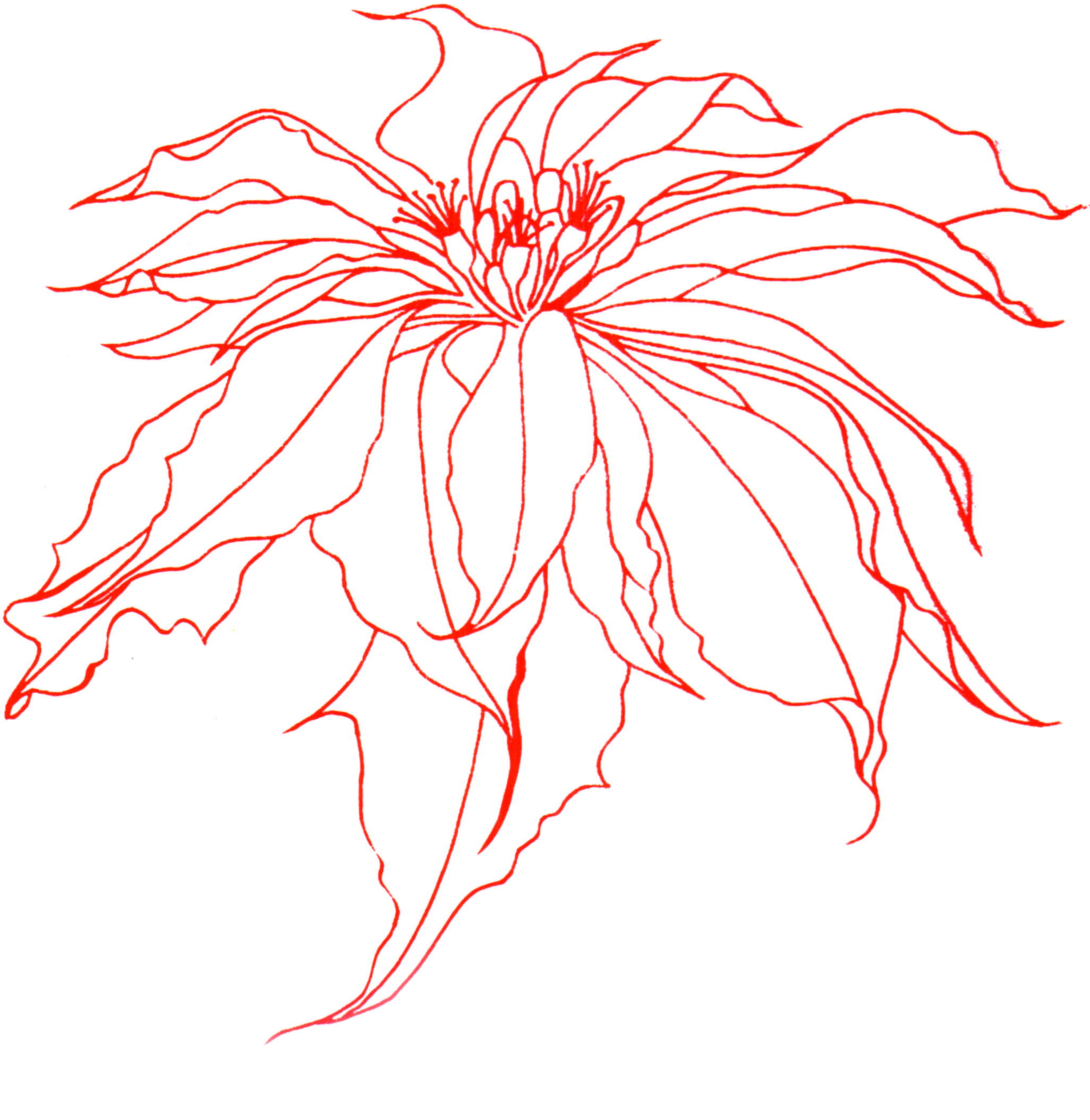

Red fir *(Abies magnifica)*

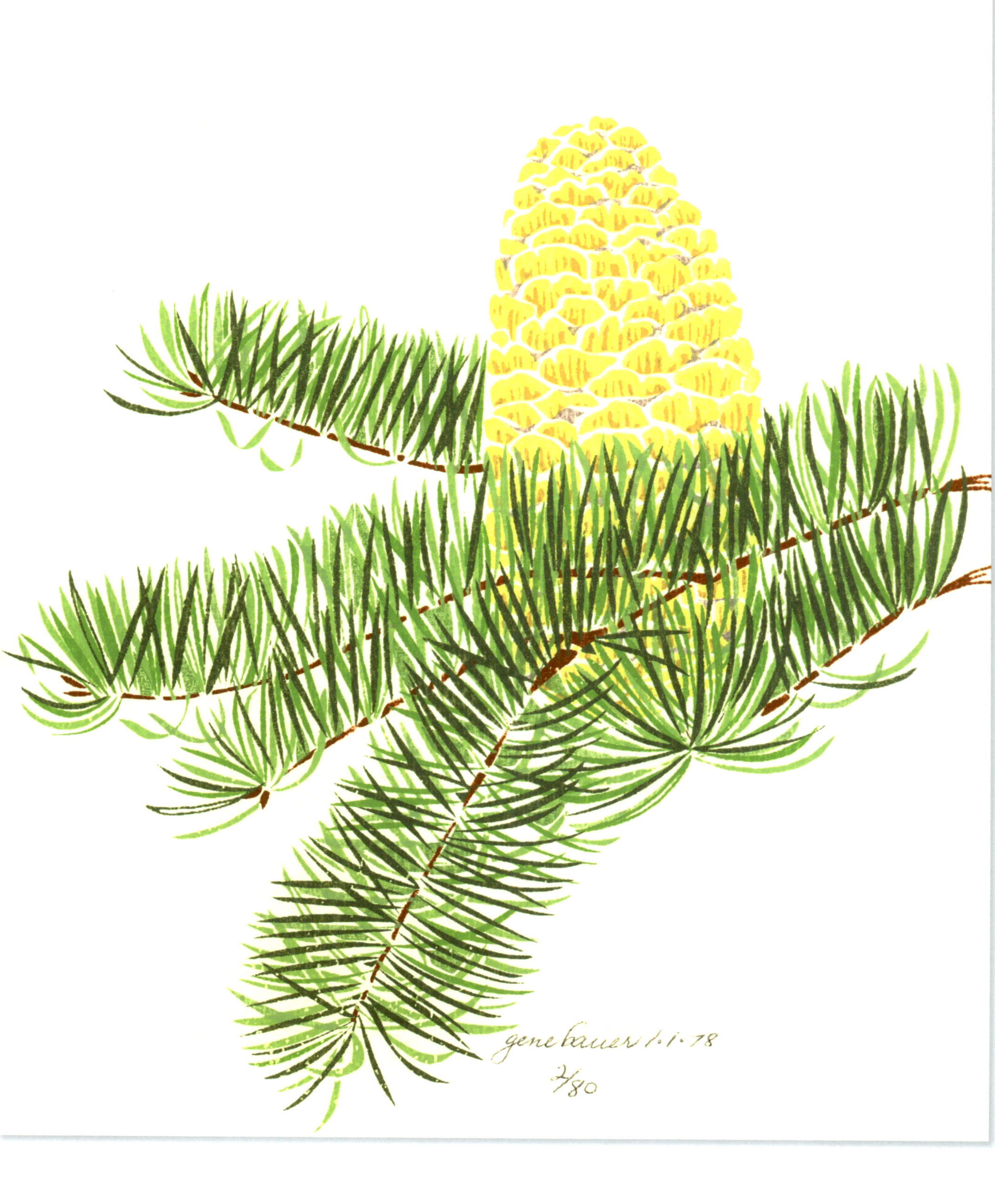
gene bauer 1.1.78

Early January 1978

Thanks to two kindly rangers of the U.S. Forest Service, I had the best excursion into what can only be described as a wilderness paradise, an area of high mountains, lakes, rivers, and seemingly endless forests of the southern Cascade Range and northern Sierra Nevada Range, including two national forests, Plumas and Lassen, and one national park, Lassen Volcanic. This is Feather River country, the place where canyon waters flow into the Oroville Dam and eventually into the homes of many thankful people, including myself, in the southern part of the state. The Red Fir Nature Walk near Gold Lake led me to the subject of my efforts here.

Such incredible silence! No automobiles, no buses, no motorcycles, no streetcars, no trains, no airplanes, no people. At least these were the conditions when I took a gentle fifteen-minute walk here. Just two tenths of a mile into the nature trail I came upon an almost pure stand of virgin (unlogged) *Abies magnifica* (red fir) trees. The average age of the trees in this even-aged stand is 275 years. There are very few *Abies concolor* (white fir) and even fewer *Pinus monticola* (white pine) growing here so this might be as near to a pure stand of red fir as is possible to find without serious trekking. And what an impressive stand it is. Wee seedlings only a few inches tall, some only a few feet tall, sterling specimens from five to fifteen feet tall, and many monarchs rising two hundred feet into the sky.

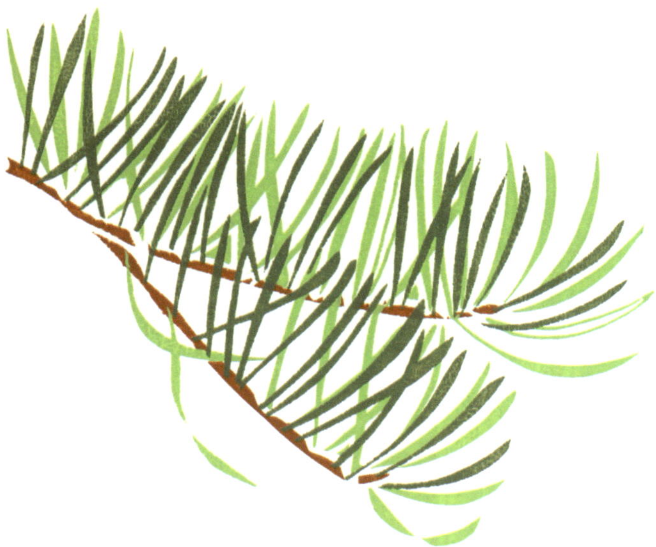

Abies magnifica

After being with this spectacular stand of trees, how could I not try to capture what I felt and saw. It is easy to understand how *Abies* became favored as the tree of choice for the yuletide season. *Abies nobilis* (noble fir) was the tree sent from the northwest to become the Christmas tree at the White House in 1977. There are about forty species in the genus *Abies,* all native to the northern hemisphere and nearly all found at high elevations. *Abies* is the ancient Latin name for the tree commonly known as fir. *Magnifica.* Aptly named, I will attest. Strikingly beautiful, exalted, sublime, exceptionally fine, eminent, distinguished—magnificent. This species is popularly known as red fir, not nearly as hyperbolic as its botanical name but just as fitting. Its bark is a red brown and its wood when cut takes on a reddish tinge.

All fir trees are graceful, especially in their youth. As pyramidal evergreen trees, they are usually symmetrically perfect. Their branches are arranged in regular whorls, often in groups of five protruding horizontally from the central trunk of the tree like spokes from the hub of a wheel. The flat sprays taper into a spire at the top. The short needles are arranged spirally around the branch but curve in such a way that they appear to grow from only two sides. Red fir needles are usually blue green, with some trees definitely more blue than green.

The cones of all fir trees are erect and always borne high in the crown. They ripen in one season and by September disintegrate while still on the tree. Cone scales and seeds are scattered on the wind and this is the reason fir cones are rarely found intact on the ground. The only cones found whole at your feet have usually been severed by the tree's dexterous interloper, the squirrel.

Few sights are more appealing than a fir tree upon whose lofty branches snow has recently fallen. As the snow nestles in among the needles and slowly melts, the branches drip with a sparkling iridescence. A tree covered with snow and enhanced by the final touch of glistening icicles is magnifica, indeed.

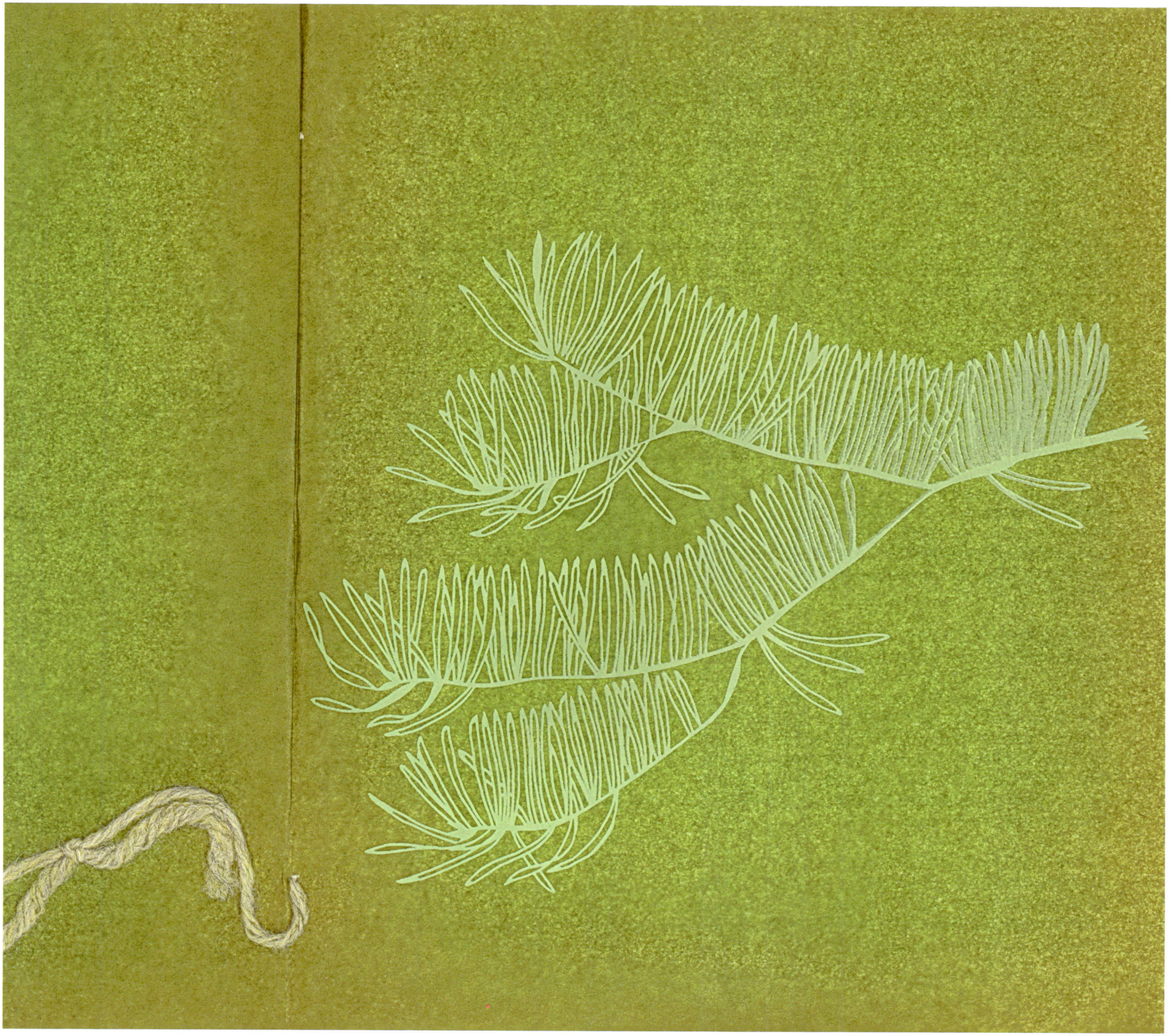

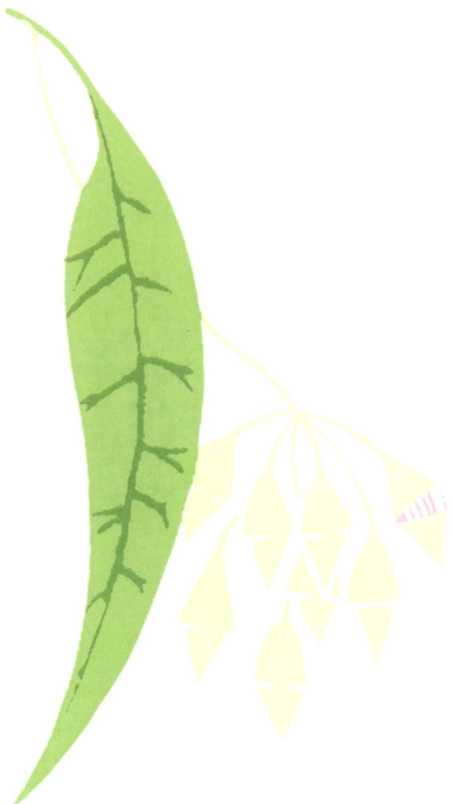

Red ironbark eucalyptus *(Eucalyptus sideroxylon)*

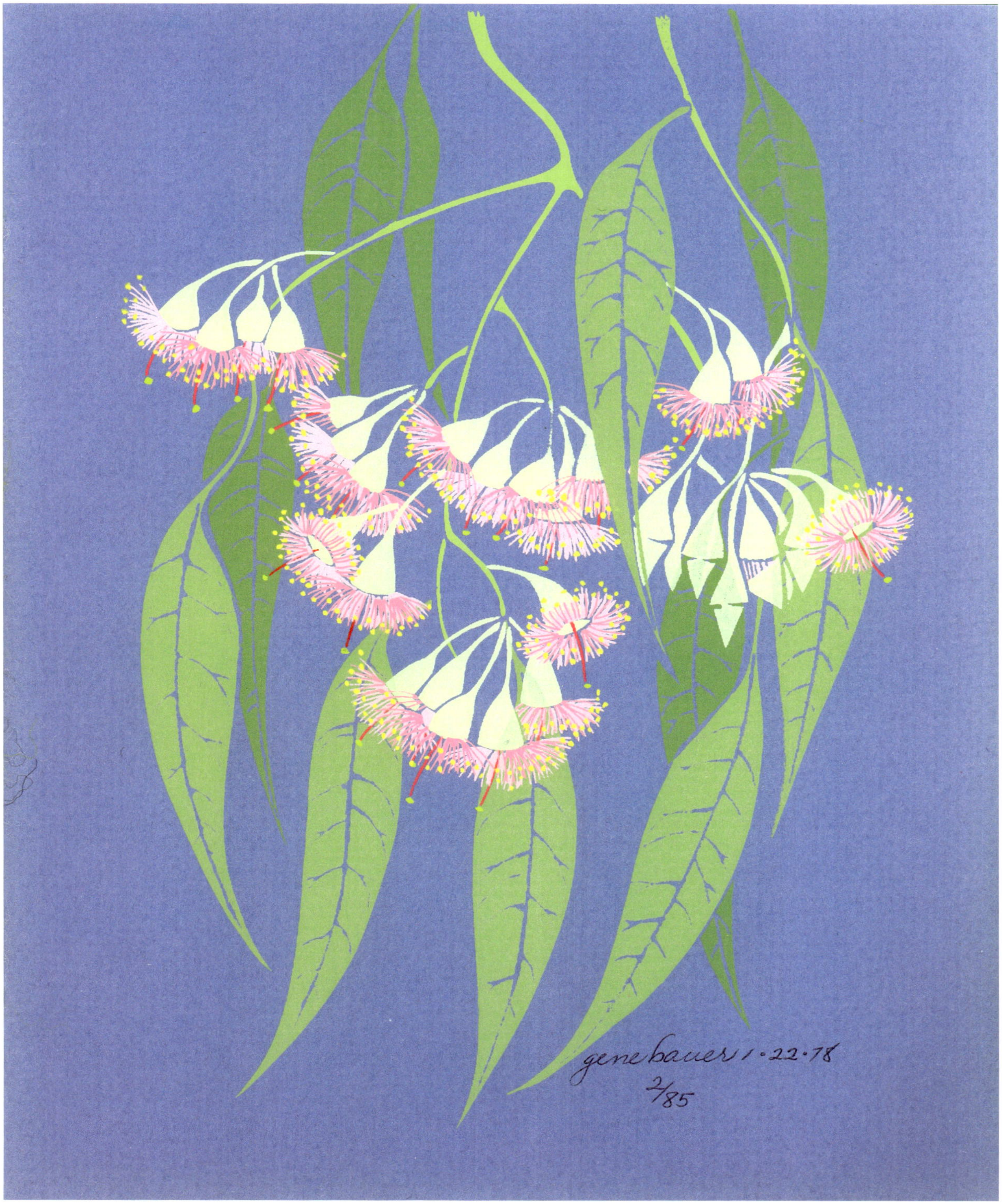
gene bauer 1·22·78
2/85

January 1978

Few, if any, public gardens in this state have a more interesting history than does the South Coast Botanic Garden, one of the world's first botanic gardens developed on a completed sanitary landfill. Its evolution might be described as a Cinderella story: the site transmogrified from the Pacific Ocean floor to a large mining operation; from that to a trash dump; and finally, from a sanitary landfill operation to eighty-seven acres of plant growth, including seven hundred genera, more than two thousand species, and a lake complete with waterfowl, a stream, and pleasant walkways. What used to be a wasteland is now a garden, an outstanding example of land reclamation. The eyes of the world continue to watch it with interest as a solution to the problem of waste disposal.

Upon visiting this garden I was captivated by many unusual flowering plants, any of which would have made an attractive cover. However, none satisfied me. But as I stood scanning the garden's skyline, suddenly it came to me … *Of course! What else?*

Eucalyptus

To think that I almost completed this series of studies without including a tree I have known all my life and to which many fond childhood memories are linked. I grew up in Orange County where the only trees I knew, other than some deciduous fruit trees and an occasional palm and pepper tree, were orange trees (grown in groves for the fruits they produced) and the many rows of eucalyptus trees (planted for windbreaks and for fuel). Warmed by eucalyptus wood, our home was infused with its scent. Like many a California native of that mid twentieth-century era, I thought the whole world consisted of only orange trees and eucalypts. In an area of Long Beach not far from the site of the present South Coast Botanic Garden was the first forest I ever knew, acres and acres solidly planted with those gangling trees with drooping leaves—it is extremely difficult for me to visualize the California landscape without eucalyptus trees.

Few other introduced plants have adapted so well to California conditions as have those of the genus *Eucalyptus.* They are so prevalent in this state that many people erroneously regard them as indigenous. As the most widely planted nonnative trees in the state, it is possible to drive several hundred miles without ever losing sight of them.

The first ones, planted here in 1856, were so successful that by 1870 they were being planted in great quantities for shade, beauty, windbreaks, and firewood. Between 1904 and 1912 thousands of idle acres were planted in an ill-advised plan for timber. They are still planted in great quantities but not for the get-rich-quick-by-timber-sales reason. Many more species are now available, and people are far more discriminating in selecting certain species for very definite purposes. They are now grown for their beauty, tolerance to drought conditions, immunity to pests, and extremely rapid pace of growth, sometimes as much as ten to fifteen feet a year.

The genus *Eucalyptus* is a large one containing more than six hundred species, all native to Australia, nearby islands, and Indo Malaysia. The name eucalyptus is from the two Greek words *eu,* meaning "well," and *calyptras,* meaning "lid" or "cap," in reference to the cap-like covering over the immature flower. Fortunately, these trees are commonly known by their botanical name, but some call them gum trees or blue gum trees. Technically, the name blue gum should be used only when speaking of one species, *E. globulus,* the Tasmanian blue gum planted in great numbers during the eucalyptus boom of 1904–1912.

Eucalypts are evergreen trees and shrubs. Nearly all feature beautiful bark and often are grouped according to the characteristics of this bark. The gums have exquisitely smooth bark that peels off in irregular patches revealing lovely, soft, subtle colors; the bloodwoods have scaly bark; the stringy barks are just that, with the bark peeling off in long fibers; and the ironbarks grow a dark, furrowed bark.

The flowers on all species are similar, differing only in size and color. They are borne in umbels or clusters and are composed of many showy stamens. Colors range from white, yellow, orange, pink, rose, and red. Some species produce extraordinarily beautiful, conspicuous flowers while other species bear flowers that are hardly noticeable. All have interesting, woody seed capsules.

Leaves are usually aromatic and range in color from silver to blue to blue green to green. Some species have round leaves, some oval leaves, and some sport long, sickle-shaped leaves. All eucalypts are fascinating trees and much is written about them; this is barely an introduction.

Eucalyptus sideroxylon

Nearly all the botanic gardens in California could boast about their outstanding collections of eucalypts, and South Coast Botanic Garden is no exception. It was awfully difficult to choose one species for my drawing. That day in October many were blooming and I knew many would continue blooming throughout the winter months. It seems eucalypts bloom all year long, as if they haven't quite made that final adjustment from the southern hemisphere to the northern. The specific name of the one I chose, *E. sideroxylon,* means "red ironbark." Other species bear showier flowers and perhaps finer leaves, both in color and shape, but to me, *E. sideroxylon* epitomizes the eucalyptus tree.

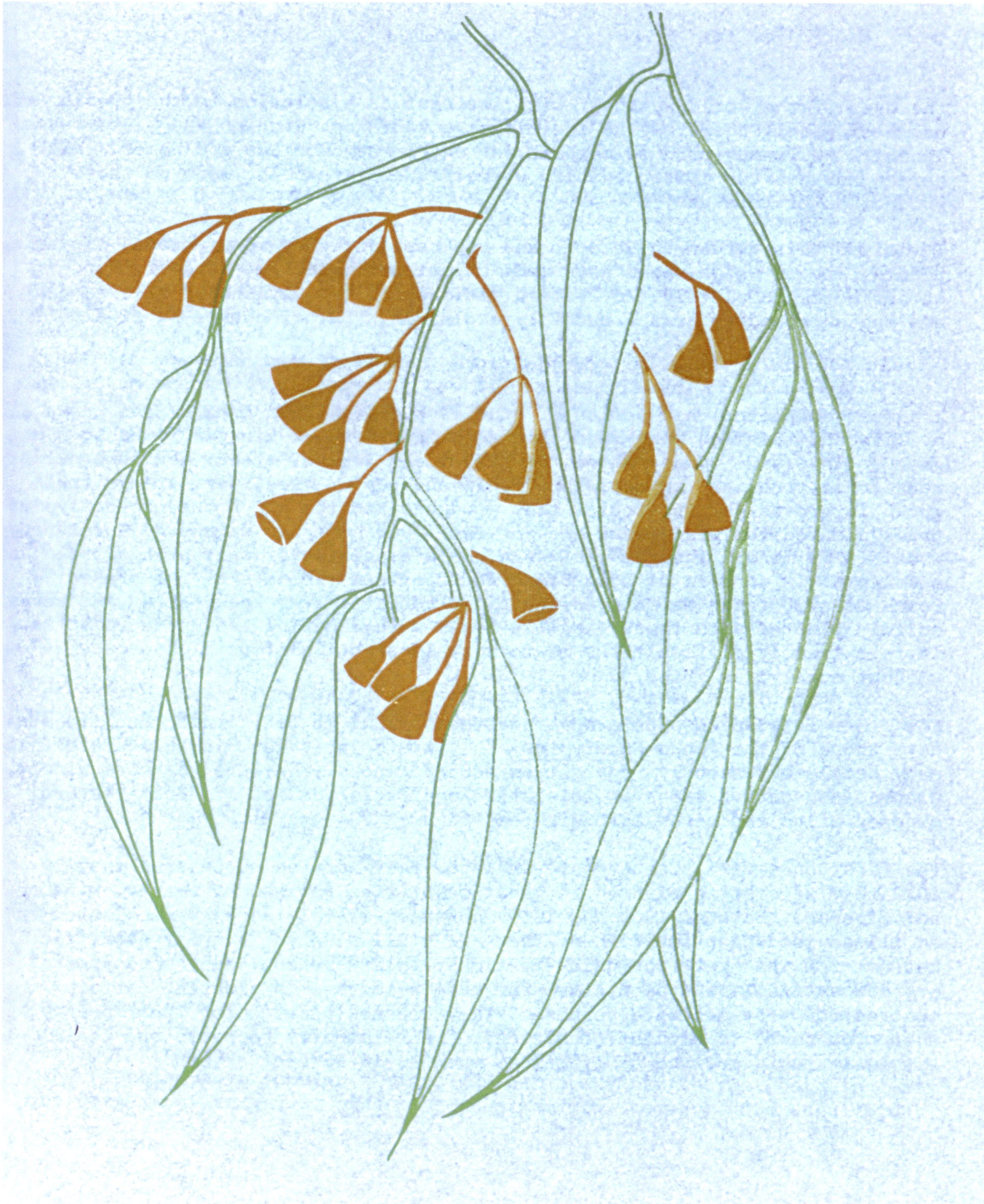

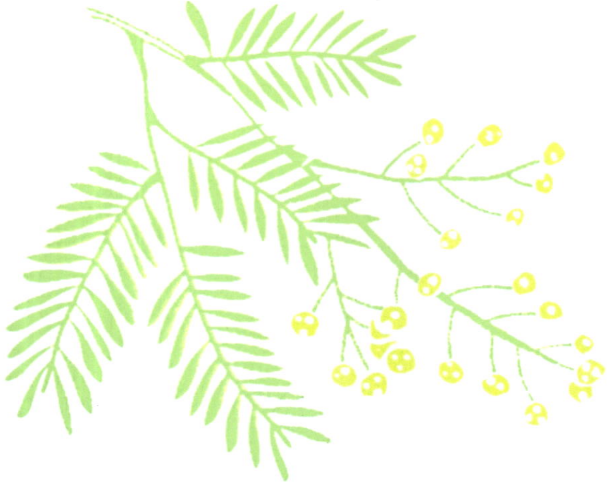

Bailey acacia *(Acacia baileyana)*

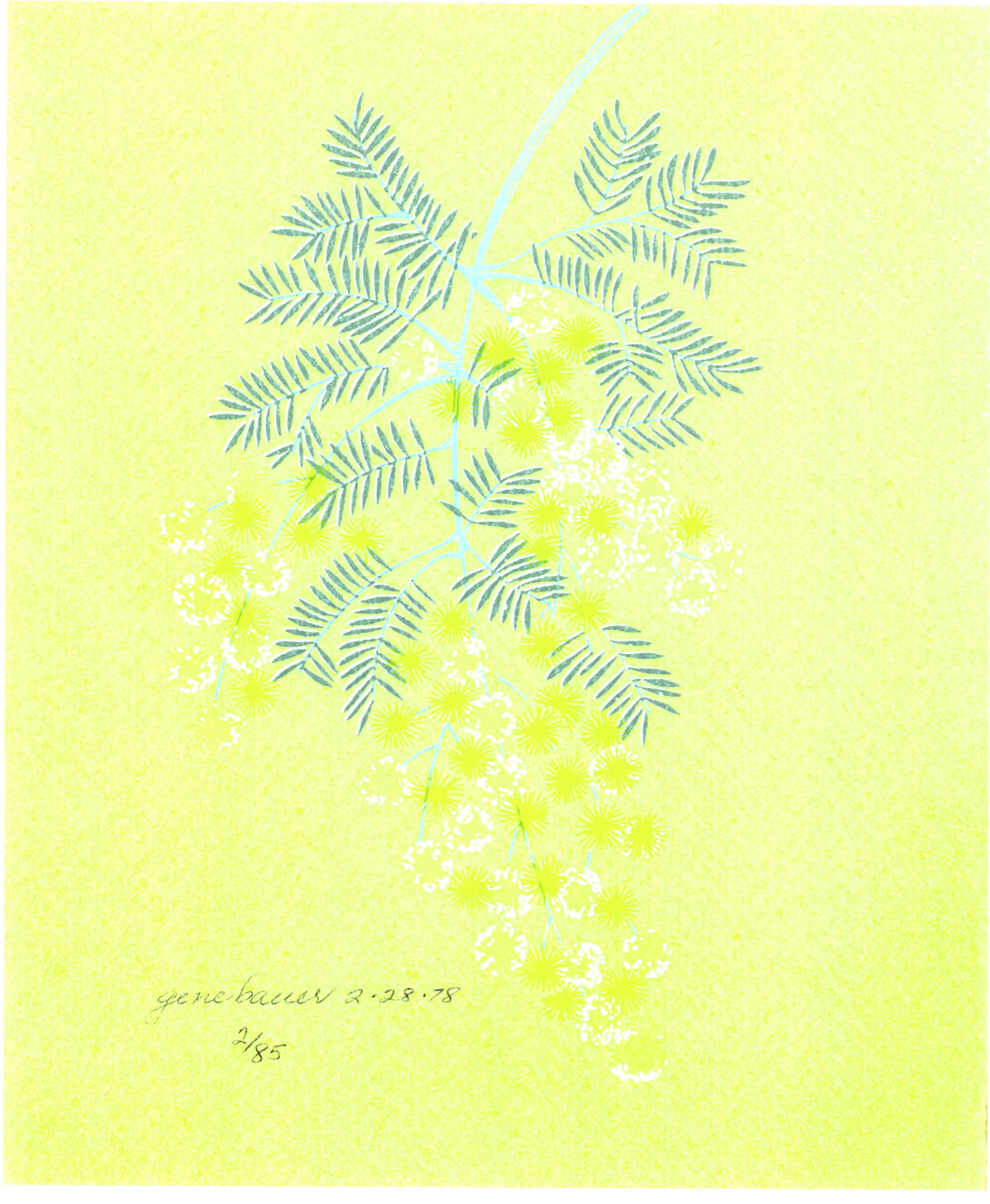
gene bauer 2·28·78
2/85

February 1978

The eighty-three acres of the Rancho Santa Ana Botanic Garden located in Claremont is devoted entirely to the cultivation and study of plants indigenous to the state. Entering the garden through a breezeway, you will be greeted by a magnificent vista that *is* California. My visit here in early February offered a particularly breathtaking panorama of the San Bernardino Valley. Snow-clad Mount Baldy was clearly visible, with peaks of lesser stature of the San Gabriel Mountains in the foreground, all clothed in rich deep blue hues. It formed a dramatic background for this garden: on the left were tall, open, lacy, blue-gray Digger pines; on the right, the denser, tighter, greener Monterey pines. In between these pines were vast pools and clumps of the varying heights, hues, and textures of these colorful Californians: manzanita, *Ceanothus, Fremontodendron,* toyon, madrone, *Dendromecon,* and *Calycanthus.* In the distance, jutting out of these myriad green, blue-green, and blue masses, the white bark and handsome structural form of the deciduous California black walnut *(Juglans hindsii)* repeated in a staccato silhouette the sparkling whiteness of the far distant Old Baldy. In the immediate foreground, many clumps of wildflowers surrounded a small pool of water. Mirrored at the water's edge were the first blooms of the first flower of this state, *Eschscholzia californica.* Though only one of the many vistas in this garden, it seemed a grand summation of the beauty to be found in this state, from snow-capped mountains to reflections of California gold.

Panoramas in this garden, as in all gardens, change with each season. Perhaps the spring months, usually March, April, May, and early June, are the most exciting with the many wildflowers in full bloom as well as shrubs and trees.

Situated on thirty-nine acres of rugged, hilly terrain, the University of California at Riverside Botanic Gardens afford spectacular vistas that are entirely different from gardens located in other parts of the state. Included here are two prominent arroyos with steep slopes that vary in elevation from 1,100 feet at the entrance to 1,450 feet. It is possible to walk the entire perimeter of this garden and see stunning views of the town of Riverside below and across to portions of the adjacent Box Springs Canyon, covered with its enormous, weathered granite boulders. In the beginning, lack of funds dictated that the gardens would be incorporated with the general landscaping on this campus. Outstanding specimens of many unusual species, clearly labeled, may be seen around the buildings. Sometime in the early 1960s the decision was made to have a separate and complete botanical garden here, but it wasn't until 1978 that the plantings begun in 1963 took shape into more than two thousand different species.

One of the most outstanding areas among many here is the Australian section. Meandering around there I almost forgot I was really in Southern California. Engulfed as I was in such a fantastic array of leaves, unusual both in color and form, and flowers, flower buds, seed capsules, and aromas, I would not have been surprised had a kangaroo hopped across my path. Included in the Australian section are many species from these genera: *Hakea, Grevillea, Callistemon, Eucalyptus* … and *Acacia.*

Acacia baileyana

Fountains of golden sunshine in midwinter! How startling to see trees whose rounded crowns are one solid mass of gleaming yellow in December, January, and February. Although it is during these months that the most spectacular color display is unveiled, it is possible by careful selection of species to have *Acacia* bloom every month of the year.

The genus *Acacia* is a large one containing about eight hundred species native to the tropics and warmer regions of the world with about half of them native to Australia. In Australia, they are second in importance only to the *Eucalyptus. Acacia* is from a Greek word *akis,* meaning "a sharp point," and is in reference to some species that have very formidable sharp thorns. In California they are known commonly by their botanical name. In Australia they are commonly known as wattles, a name derived from the fact that early settlers used *Acacia* branches for temporary "wattle and daub" dwellings. The branches were interwoven to construct walls that were later plastered with mud.

Acacias are evergreen trees and shrubs whose foliage differs widely both in color and form. Leaf color ranges from silvery gray to deep waxy green. Leaf form ranges from feathery, finely divided leaves to flattened leaf stalks that serve the purpose of leaves. Flowers are similar on all species differing only in color (yellow, orange, or white). They are composed of many showy stamens gathered into globular heads appearing as small fluffy balls in large panicles.

Named for F. M. Bailey (1827–1915), a colonial botanist of Queensland, this tree is commonly known in the United States as Bailey acacia. It is native to New South Wales and is known there as Cootamundra wattle. When in bloom, the silvery gray foliage of this small evergreen tree is almost hidden by the sparkling clear yellow flowers. The tree is so heavily laden with astonishingly profuse flowering that it appears to be dripping with these airy little yellow puffballs. This drought-resistant tree has been widely used in freeway plantings in California where it tolerates much neglect. On drab, dreary winter days, it is greatly admired and appreciated when it bursts open as a bright and cheery fluffy cloud of yellow amidst a sea of somber green plants.

Japanese flowering cherry (*Prunus serrulata* 'Kwanzan')

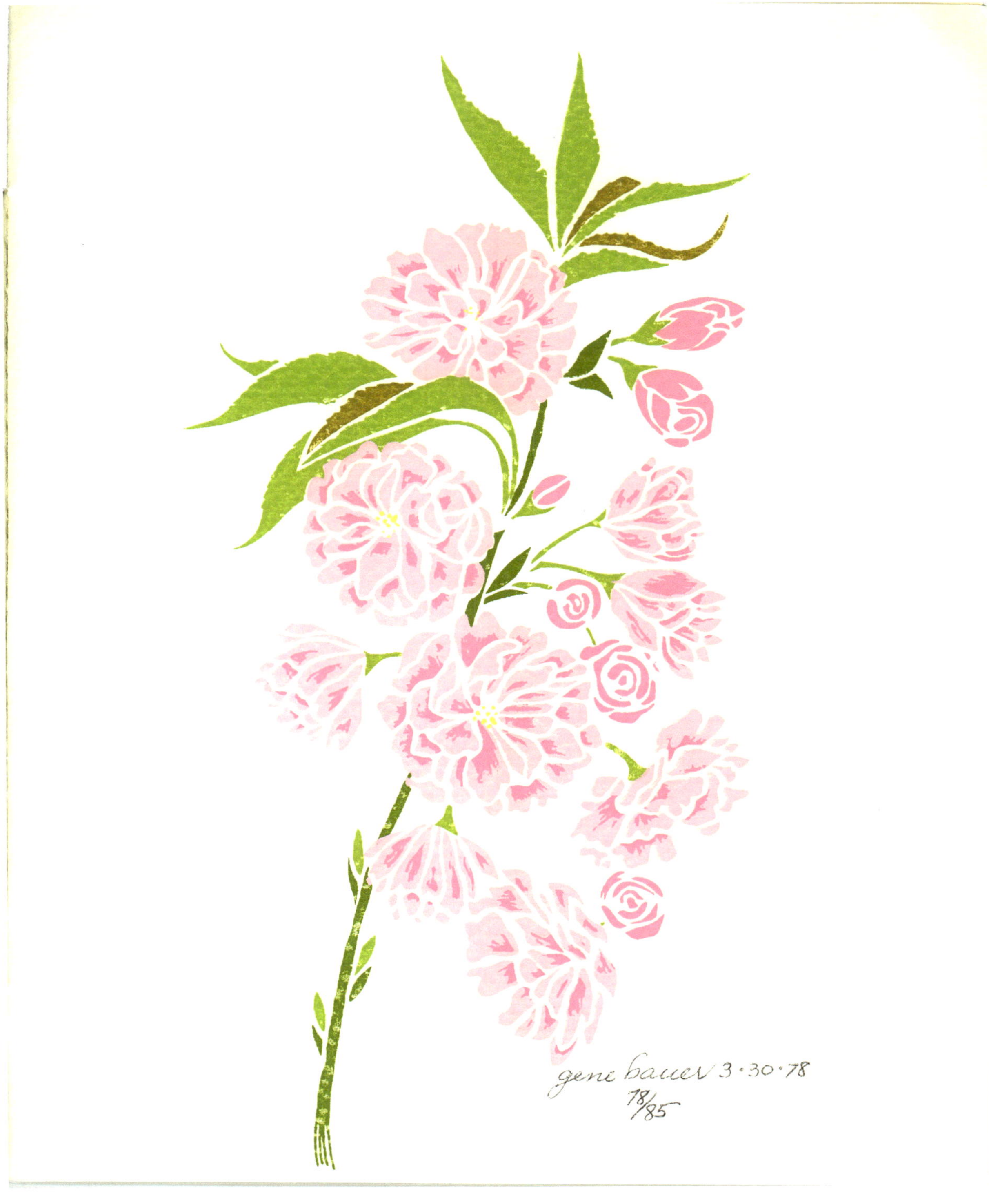
gene bauer 3·30·78
78/85

March 1978

John Muir arrived in California in 1868 and, heading for the Sierra Nevada, he described the valley I visited 110 years later when it was without subdivisions, freeways, and shopping centers:

> "We crossed the bay by the Oakland Ferry and proceeded up the Santa Clara valley to San Jose. This is one of the most fertile of the many small valleys of the coast… It was now spring-time, and the weather was the best we have enjoyed. Larks and streams sang everywhere; the sky was cloudless and the whole valley was a lake of light. The atmosphere was spicy and exhilarating… This San Jose sky was not simply pure and bright, and mixed with plenty of well-tempered sunshine, but it possessed a positive flavor, a taste that thrilled throughout every tissue of the body.
>
> "The foothills of the valley are in near view all the way to Gilroy, those of the Monte Diablo Range on our left, those of Santa Cruz on our right; they are smooth and flowing, and come down to the bottom levels in curves of most surpassing beauty. They are covered with flowers growing close together in cloud-shaped companies, acres and hillsides in size, white, purple, and yellow, separate, yet blending like the hills upon which they grow…."

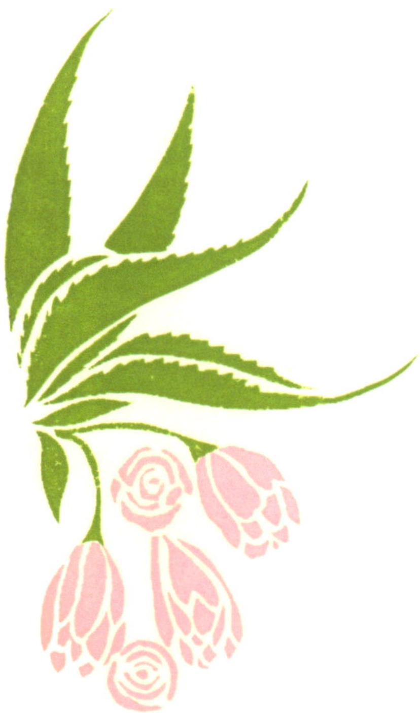

I arrived in this area in springtime, too, and if I could have erased the automobiles, noises, and houses, the Santa Clara Valley would have been exactly as he described it. The Monte Diablo Range is still there, too early for wildflowers but the most exquisite green imaginable. Mrs. Betzi Akerstrom of the Educational Park Library called them "my velvet hills." And so they are, and what a lovely background for the places I visited.

In 1959 Mildred Overfelt, daughter of a pioneer family, gave thirty-three acres of land to the city of San Jose with the wish that it offer people both an aesthetic and educational experience. The Overfelt Gardens opened to the public in 1966, and notably, developed a fragrance garden for the blind. Its three lakes provide food and shelter for waterfowl, including many migratory birds, and also function as percolation ponds helping to maintain the water table in this valley. Not terribly far from there, in the Saratoga foothills, is the Villa Montalvo Arboretum, whose grounds were planned by John McLarren, the landscape designer who also designed San Francisco's Golden Gate Park.

My inspiration came from two gardens featuring what traditionally ushers in spring for the Japanese, the delicate pink blossoms of the cherry trees. I saw them at the Japanese Friendship Garden, patterned after San Jose's sister city's Okayama Korakuen Garden, and again in Saratoga's Hakone Gardens, designed by Mr. N. Aihara, son of the head gardener at the Imperial Palace in Tokyo. Japanese gardens are lovely every day of the year primarily because of the variety of choice evergreens that are selected. Yet in spring the flowering cherry trees burst into bloom, glorifying the world with their beauty for an intensely short period of time.

Prunus serrulata 'Kwanzan'

Among the most spectacular species in the genus *Prunus* are the small trees known as Japanese flowering cherries, which include *P. serrulata, P. subhirtella,* and *P. yedoensis,* along with the many cultivars derived from them. Flower color may be white, pale pink, or deep pink, and flower form may be single, semidouble, of full double. Perhaps the most popular of the flowering cherries is *P. serrulata* 'Kwanzan'. It grows from twenty-five to thirty feet tall with stiffly upright branches that are almost hidden by the massed clusters of deep pink, double flowers.

The species are indigenous to the mountains and forests of Japan where they are much cherished by the Japanese people as a symbol of their philosophy of life. A poem by Taiko Hideyoshi written four hundred years ago expressed it this way:

> "Born into this world like a dewdrop,
> The present life is but a dream in the dream
> And soon vanishes into nothingness."

Southern magnolia *(Magnolia grandiflora)*

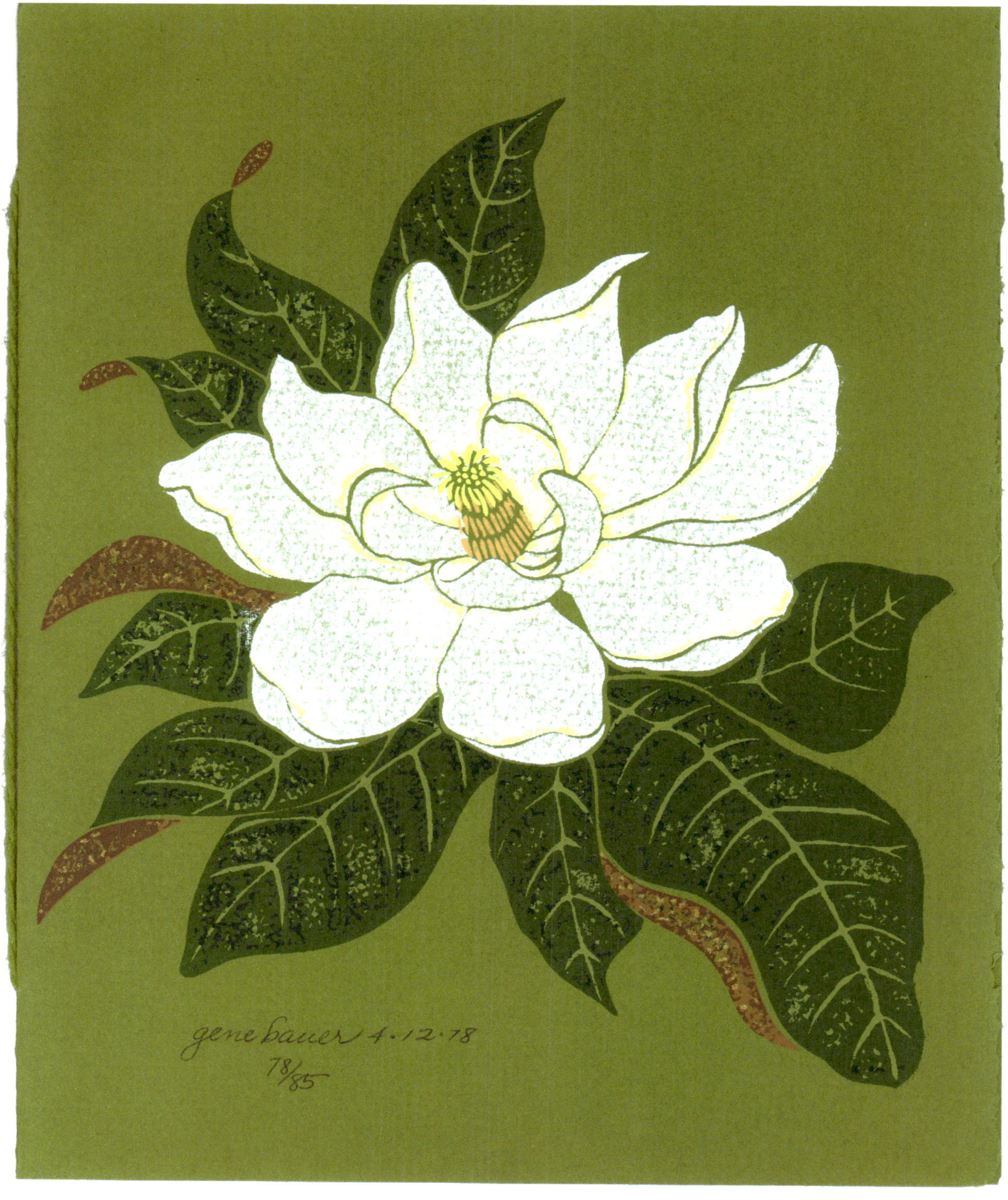
gene bauer 4.12.78
78/85

April 1978

Some of the finest tree specimens in California can be found on the grounds of the state capitol in Sacramento. Many of the trees on these forty acres are from the original plantings of 1870, and they include examples from every continent and climate in the world. With so much to choose from, I was experiencing some difficulty in deciding until I realized that an ending should be no less grand than a beginning. I began the Golden Botanical Gardens series with the much-admired *Liriodendron tulipifera* as the cover of the first booklet, so it seemed appropriate to approach the end of the series with a plant equally as magnificent.

Magnolia grandiflora

You know this tree. It is the state tree of Mississippi, much loved in the South, but it has always been and continues to be beloved by Californians as well. Among the most respected introduced trees in the state, it lines streets in both commercial and residential areas, and many can be seen on the capitol grounds. In California, this tree is known simply as magnolia or southern magnolia, but in its native habitat, the southeastern United States, it is called bull bay, possibly because of its habit of taking over wherever conditions are favorable. A noble evergreen capable of growing taller than one hundred feet, in elegance of stature *Magnolia grandiflora* virtually stands alone in its genus.

The genus *Magnolia* does include about eighty other species, native to the southeastern United States, Mexico, Central America, and Asia. The genus includes both evergreen and deciduous trees and large shrubs, all with beautiful foliage, spectacular flowers, and attractive fruit. Yet *grandiflora* is still incomparable even among such lovely relatives. Its specific name means large-flowered, and the leaves are large, too, green and glossy so that they form a perfect background for the white, waxy flowers. The heavily lemon-scented blooms begin to open in May and continue throughout June and early July. As the petals fall, the center of the flower develops into a cone-shaped seedpod. After the pod matures it bursts open to reveal the jewel-like seeds in a brilliant red.

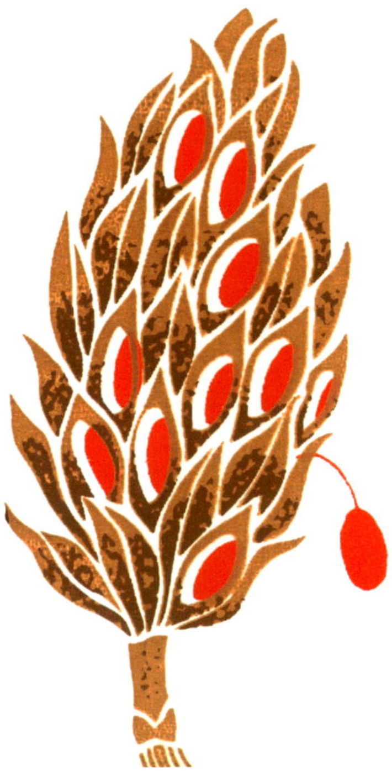

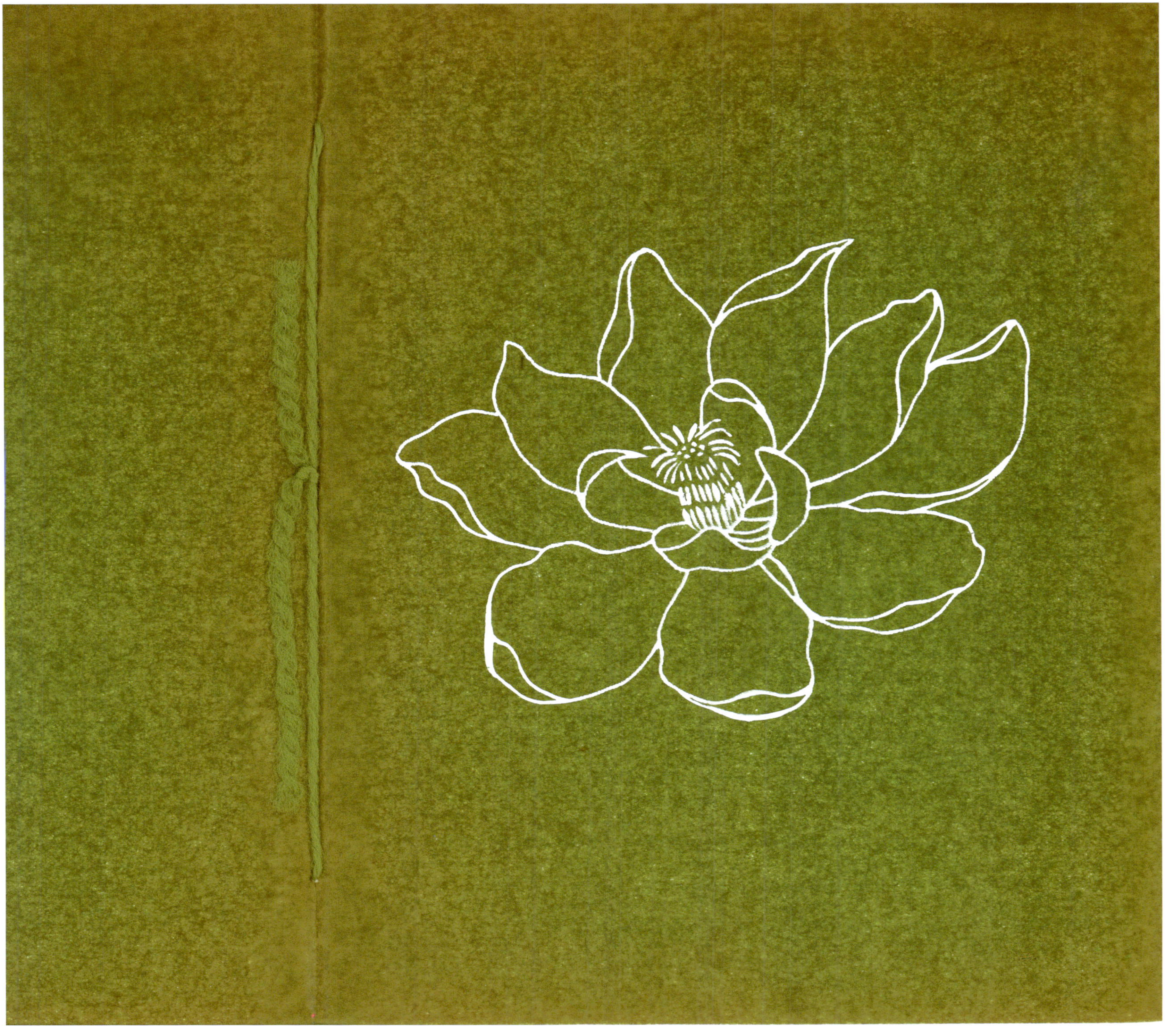

Early May 1978

In late April of this year, a stranger was admiring our mountainside and asked me, "Is this your occupation or your hobby?" It is neither. It is my love.

Spring of each year is very special to me, and spring 1978 is particularly special. It is the twentieth anniversary of my first purchase and planting of spring bulbs. Grand and glorious festivities were being contemplated, but then came the abominable weather (five continuous days of clouds or fog, rain, hail, sleet, snow, wind, and two days of sunshine then a repeat of five days of who-knows-what). Even today—on May 1—it is snowing. This kind of weather, along with the road conditions that accompany it, has prevailed since late December of last year, so I have abandoned all plans for any kind of outward celebration. Inwardly, however, I rejoice in the knowledge that I found *the* flower that made my dream world a reality.

Twenty years ago I saw a small group of tulips and daffodils (probably one dozen of each) growing in a yard near our mountain property. I asked the owner what time of year she planted them and she replied, "In September and October." So that fall I went to a nursery to purchase my first bulbs, selecting them according to the pretty photographs accompanying their boxes. I clearly remember the clerk trying to sell me some hyacinths. Hyacinths? What did they do?! They were terribly expensive, I had never seen one, and the photographs didn't excite me, so no hyacinths that first year. Tulips and daffodils were my first purchase. I paid for the bulbs, adding in my complete ignorance, "Oh yes, and which way is up?"

I've planted three thousand new bulbs each year for the last two decades, so I guess I know a few more things by now (see page 238). Every year I add more drifts as well as some of the more expensive single bulbs. Right now I am planning where the next and continuous sweeps of color should be located and also what colors they should be, either all yellow, yellow/orange, all white, white/yellow, or that exquisite soft sulphur-lemon yellow of which I am especially fond.

Narcissus 'Hawera'

With hundreds of different flowers to choose from, including trumpet, large-cut, small-cup, double, *triandrus, cyclamineus, jonquilla, tazetta,* and split-corona *Narcissus,* how could I possibly make a choice of what to draw? Some of my favorites are 'Daydream', 'Fawnglo', 'Rushlight', 'Binkie', 'Stint', 'Amberjack', 'Euphony', 'Rainbow', 'High Note', 'Pipit', 'Oryx', 'Old Satin', 'Arish Mell,' 'Tuesday's Child,' and 'Ceylon'.

I chose an older cultivar, 'Hawera'. It is a miniature *triandrus* hybrid growing about ten inches high and producing two to three flowers on each delicate stem. The small, nodding flowers are beautifully proportioned and a soft shade of yellow. I couldn't resist its graceful elegance.

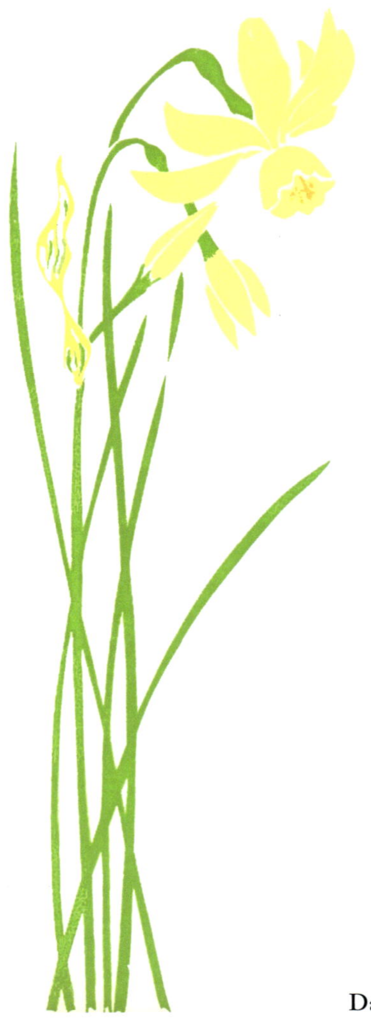

Daffodil (*Narcissus* 'Hawera')

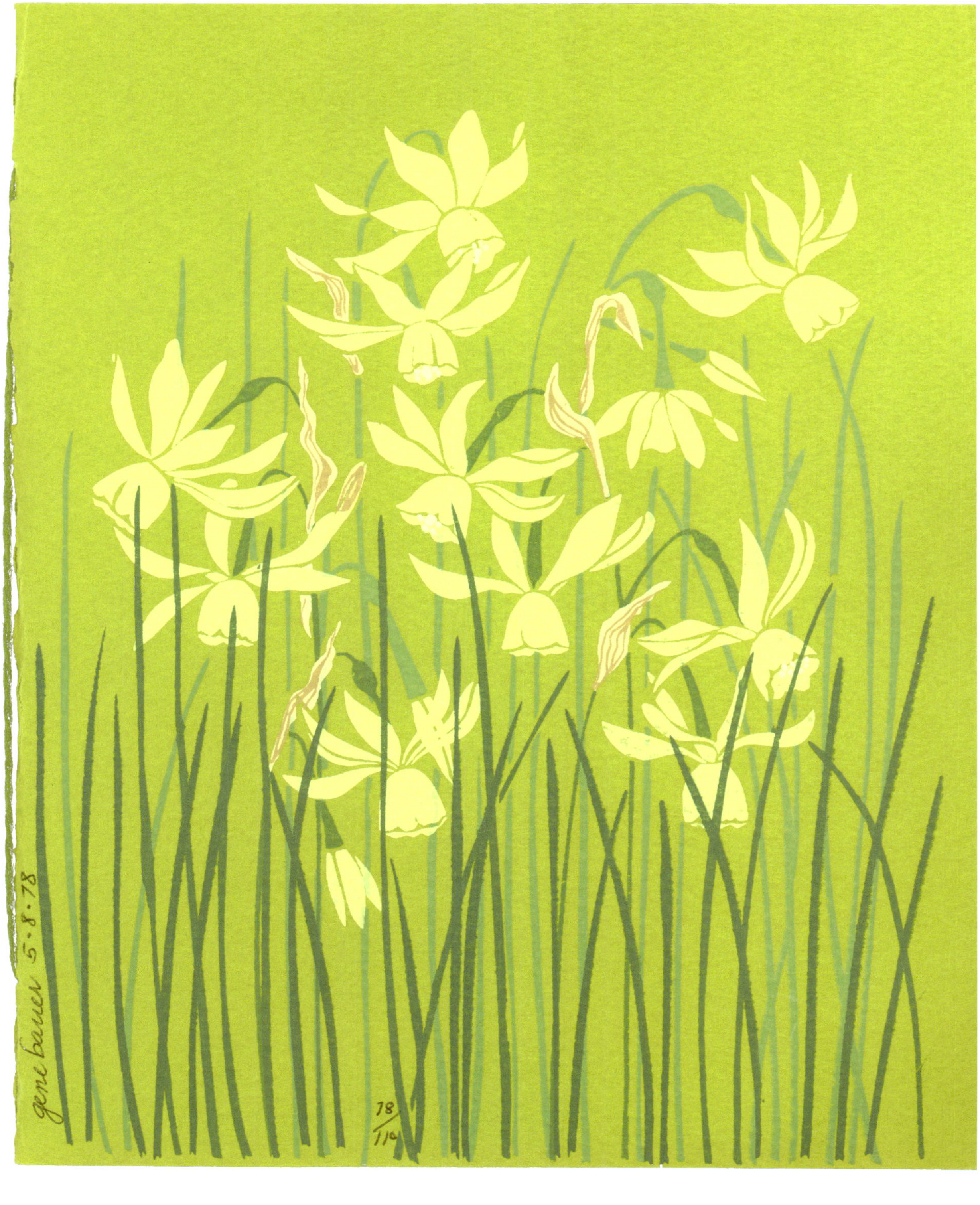
gene bauer 5·8·78
78/110

White geranium *(Pelargonium hortorum)*

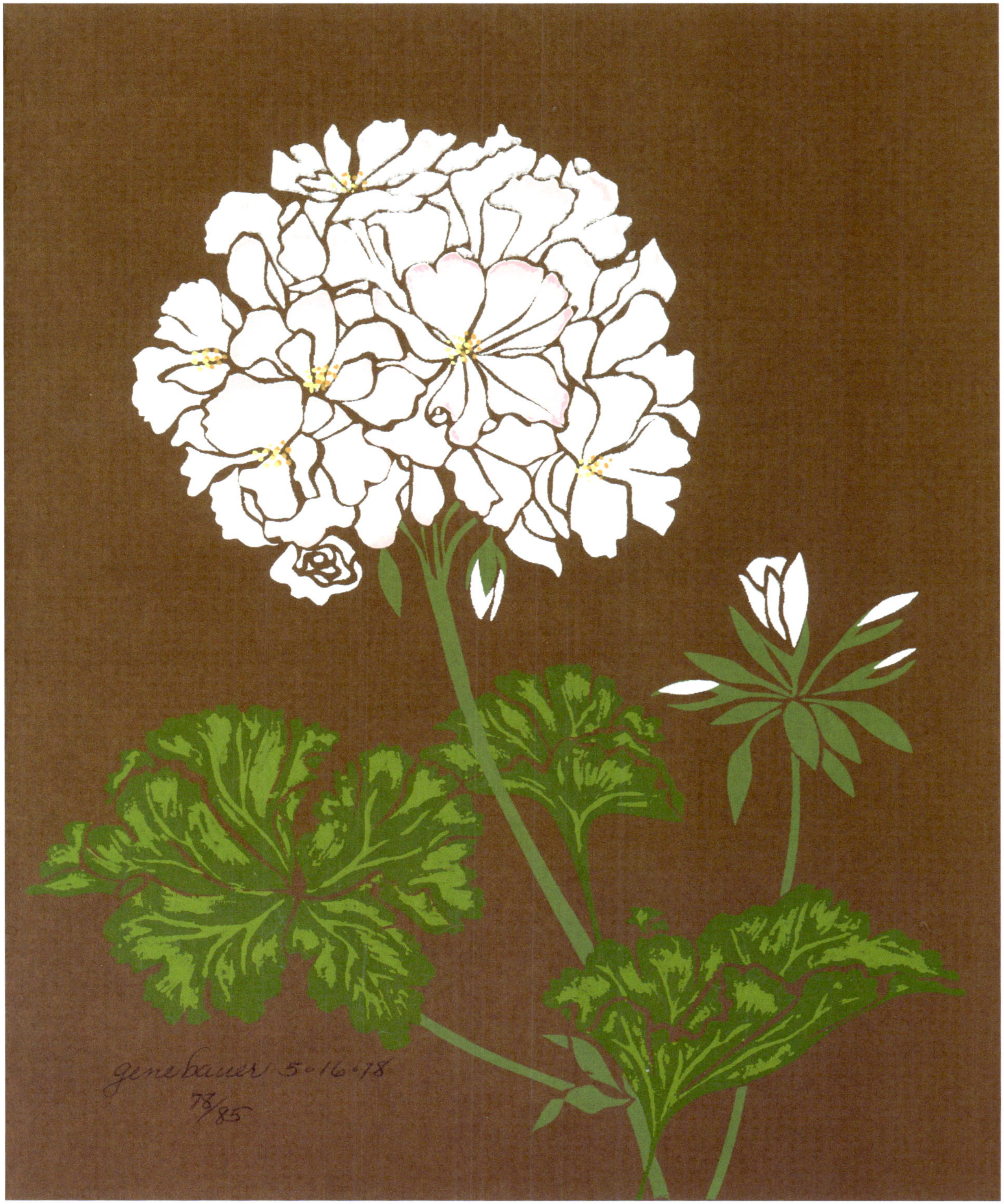
gene bauer 5·16·78
78/85

Late May 1978

In the past two years, I personally visited fifty-six public gardens in the state of California—arboreta, botanic gardens, mission gardens, rose gardens, oriental gardens, and interpretive nature trails. But to draw the next and last plant, all I had to do was stay home.

Pelargonium hortorum

This final entry is a very special plant. This one I chose especially for my husband, Dale. He is not a connoisseur of geraniums, but he is very selective about those he chooses to grow. Dale grows geraniums only in containers and has limited himself to those cultivars that bloom freely and have good color, well-proportioned stems, and attractive foliage. He keeps the containers outside at his office in frost-free Redlands during the winter and brings them home where it is always ten or fifteen degrees cooler in the summer. He grows a stunning red cultivar ('Mexico') as an upright plant, and a lovely lavender, ivy-leaved cultivar ('Santa Paula') in hanging baskets. For me, because I adore its salmon-orange color, he also grows 'Maxim Kovalevski'.

He had tried several different white cultivars but each had some fault, such as a too-long stem, too-small flower, or unattractive foliage. One day he brought home another white cultivar, this one named 'Modesty'. About all I could say was, "Modesty? I can certainly think of many more charming names than that for a white geranium."

The plant did not have a flower at that time, but soon one developed. It was large, full, pure white on a nicely proportioned stem with attractive foliage. *Good,* I thought. *At last he has found a nice white.* I looked at it each day and after about four days I could hardly believe my eyes. The most beautiful soft pink appeared on each petal. Each day the pink grew deeper and lovelier. Of course—it was blushing. No other name could possibly be as appropriate. Indeed, 'Modesty' is a very special plant.

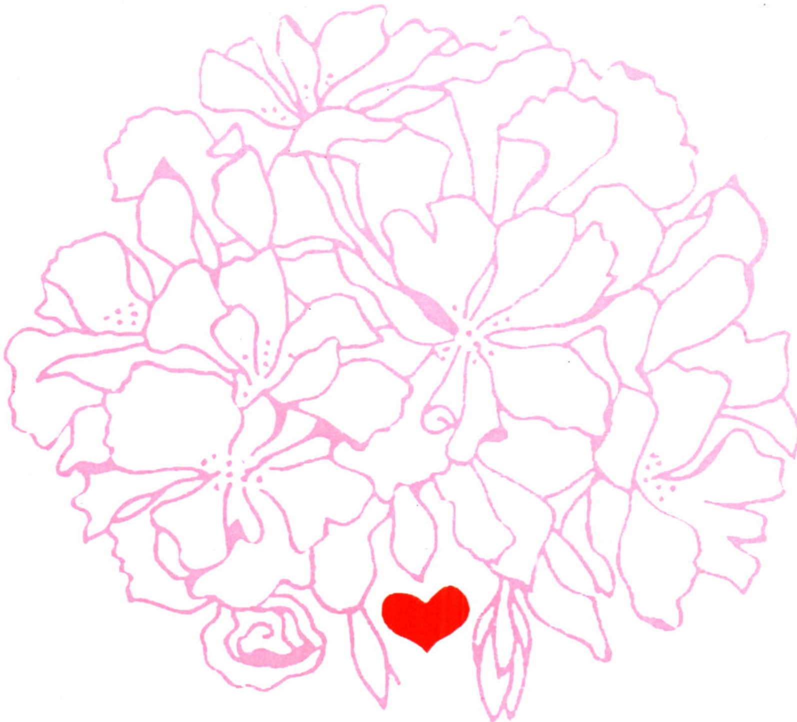

Collection 3

Living with Nature

AS A NATIVE CALIFORNIAN I grew up with many living things to appreciate around me. My childhood was spent at a time when flora, both indigenous and cultivated, was still part of my hometown in Southern California. We spent summers at the seashore in nearby Newport Beach, but home was on an orange grove just inland from there. This interior of Orange County then was known for its rich, fertile agricultural land and produce and of course for its excellent citrus groves, from which my hometown Garden Grove derived its name. The name made sense then—nearly everyone had a grove of some kind and a garden in the 1930s and '40s. Even by the 1970s when I began my drawings, few if any groves were left and the town was composed of two elements—houses and people. It seems incredible that in my youth I rode a horse on the isolated country roads by the walnut groves of neighboring Anaheim—in the exact spot where now stands the world-famous Disneyland, with accompanying motels, shopping centers, and glittering lights.

Almost every part of the state of California—coastal, coastal mountain, valley, foothill, high mountain, desert—is or was at one time richly endowed with indigenous plants. By exploring this state it is possible to see these fascinating native plants in the exact place where they were intended, mingling with complementary plants similar in their requirements for survival. Unscarred land is rapidly disappearing, but it is still possible to find areas that have escaped the ravages of the bulldozer. These natural areas are thrilling to behold and one can only marvel at the superb textures, forms, and colors found here. It seems all the plants exist happily without the interference of man.

Plants present themselves so appealingly in wilderness areas that people are often tempted to relocate them to their personal garden. In addition to being illegal (California Penal Code, Section 348a protects all native plants), chances of their survival in a cultivated garden are very slight. California native plants do not like to be grown in containers in a nursery where they are routinely watered, as are exotics, and they frequently, and sometimes quickly, succumb to this treatment. It takes knowledge to grow natives under cultivation as well as a good selection of California native plants in containers from which to make a purchase. The Theodore Payne Foundation for Wildflowers and Native Plants provides both.

Theodore Payne arrived in California in 1893 as a young man, became enchanted with the native flora of this state, but even then saw it beginning to disappear from the landscape. He believed if he could grow the native plants from seed, learn their cultural requirements, and offer them for sale to the public they would be saved from eventual extinction. He opened a nursery in 1903 and along with other plants began growing natives. By 1922 he was devoting nearly all his time to the growing of native plants until his retirement at the age of ninety. Continuing the work of Mr. Payne, the foundation created in his name in 1960 now occupies twenty-two acres situated in a natural canyon bottom with surrounding hillsides and terraces overlooking the San Fernando Valley. A nonprofit organization, its primary function is to make available to the public seeds and plants of natives that are difficult to obtain, and to offer information on natives—in this way helping to preserve rare and endangered species by cultivation.

TREES

January 1973

One day in winter I discovered a particularly handsome and interesting stump of a ponderosa pine *(Pinus ponderosa)* that became the inspiration for this serigraph. Located near my home, the tree was destroyed in a recent forest fire and cut shortly thereafter, revealing all the strains and stresses it had endured for more than two hundred years. When a tree is cut, not only can its age be determined by the growth rings, but also its entire history. Wet and dry years and natural catastrophes, for example, are recorded in the rings. Through careful observation one may learn many things.

Awareness and appreciation of trees need not wait until it's too late to preserve them. The American Forestry Association has been locating the largest living specimens of American trees since 1941, designating the largest species of a tree as champion. Three factors are weighed in sizing up whether a tree is champion material: stem circumference at 4½ feet above ground, total height in feet; and one quarter of the crown spread footage. The three measurements are added up to attain a single score; the highest score is the champion. The list of champions is AFA's Social Register of Big Trees. Some trees become victims of disease, pollution, the elements, commercialism—and others can be dethroned whenever a larger specimen of their species is discovered—so the list is constantly changing. By 1973 champions had been discovered for only half of the 865 species of trees native to the United States, a number that includes a few introduced species that have become more or less naturalized.

California has been the home of the world's tallest, largest, and oldest tree: a 368-foot coast redwood *(Sequoia sempervirens);* the giant sequoia *(Sequoiadendron giganteum),* containing 600,120 board feet of wood; and a bristlecone pine *(Pinus aristata)* estimated to be 4,600 years old. The state abounds in handsome, majestic trees, many species of which I am glad to say find the area where my husband and I live as amenable as we do. Our home is located at an elevation of 5,500 feet on the south slope of the San Bernardino Mountains in the southern part of the state. Annual precipitation during the months of November to May varies from twenty-five to fifty inches usually (though sometimes as much as eighty-five inches), with some falling as snow. Infrequently, the winter low temperature may be 15 degrees. Summers are warm and rainless and, while in the valley below summer temperatures can rise higher than 120 degrees, here it is rarely above 90.

The predominant indigenous trees in our area include seven species of cone-bearing evergreens: four species of pines *(Pinus ponderosa, P. coulteri, P. attenuata, P. lambertiana),* one species of true fir *(Abies concolor),* one "false" fir *(Pseudotsuga macrocarpa),* and one species of cedar *(Calocedrus decurrens).* Also native and in abundance here are two species of oak, one deciduous *(Quercus kelloggii)* and one evergreen *(Quercus chrysolepsis).* In protected canyons grow big-leaf maple *(Acer macrophyllum),* California sycamore *(Platanus racemosa),* California bay *(Umbellaria californica),* and along watercourses, white alder *(Alnus rhombifolia).* In very favorable locations—usually somewhat protected and with moist, rich soil—I find that glorious tree, western or pacific dogwood *(Cornus nuttallii).* About seven miles away, the canyons of the north-facing slopes are densely covered with this spectacular tree.

Some chaparral plants spill over into our area, including several species of manzanita *(Arctostaphylos),* buckthorn *(Ceanothus),* and coffeeberry *(Rhamnus californica).* Wildflowers are abundant here, including annuals, biennials, perennials, bulbs, corms, rhizomes, and even saprophytes. We are not lacking in tree squirrels either, along with coyotes, raccoons, and deer, and when snow lingers on the ground for a prolonged period of time we see the occasional fox and bobcat. Stories of bears persist, but we have not seen one. Many species of birds live in this area, and the annual visits of migratory ones are much enjoyed. For one thing, we share a mutual appreciation of trees with the birds, so how could we not feel kindred?

I'm always delighted to watch the birds scampering around, in early spring their bills packed with scraps of this and that, hurriedly constructing their nests. By careful observation it is possible to learn the exact location of the nest, often located in the lower branches of a tree and sometimes in shrubbery. Now and then, horror of all horrors, the nest is made on the ground, where there are all kinds of marauders, four-legged, two-legged, and no-legged slinking around. In past years I have observed Steller's jays, juncos, towhees, and grosbeaks building their nests, laying eggs, incubating, hatching, feeding, and fleeing the nest.

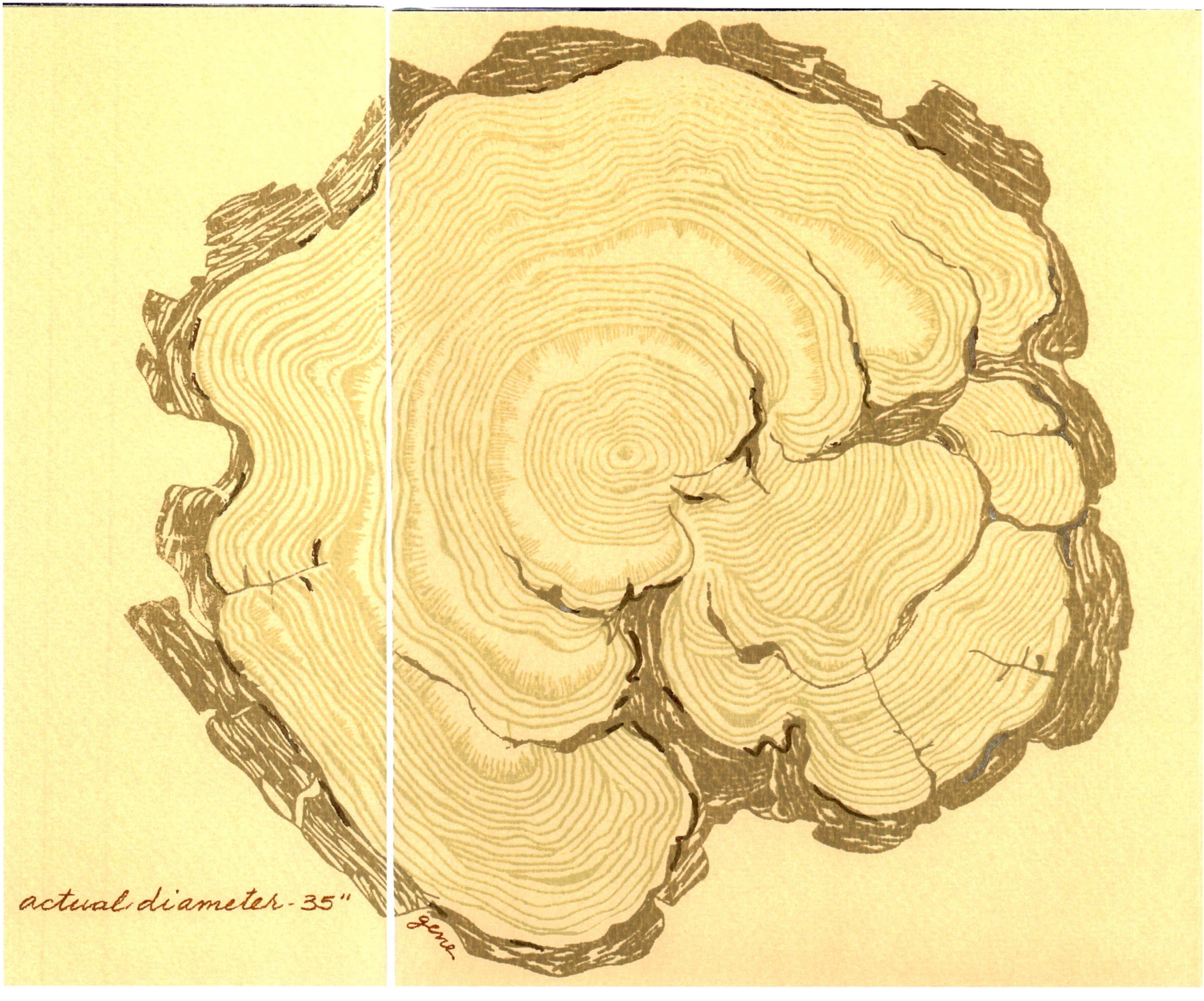
actual diameter - 35"
gene

THE YEAR OF THE QUAIL

July 1979

In March 1979 I heard and caught glimpses of an unusually large population of adult quail not far from our house. This was, of course, the mating season and I suppose all kinds of acquaintances were being made. On April 6, as I was tidying the area for the daffodil season, I noticed a large clump of unattractive perennial grass near a drift of blooming flowers about eight feet from a path and decided it must be eliminated. Kneeling down, grabbing it with my gloved left hand, ready to chop it out with the hand cultivator in my right, I pulled the grass back and discovered what had been hidden within: a nest with one egg, quite alone. I knew immediately it was a quail egg.

Of the many resident birds here, none is more admirable to me than the mountain quail *Oreortyx pictus eremophilus.* In its form, bearing, and exquisite coloring, the mountain quail is aristocratic and distinctive. Its protective nature and well-disciplined families, along with its many different calls, are a joy to behold. The quail is a ground bird and although capable of flying it will do so only when disturbed. Yet even in flight the quail gives off a distinctive whirring sound unlike any other bird. It usually retreats to lower elevations in winter, but now and then a hardy soul makes an appearance when snow is on the ground. Even then it will walk rather than fly. The trail made in a foot of snow as the rounded, plump body waddles through it is most uncommon. There are no footprints, simply a smooth, beautifully sculpted half-tunnel that seems endless. To someone unfamiliar with the habits of the quail it would appear that a one-legged, footless abominable snowman was roaming the countryside.

From the day I found it, I watched the quail nest and diligently recorded all the happenings. I never did learn whether more than one female was involved in the egg-laying, but when two eggs were laid on three different occasions that seemed a probability. Nor did I ever learn if the male was involved with the incubation of the eggs. I did learn, however, that there never was more than one bird sitting on the nest at one time.

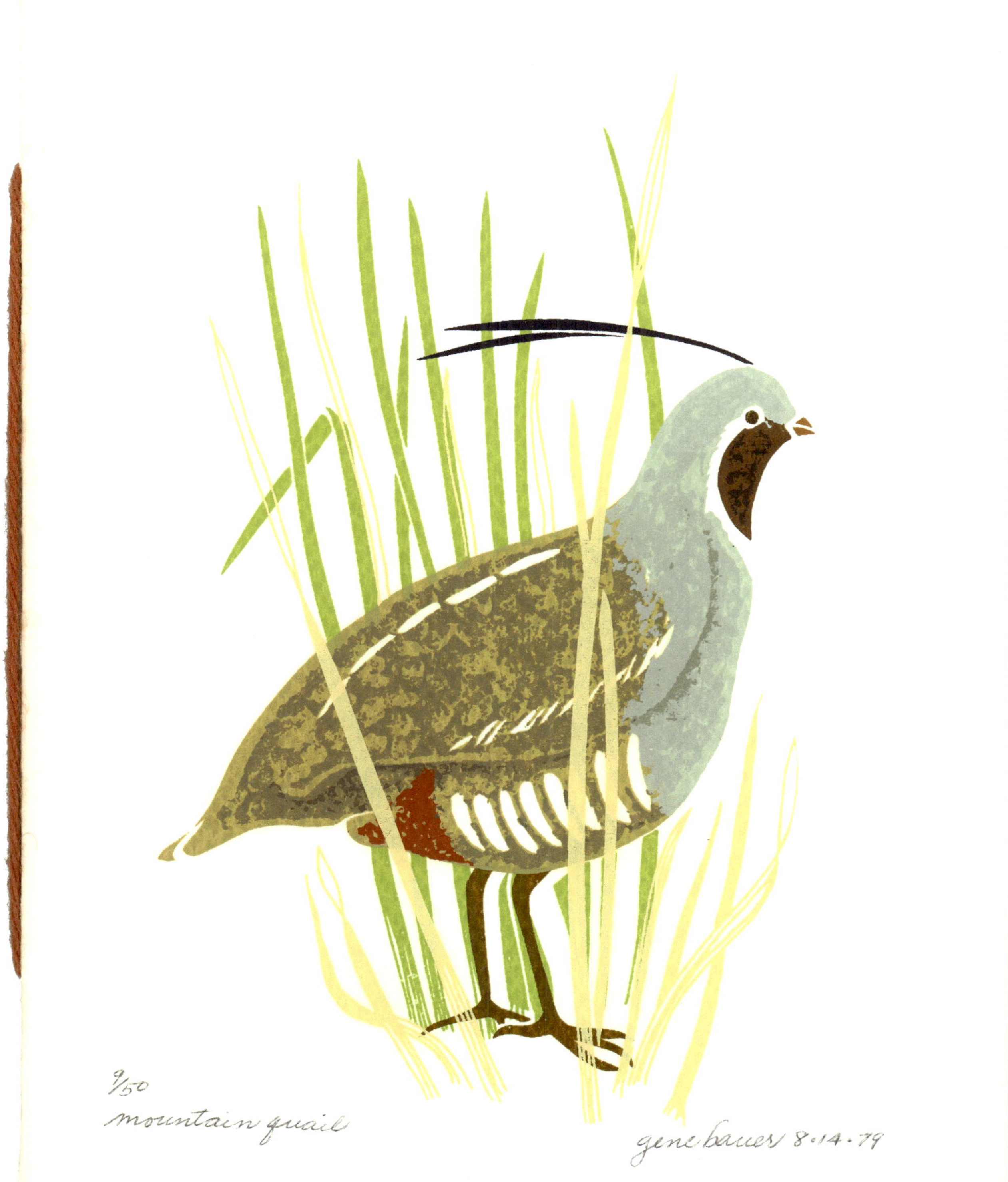
9/50
mountain quail
gene bauer 8.14.79

These are my records of the daily occurrences on and around the nest sixty-eight days from its discovery on April 6, 1979, to the grand and glorious day of June 12, 1979.

The eggs were definitely not being incubated before May 6. Then, out of state, I had no idea what transpired during the next eleven days. But on my return on May 17 someone was sitting on the nest. During this incubation period I visited the nest at least three times a day, early morning, midday, and evening. Some days I came more often. On several occasions the bird was off the nest but returned within fifteen minutes. I was able to inspect the eggs closely and found they were in a single layer, unlike they had been earlier, no doubt so that all eggs could receive the same warmth. The adult bird on the nest made for a reassuring presence. She (or he) was always fluffed up and flattened out like a pancake to cover all twenty eggs. The head tucked down into the body, the rich colors blending one into another, the fluffed feathers almost hid completely the quail's dark plume. It appeared as a lovely, soft, watery watercolor painting.

Small birds require fourteen days for incubation while quail, rather large birds, take twenty-four days. In learning that, I read something disturbing: quail are notorious for sitting on infertile eggs and will continue to do so without the eggs ever hatching. Days and days then weeks went by and I began to believe what others were warning me, that they would never hatch. My growing despondency almost lifted when on June 9 I inadvertently disturbed a large covey of newly hatched chicks and their parents in the adjacent vicinity. But they were not mine.

When quail hatch they are ready to leave the nest, so I kept my eye on them. When I checked the nest on the evening of June 11, the adult bird was off and all twenty eggs were there. The next day I had to be in town very early and did not check the nest in the morning, but when I returned home at 2 PM they had hatched! Peering into the nest I could see only neatly broken shells. Then I spotted them, about ten feet away under a manzanita bush, an exhausted adult and the spritely chicks. Sixteen of the twenty eggs had hatched into healthy chicks to freely roam the countryside. But where were the other four?

I returned to the nest to look at the shells, so cleanly opened, not at all roughly jagged as most newly hatched eggs are depicted. And then I saw feathers underneath them and heard peeping. The baby bird, obviously crushed under the weight of all the others, didn't last long but I felt its beating heart in the palm of my hand before it expired. Then, carefully removing the shells to see if there were others, I heard peeps coming from one of the three eggs left and realized one egg was cracked. It was peeping! Removing it from the nest, I carefully placed the egg in a small container lined with a paper towel. It kept peeping and the crack kept enlarging until two hours later, first the head popped out and finally the body emerged. Unfortunately, though its eyes were open, this chick seemed seriously lacking in vitality, unable to hold up its head or stand on its feet. It lived through the night but, still peeping, expired the next morning. The other two eggs never hatched.

During the sixty-eight days of this bittersweet adventure the clump of grass grew from one foot high when first noticed to three feet high with mature seeds. I cannot bear to remove it, and it is very lonely now when I walk past it.

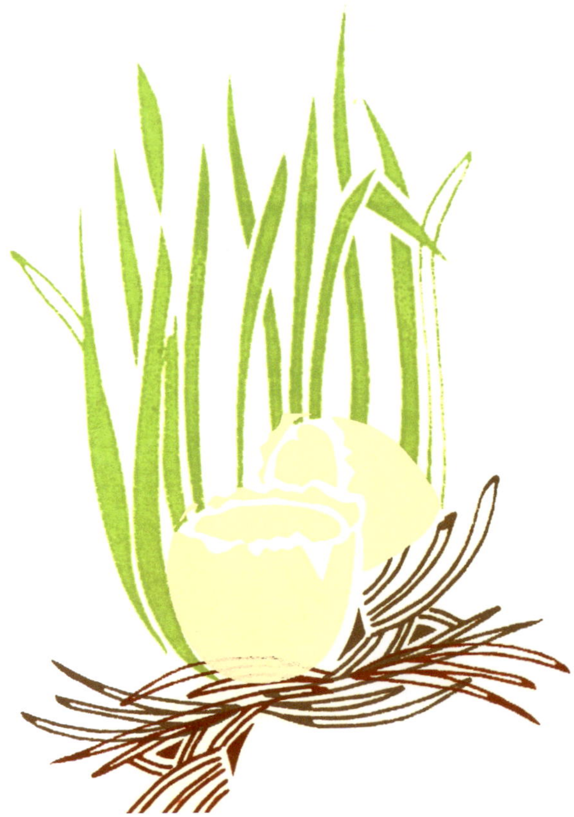

QUAIL LOG

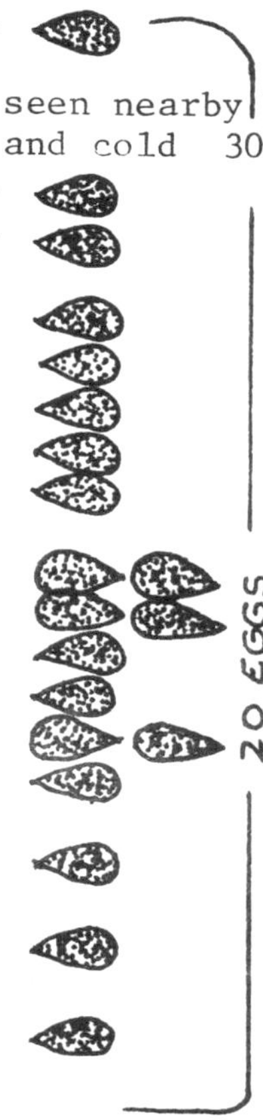

April 6 ......
7
8 Quail seen nearby by Mrs. Hill
9 Foggy and cold 30°
10 ......
11 ......
12
13 ......
14 ......
15 ......
16 ......
17 ......
18
19 ...... LAID BET. 6-9am
20 ...... LAID AFT. 12 pm.
21 ......
22 ......
23 ...... LAID BET. 2-4pm
24 ......
25
26 ......
27
28 ......
29
30 ......
May 1
2
3
4
5
6
7
8
9
10
11 Snow and cold 30°
12
13
14
15
16

May 17......Incubation <u>FOR SURE</u>
18
19
20
21
22
23
24
25
26
27
28
29
30
31
June 1
2
3
4
5. 3' rattlesnake near house
6. 1' rattlesnake on porch
7
8
9
10
11
12 O Joyous Day!

MOUNTAIN CHICKADEES

January 1972

I chose *Parus gambeli,* the mountain chickadee, for a booklet cover because it is my favorite bird. It is certainly not the most beautiful or the most striking to look at; it does not lay the most beautiful eggs or sing the most beautiful song; it is neither the largest bird nor the smallest. It possesses a very simple virtue that endears it to me, its friendliness.

Of all the birds who make their home here in the mountains, including the migratory birds as well, the chickadee is the only one that will come and eat from my hand. If I am sitting down a group of them will come; one will sit on my shoulder, one on my knee or foot, each patiently waiting its turn to come to my hand and eat. It is such a trusting little bird that it will also eat from the hand of a visiting stranger. These birds expect me to have food with me at all times when I am outdoors. They will hover very close and follow me around ever ready for a handout.

When we purchased our land here in the mountains in 1955 I did not know one bird from another. But a most distinctive and insistent call was always present and, though it took several years, I finally put call and bird together. Everyone here is familiar with the song, though few have the time to really know the bird. Its call consists of four notes, the first three going down the scale with the fourth note on the same level as the third. Because it has four "syllables" we fondly refer to it as the "*I love Gen-ie* bird." At times there are slight variations in the call, yet it is so simple and distinctive that my husband and I often use it to summon each other by whistling it.

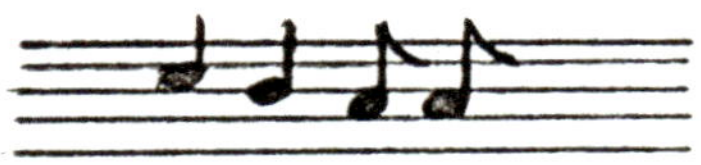

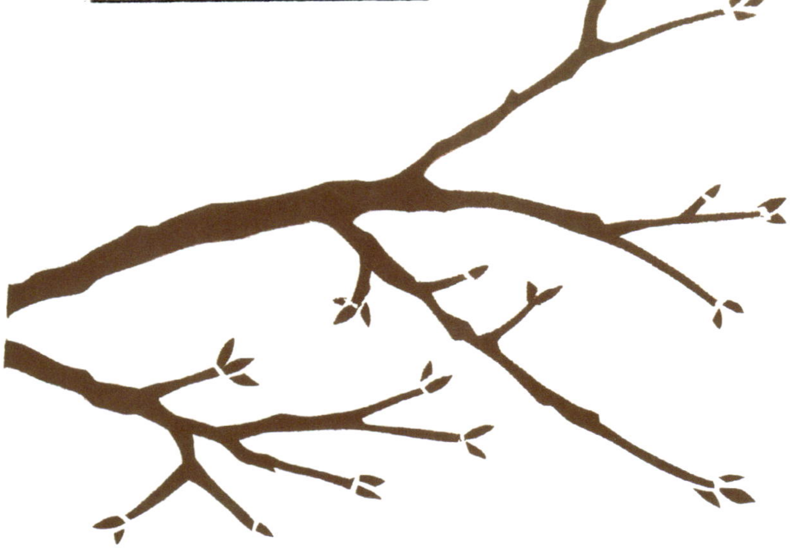

The chickadee is a member of the family Paridae, the same family as that of titmice, verdins, and bushtits. All of the birds in this family are small, ranging in length from four to six inches, but they are hardy. Chickadees rank especially high even among these hardiest of birds, ranging to the farthest limits of forest growth in the Far North. The mountain chickadee is distinguishable from the many other species of chickadees by its white eyebrow. While I was drawing this bird, on December 27, 1971, it occurred to me that the mountain chickadee coloring and markings are similar to another favorite of mine, this one a furry friend, the raccoon. It snowed all that day. The chickadees, along with nuthatches and juncos, were spending the entire day at the feeder, always eight to ten birds perched on the oak branch patiently waiting for their turn at the peanut hearts. Fluffed up most of the time, they looked like chubby little puffballs. The snow gathering on their feathers did not deter them. They were marvelous models.

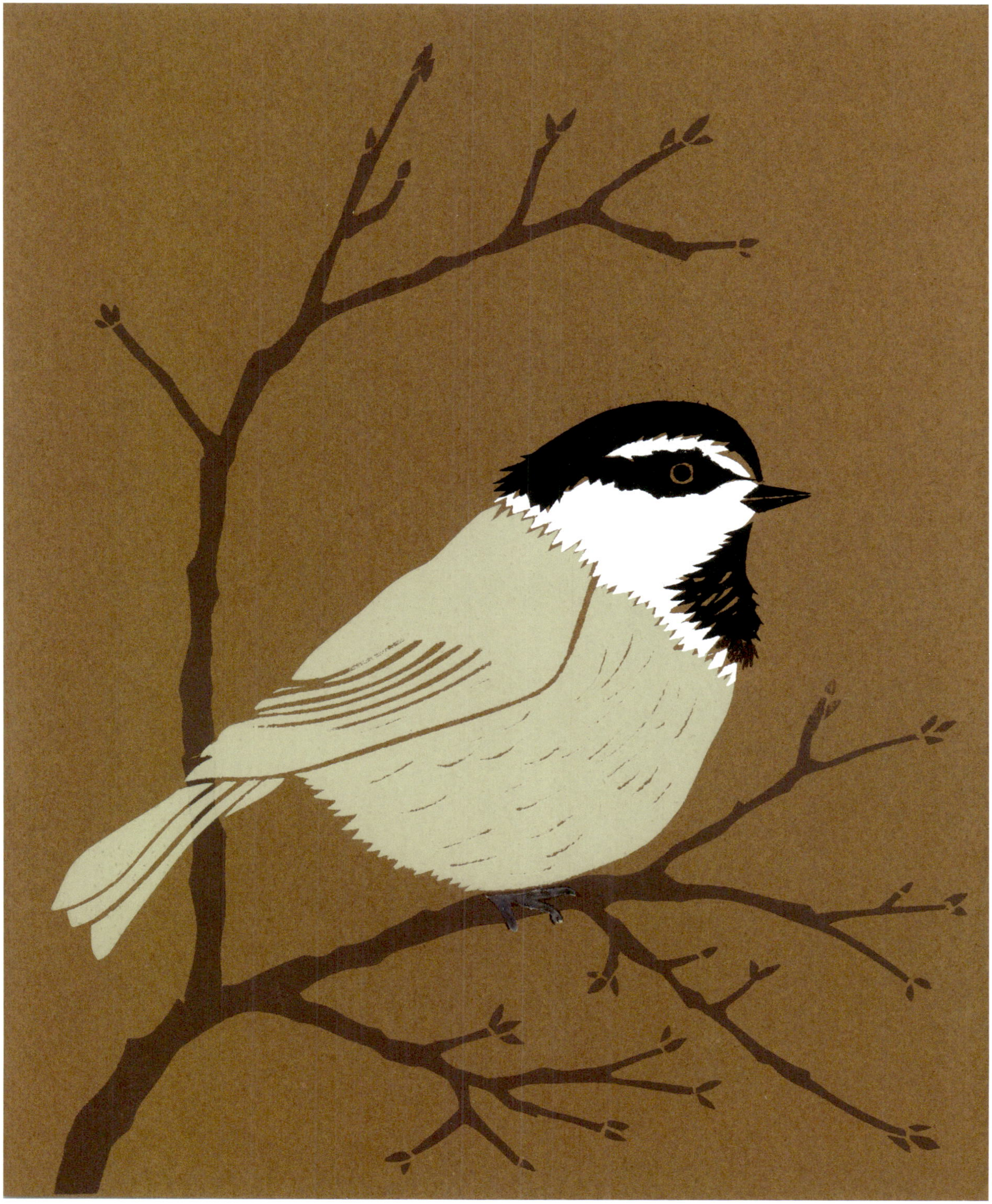

I, too, am not deterred by snow. In fact, I love when each season brings its most traditional characteristics with it. Much of the state of California enjoys year-round sunshine, to many people an ideal climate. To me however, 365 days a year of sun can be a bit dull and monotonous. I was born and lived more than thirty years of my life in this kind of environment, and even as a child I found the mild weather dreary. Because we lived on an orange grove, we paid rapt attention to the 8 PM nightly fruit frost warnings, and I learned very early in life what "above 32" meant. Of course, my parents were pleased when night after night the forecast for our area was above 32, but secretly I yearned for it to drop to 25 degrees, even to 20. Dropping into the teens was unheard of, but I could dream.

The two areas of the plant world that I am most interested in reflect the things denied me in those early years. I adore the turning of the seasons. When it is winter I want it to be cold, the colder the better. I consider snow falling from the sky to be one of the most beautiful of all phenomena. I can think of no sight more stunning than afterglow, those very brief few minutes when the sun descends below the horizon and the sparkling white world of snow-clad trees turns a vibrant, rosy hue. I want spring to be an awakening time. I want to watch the buds on trees swell and see clean, fresh, new leaves unfurl. I want to see flower buds burst open spreading warmth and sunshine after the chill of winter. Summer is a warm and relaxed time, but as fall approaches I want the air to be crisp and clear and the leaves on deciduous trees and shrubs to turn those exquisite colors known as autumnal tints.

Those scenarios are not typical of most of California. However, at the high elevation where we live we have fully the four seasons to enjoy, and my interests in the gardening world reflect two of them, spring for the daffodils and fall for the maple. It seems fitting to talk about fall foliage first because in the gardening chronology of plant life cycles trees come first—in order to enjoy their turning leaves in the autumn, I have to prepare in the spring for their planting.

The only native tree in our immediate area to produce fall colors is *Quercus kelloggii,* California black oak. In a good year the leaves turn a lovely yellow, but in other years only a drab yellow brown. I have introduced many trees to add to our deficiency in fall colors. Among these are tupelo *(Nyssa sylvatica),* northern red oak *(Quercus borealis),* pin oak *(Quercus palustris),* scarlet oak *(Q. coccinea),* flowering dogwood *(Cornus florida* and *C. florida rubra),* western dogwood *(Cornus nuttallii* 'Goldspot'), Chinese dogwood *(Cornus kousa chinensis* 'Milky Way'), and Chinese pistache *(Pistacia chinensis).* More rewarding are several species of the genus *Acer:* vine maple *(Acer circinatum),* a California native but only in the northern part of the state; sugar maple *(A. saccharum);* paperbark maple *(A. griseum);* amur maple *(A. ginnala);* and most rewarding of all, the selected forms of *Acer rubrum*—'October Glory' and 'Red Sunset'.

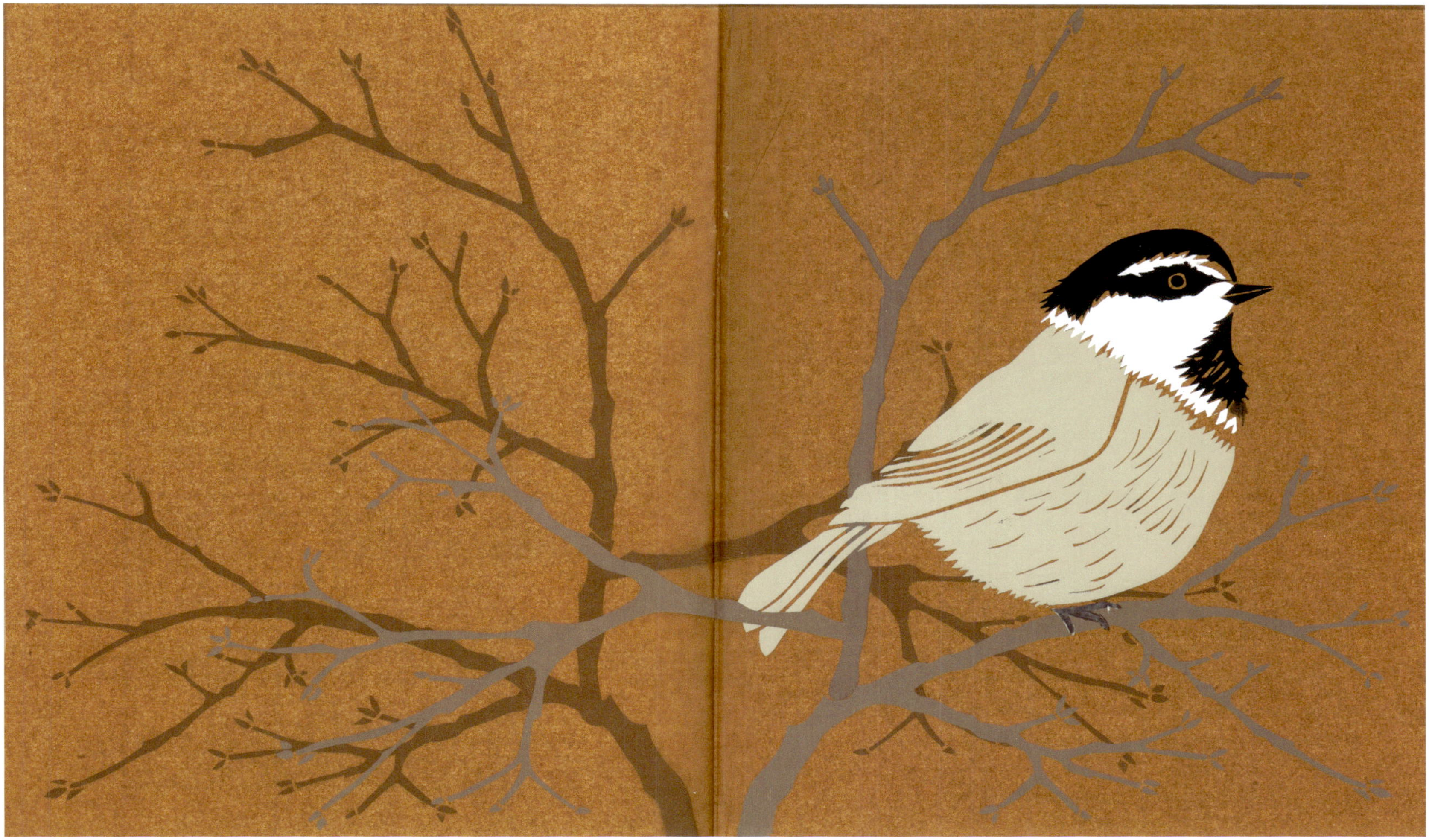

THE MAPLERY

May 1976

I made the original line drawing for this serigraph on April 12, 1976, during the month I started growing from seed more than five hundred trees of *Acer grandidentatum.* It takes time, but I am gradually getting the wee two-year-old canyon maples in the ground. It's been a joy watching them grow from seed to a baby tree. This one especially astounded me in the way it developed the many characteristics shown here within only seventy-two days.

First the seedling popped up; it was about 1¼ inches tall when it pushed its shell off and flung out its seed leaves. You will notice the two seed leaves, having served their function, have dried and turned brown. Soon they will simply fall from the little stem trunk. Immediately after the two seed leaves appeared, it sent out its first true leaves. I was surprised to see that these first leaves were not maple-shaped: their edges were finely serrated and the leaf itself was fuller, more rounded and heart-shaped, not at all deeply lobed or cut as the typical maple-shaped leaf.

The baby tree grew almost another one half inch and sent out another pair of leaves, 90 degrees around the stem from the first pair. Looked at from the top it appeared as a cross or a plus mark.

These leaves were the typical maple shape. After that, the tree grew another half inch and sent out another pair of leaves, again maple-shaped and 90 degrees around the stem from the last pair, and they lined up exactly with the first pair of true leaves. So far all its leaves have been produced in pairs.

Pulling out my granny glasses I was astonished to see that the tree had even formed wee buds in the little crevices where the petioles (leaf stems) joined the tiny trunk. I could even detect little spots on the trunk that would eventually develop into furrows—to become the textured surface of the trunk as it grew older—in other words, the bark. So many adult characteristics appeared within little more than ten weeks of my placing the 1½-inch rootlet of this seed into a Jiffy-7 peat pellet. Only about a year before, it was an inconspicuous flower on a maple tree in Blacksmith Fork Canyon in Cache County, Utah, a yellowish flower merely a quarter-inch in diameter. She was the fruit of a prime specimen of her species, a thirty-five-foot canyon maple *(Acer grandidentatum)* with a twenty-inch trunk diameter (measured at human breast height or 4½ feet above ground).

Just like corn and beans and many good and plentiful food crops begin, the fruit harvested from this canyon maple was selected, hand-picked, and labeled with an accession number, a seed's most important single identifying feature. By knowing the accession number, it is possible to trace back to all the recorded data about the seed.

As I write this, the new tree is sending out another pair of leaves directly over the second pair. (Those are the small leaves at the top of the serigraph.) Right now from soil level to the top, the tree is about three inches tall and (in its milk carton) weighs five pounds. I have eighty-four of these in milk cartons in our sky room. Too heavy and inconvenient to carry in and out of the house, they will stay here until the weather is more stable, perhaps mid-June. I am pleased with the progress of my maple trees. As soon as the weather warms up for the summer I will put all the rest of the five hundred three-inch peat pots in the milk cartons, where they will remain outdoors under a shaded canopy, regularly watered, cared for, and admired.

I will spend the summer preparing the ground in preparation for planting the small trees in their permanent location. As soon as cool weather arrives in October, the little trees will drop their leaves and then be dormant for the winter. Then I will be able to plant the trees and truly mark the beginning of our shady grove of canyon maples, and hopefully, someday our mountainside will be as stunning in the fall as it is with the daffodils in the spring.

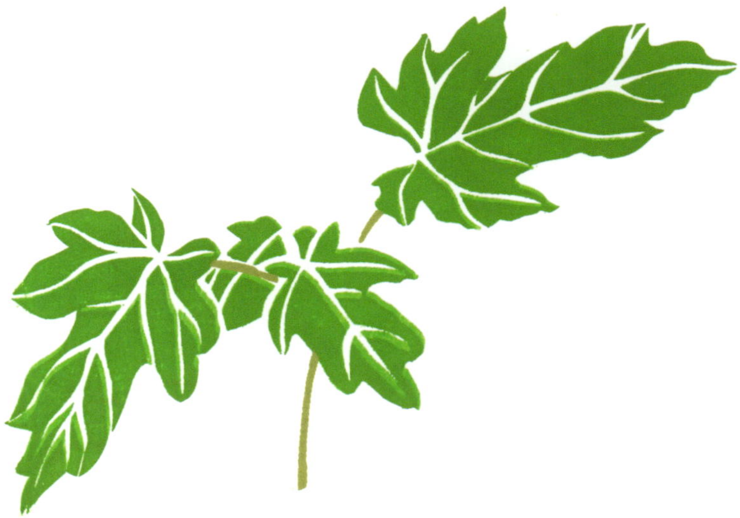

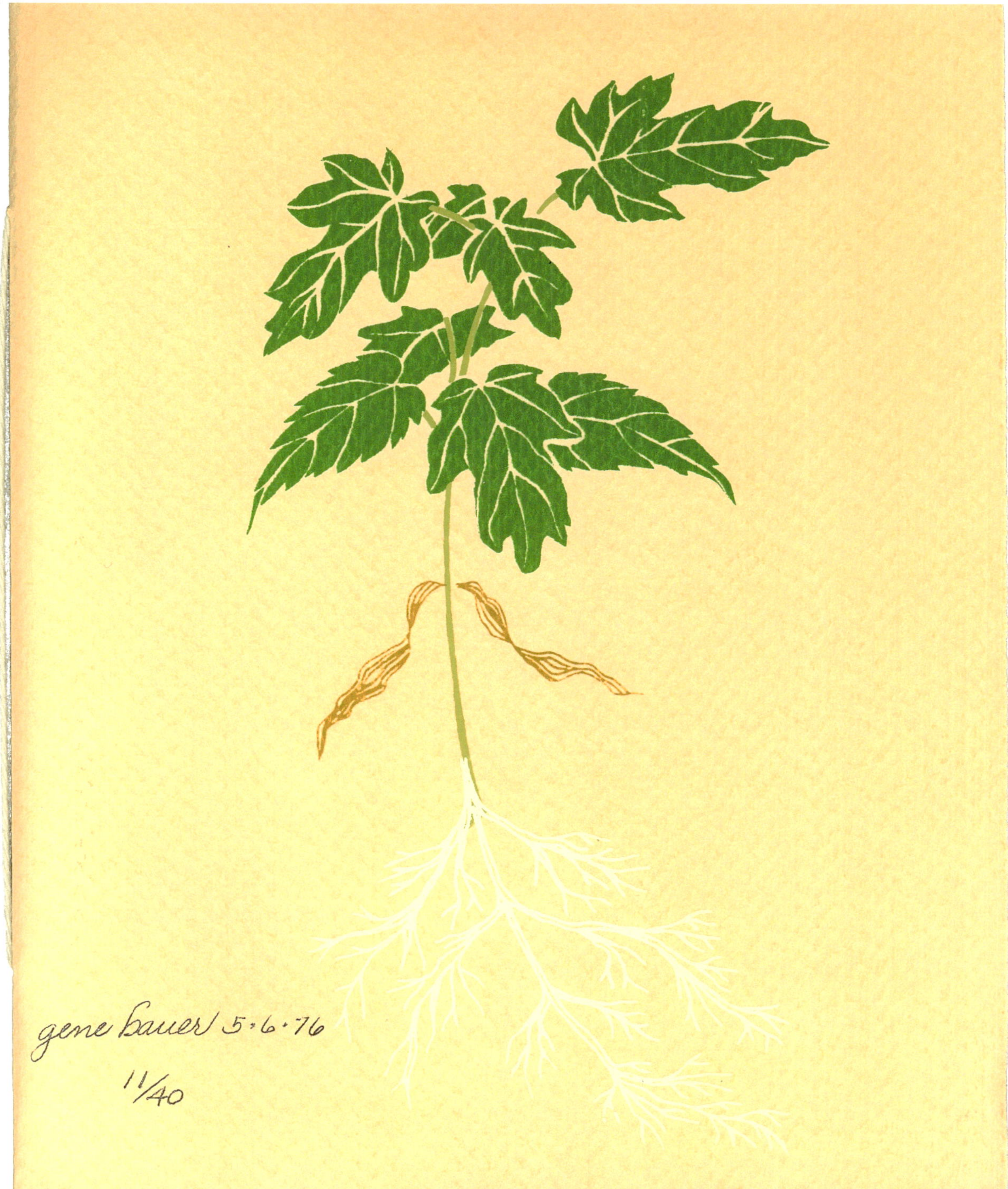
gene bauer 5·6·76
11/40

DANCING DAFFODILS

April 1975

Since 1958, I have become totally addicted to spring-flowering bulbs, especially the big three: tulips (form, colors), hyacinths (heavenly fragrance), and daffodils (dependability and durability and so much more). I will always grow some tulips but usually in containers and treated as annuals. Some of the tulip species do very well for me, retaining their quality blooms for many years, but soon disappear if a gopher finds them before I find the gopher. I am enchanted with tulips, but at times they seem a fleeting extravagance.

Daffodils are more practical. Their bulbs are toxic to gophers and they don't eat them, leaving the daffodils to become cherished friends that make their yearly appearance in good health. Some are the original bulbs I purchased in 1958, part of a collection that twenty years and 25,000 bulbs later produces from three to twenty flowers per bulb. Most are planted in drifts of fifty to fifteen hundred bulbs, each drift composed of only one cultivar. Therefore all the plants in a drift are the same height, shape, color, and they bloom at the same time, appearing as the natural evolution of one bulb. I have more than three hundred different cultivars, many of which are one-of-a-kind, expensive bulbs. I consider daffodils to be a truly remarkable bulb under my conditions; never artificially watered or fertilized; never dug, divided, and replanted (how could I possibly have managed that little chore?). I purchase new bulbs each year to add to the collection.

I know exactly how many bulbs have been put in because I planted every one of them myself [in 2010, one million]. It is not the actual planting that is time-consuming but the preparation of the soil. I decide where the drift will be located, and then with a shovel, turn the soil over to about a one foot depth, removing rocks, weeds, branches, and other debris, then rake it and turn it over once more and rake it again. Then I place the bulbs where I want them and start planting. I do this all by hand, scooting up to sit on the bulbs I've just buried in order to plant the next above them on the hillside. Of course, mechanical devices might be less energy-consuming in preparing the soil, but machinery does not recognize the wild iris or corms of *Bloomeria, Brodiaea,* and *Calochortus* that must be relocated.

I consider that on our mountainside I have created not a garden but rather a living and breathing three-dimensional painting. Why would anyone want a garden in an area blessed with such a bounty of natural beauty? Never do I consider that I have a garden or even a yard. I believe we live in nature's wonderland and I do not care to change or destroy it, only protect and preserve it, and where possible, enhance it.

Our mountainside becomes a soft, green tapestry about mid-January and with each day passing its richness increases. It is like an enormous pointillism painting except that its depth is achieved not by dots but by vertical strokes. My daffodils in winter appear as large drifts of strong-bladed grass flowing across the hillside. Some drifts are intertwined with others and some are isolated. Some drifts have a slightly bluer foliage than others, adding an interesting and subtle distinction. The tapestry grows more luxuriant each day, until suddenly in late February or early March, several of the drifts burst open with the sun-kissed yellow faces of 'February Gold'. From that time on, other drifts develop color until early May (maybe June this year). Even after flowering, the hill returns to a rich green tapestry until the foliage has matured, and by then large clumps of wild iris bloom—not planted by but protected by me—followed by other wildflowers and eventually lush bracken or brake fern.

In opening the daffodil garden to the public each spring, I've opened myself to many interesting questions and comments about the flowers these past few decades. Even the United Parcel Service delivery man whose usual vocabulary consists of three greetings—"Hello" (if I'm lucky), "Sign here," and "Good-bye" (if I'm lucky)—became euphoric when he saw them for the first time this year. "These flowers," he began, "I went way past your place just to see where they went, and they go on and on and on. Are they all yours? Are they wildflowers? They don't grow anywhere else. Man, oh man, are they ever a show."

I was really quite thrilled that he was moved to express himself in something more than the two-syllable greetings I am accustomed to hearing. During another encounter several years ago, with a visitor from the public, I was more perplexed than delighted: "They're so gorgeous," the person said, "did they...did they... [groping for words]... just happen?"

Never have I appreciated Wordsworth's poem about the dancing daffodils more than I have this year. Days and days we have been enveloped in clouds, our only relief the enhanced beauty of the daffodils against the clouds coming up through the ravines and silently floating by. I love these flowers in late afternoon with the sun streaming through them and later just as the sun sets. On full moon nights they literally glow, especially the white ones. But I alone experience all these phases of their development. I'm fairly sure the poet saw only one species of daffodil, the yellow *Narcissus pseudo-narcissus.* What might he have written had he seen what I witness every year in every light, the drifts of yellow; yellow orange; soft, luminous sulfur-lemon and white?

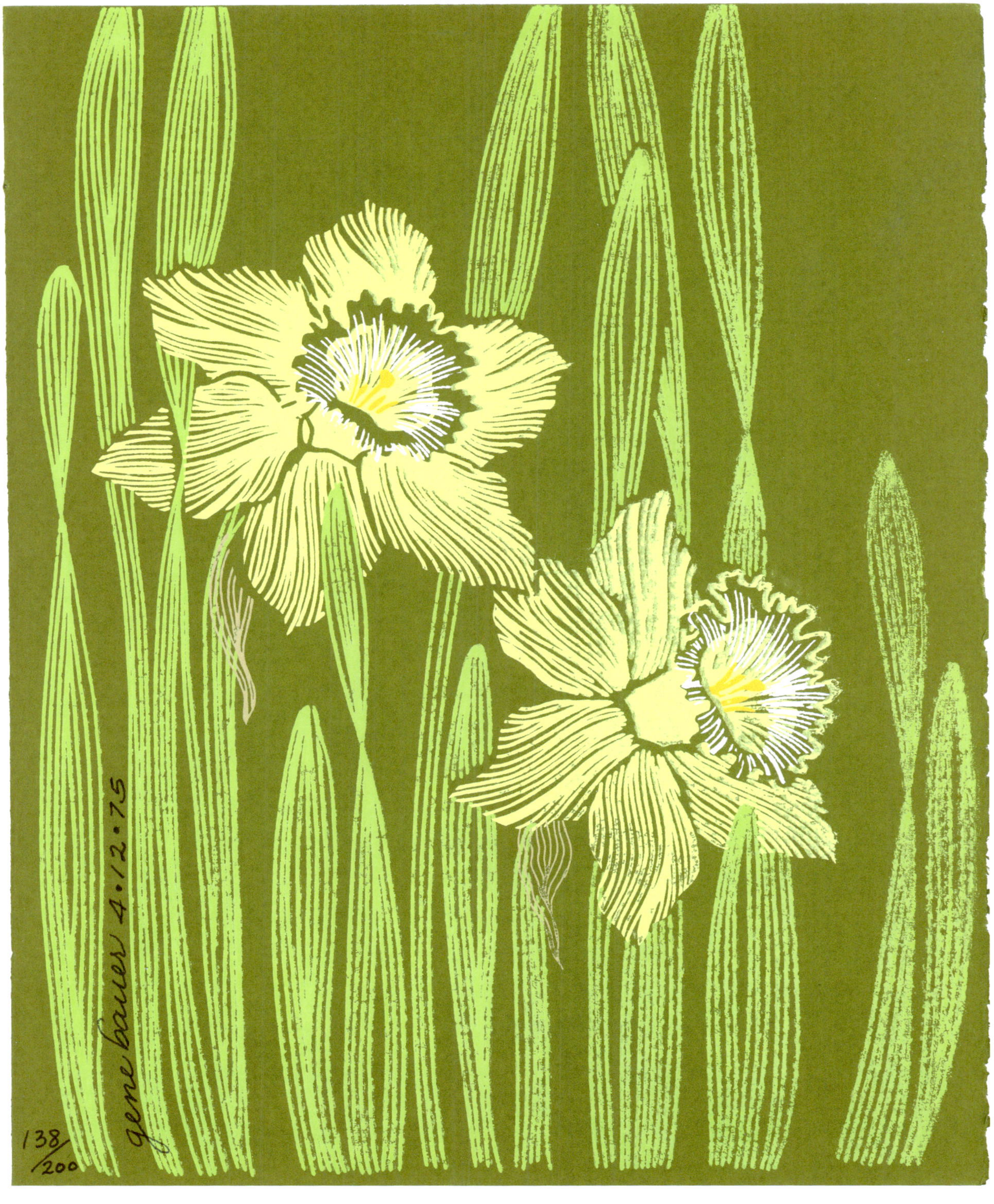
gene bauer 4·12·75
138/200

Public and private collections

Sets of Gene Bauer's original serigraphs are archived in the permanent collections of the following institutions:

The Hunt Institute for Botanical Documentation Carnegie-Mellon University, Pittsburgh, Pennsylvania

The Missouri Botanical Garden, Saint Louis, Missouri

The Frelinghuysen Arboretum, Morristown, New Jersey

Strybing Arboretum, San Francisco, California

The Sherman Foundation [Sherman Library and Gardens], Corona del Mar, California

The following exhibitors showed a permanently mounted display of Gene Bauer's original serigraphs on a limited-time basis in the late 1970s:

Arnold Arboretum, Harvard University, Jamaica Plain, Boston, Massachusetts

Brooklyn Botanic Garden, Brooklyn, New York

Planting Fields Arboretum State Historic Park, Oyster Bay, New York

Pennsylvania Horticultural Society, Philadelphia, Pennsylvania

The Frelinghuysen Arboretum, Morristown, New Jersey

About the author

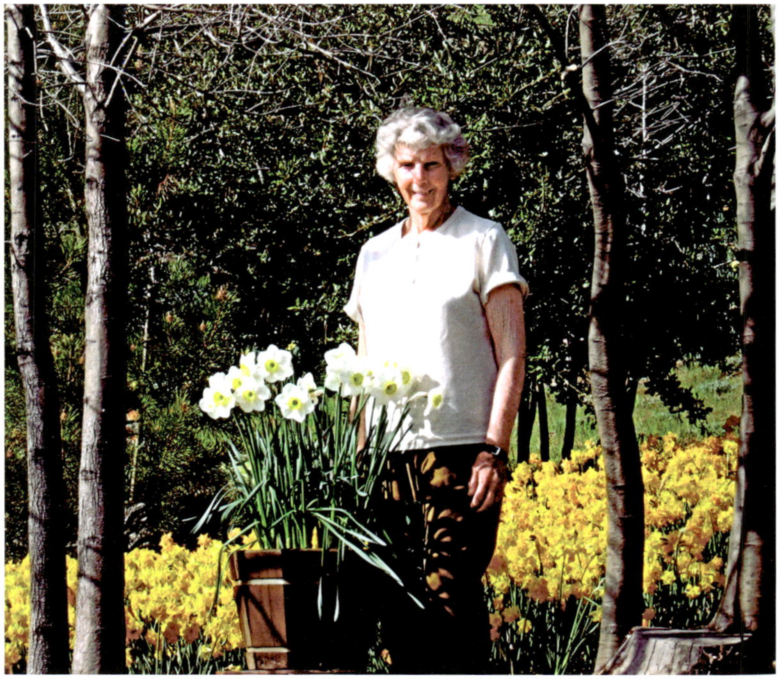

Gene Bauer, a native of California, first developed an interest in local flora among the flowers and trees of her childhood town of Garden Grove. While earning her degree in art education at the University of California, Los Angeles, she learned serigraphy, one of the oldest forms of printmaking, in order to create posters and other teaching materials. After teaching art to junior and senior high school students, Bauer moved to the San Bernardino Mountains where she became involved in a number of garden clubs and the prodigious daffodil planting (one million bulbs to date) that has become so memorable to her community. In the 1970s, her love of plants inspired her to handcraft and issue sixty-one unique booklets of serigraphs and writing reflecting on her observations of California's many public gardens.

Bauer served as native flora chairman (1972–1974) and arboreta and botanical gardens chairman (1976–1978) of the California Garden Clubs, Inc. In 1972, she was honored by the American Forestry Association for her extraordinary efforts in reforestation and conservation. Bauer continues to explore the areas around her home in the San Bernardino Mountains, the site of her magnificent drifts of daffodils. In 1997, a forest fire destroyed her home and surrounding mature native trees. Since then, Bauer has spent her time repairing the damage by planting a forest of deciduous trees.